T0199430

Series on

Reproduction and Development in Aquatic Invertebrates

Volume 4

Reproduction and Development in Annelida

T. J. Pandian

Valli Nivas, 9 Old Natham Road

Madurai-625014, TN, India

CRC Press

Taylor & Francis Group

Boca Raton London New York

CRC Press is an imprint of the
Taylor & Francis Group, an **informa** business

A SCIENCE PUBLISHERS BOOK

Cover page: Representative examples of annelid species. For more details, see Figure 1.1

CRC Press
Taylor & Francis Group
6000 Broken Sound Parkway NW, Suite 300
Boca Raton, FL 33487-2742

First issued in paperback 2021

© 2019 by Taylor & Francis Group, LLC
CRC Press is an imprint of Taylor & Francis Group, an Informa business

No claim to original U.S. Government works

Version Date: 20181120

ISBN 13: 978-0-367-78032-6 (pbk)
ISBN 13: 978-0-367-18745-3 (hbk)

Visit the Taylor & Francis Web site at
http://www.taylorandfrancis.com

and the CRC Press Web site at
http://www.crcpress.com

Preface to the Series

Invertebrates surpass vertebrates not only in species number but also in diversity of sexuality, modes of reproduction and development. Yet, we know much less of them than we know of vertebrates. During the 1950s, the multi-volume series by L.E. Hyman accumulated bits and pieces of information on reproduction and development of aquatic invertebrates. Through a few volumes published during the 1960s, A.C. Giese and A.S. Pearse provided a shape to the subject of Aquatic Invertebrate Reproduction. Approaching from the angle of structure and function in their multi-volume series on Reproductive Biology of Invertebrates during the 1990s, K.G. Adiyodi and R.G. Adiyodi elevated the subject to a visible and recognizable status.

Reproduction is central to all biological events. The life cycle of most aquatic invertebrates involves one or more larval stage(s). Hence, an account on reproduction without considering development shall remain incomplete. With the passage of time, publications are pouring through a large number of newly established journals on invertebrate reproduction and development. The time is ripe to update the subject. This treatise series proposes to (i) update and comprehensively elucidate the subject in the context of cytogenetics and molecular biology, (ii) view modes of reproduction in relation to Embryonic Stem Cells (ESCs) and Primordial Germ Cells (PGCs) and (iii) consider cysts and vectors as biological resources.

Hence, the first chapter on Reproduction and Development of Crustacea opens with a survey of sexuality and modes of reproduction in aquatic invertebrates and bridges the gaps between zoological and stem cell research. With capacity for no or slow motility, the aquatic invertebrates have opted for hermaphroditism or parthenogenesis/polyembryony. In many of them, asexual reproduction is interspersed within sexual reproductive cycle. Acoelomates and eucoelomates have retained ESCs and reproduce asexually also. However, pseudocoelomates and haemocoelomates seem not to have retained ESCs and are unable to reproduce asexually. This series provides possible explanation for the exceptional pseudocoelomates and haemocoelomates that reproduce asexually. For posterity, this series intends to bring out six volumes.

August, 2015   T. J. Pandian
Madurai-625 014

Preface

Annelids are known for the unique spectacular epitoky, regeneration and clonal reproduction. These features have attracted attention more from academic interest. An objective of this book is to elevate annelids from academic interest to economic importance. Many books are authored or edited on annelids but are limited to a taxonomic group or a specific theme like phylogeny. This book is a concise, informative elucidation of reproduction and development in annelids covering from *Aeolosoma viride* to *Zeppelina monostyla*.

The book is structured in nine chapters. In view of their importance, two chapters are devoted to regeneration and clonal reproduction, a chapter on epitoky and another on vermiculture. Some 81, 12 and 7% annelids are marine, freshwater and terrestrial inhabitants. Comprising 17,000 species, annelids are found mostly in aquatic habitats but a few in terrestrial habitats. Vertically distributed between 4,900 m depth and 2,000 m altitude, they are also found in unusual habitats like hydrothermal vents, subterranean aquatic system and migrating between the nutrient-rich anoxic and oxic zones in the sediments. For the first time, information on gutless oligochaetes and polychaetes, osmotrophism and anaerobiosis in some annelids is highlighted. In the absence of exoskeleton to escape predation, 42–47% polychaetes brood their eggs.

The second chapter deals with sexual reproduction. Polychaetes are gonochores. But clitellates are hermaphrodites characterized by internal fertilization and laying cocoon enclosing a few eggs. Updating has revealed the incidence of hermaphroditism in 207 polychaete species from 27 families. Hence, only 74% of annelids are gonochores. A directory is generated and lists 75 parthenogenic annelid species, of which 56 are earthworms. The first estimate has revealed the incidence of external fertilization in 54% polychaetes. The existence of poecilogony with triple morphs and simultaneous sex change between mating partners are projected.

Devoted to regeneration and clonal reproduction, the third and fourth chapters have brought to light a whole range of new findings. In oligochaetes, inadequate reserves in the chloragogue temporally separate the incidence of regeneration and reproduction. However, the sedentary polychaetes, which do not possess the chloragogue or its equivalent, undertake them together, of course, at the cost of reduced reproduction. The number of species capable of anterior, posterior and anterior cum posterior regeneration is 149, 206 and 143, respectively. When these numbers are considered as fractions of 13,012 polychaete and 3,175 oligochaete species, the percentage values (1.57, 1.80, 1.42) indicate that the potency is 1.5–2.0 times more prevalent in oligochaetes than the respective ones (0.88, 1.22, 0.85) in polychaetes. The earlier loss of the regenerative potency in 'older' anterior segments than in 'young' posterior segments located adjacent to the generative pygidium may be a reason for the observed less prevalence of anterior regeneration.

Clonal reproduction can sustain a species for > 30–60 years. When stressed or induced, sexual reproduction is manifested, as Primordial Germ Cells (PGCs) are transmitted up to 1,000–3,000 clonal generations. Abundant food supply and low density trigger clonal reproduction but not intense predation; for example 290 tubiculous sabellids species, only 17 are cloners. Of 100 and odd annelid families, the incidence of clonal reproduction occurs only in 12 polychaete families and 5 oligochaete families. It ranges from 2% of spionids to 54% of naidids. Further, it occurs in 79 polychaete species but to as many as 111 oligochaete species. Clonal reproduction is considered to have been derived from regeneration. However, this view is not correct, for (i) in as many as 111 oligochaete species, cloning does obligatorily require the presence of neoblasts and (ii) even with anterior cum posterior regenerative potency, 34 out of 63 polychaete species, do not reproduce clonally. In most polychaetes, the stem cells responsible for cloning are located at the posterior end and also mid-body in a few.

The epitokes are divided into semelparous epigamics and iteroparous schizogamics. For the first time, a directory is documented listing epigamy in 61 species from 12 families and schizogamy in 45 syllid species. Again for the first time, the assembled information on vertical distance traveled by 28 epitokous species reveals that the larger glycerids, nereidids and eunicids use muscular energy to climb up < 50 m but the smaller phyllodocids and ctenodrilids may engage reduced buoyancy to climb the vertical distance of up to 4,000 m.

The sixth chapter deals with sex determination by genes harbored on chromosomes. Karyotyping and breeding experiments have found heterogametism in six polychaete species only. A directory is assembled for the chromosome numbers in annelids. By selective fertilization of large eggs by X-carrying sperm, *Dinophilus gyrociliatus* have nullified the chromosomal mechanism of sex determination. In *Capitella capitata*, expression of W gene(s) is stable but that in Z chromosome is labile resulting in generation of phenotypic ZZ hermaphrodites and females.

Our understanding of endocrine sexualization in syllids and regulation of reproductive cycles in others is based on temperate polychaetes alone. A dozen neuroendocrines/hormones secreted mostly by the 'brain' regulate the reproductive cycle.

The ninth chapter on vermiculture (i) emphasizes the need for information on growth and reproduction in cultivable species, (ii) considers parthenogens and cloners, as they do not have adequate genetic diversity and cloners increases the number but may not the biomass and (iii) recognizes 'layers' as distinct from 'brooders'. There is an urgent need for research input to harvest tubificids and naidids at appropriate intervals, as it may reduce the input of nitrogen fertilizers in ricefields. The fastest growing earthworms, nereidids, tubifex and pot worms are recommended as cultivable species. For the first time, the fast growing *Branchiura sowerbyi* fed on waste paper immersed in water is identified as potential candidate species for vermiculture.

This book is a comprehensive synthesis of 737 publications carefully selected from widely scattered information from 237 journals and other literature sources. The holistic approach and incisive analysis have led to several new findings and ideas related to reproduction and development of annelids. Hopefully, the book serves as a launch-pad to further advance our knowledge on annelids.

July, 2018   T. J. Pandian
Madurai-625 014

Acknowledgements

It is with pleasure that I thank Drs. P. Murugesan and E. Vivekanandan for critically reviewing parts of the manuscript of this book and for offering valuable suggestions. In fact, I must confess that I am only a visitor to the theme of this book. However, a couple of joint publications with my student the late Dr. M. Peter Marian, my earlier book in this series and editorial service on annelidan energetics (Pandian, 1987, *Animal Energetics*, Academic Press) have emboldened me to author this book. I wish to thank my former students and associates Drs. Premkumar David (USA), G. Kumaresan, V. Mohan, P. Murugesan, S. Sudhakar and Prof. W. Westheide (Germany) for by providing copies of some publications. The manuscript of this book was prepared by Mr. T.S. Balaji, M.Sc. and I wish to thank him profusely for his competence, patience and co-operation.

Firstly, I wish to thank many authors/publishers, whose published figures are simplified/modified/compiled/redrawn for an easier understanding. To reproduce original figures from published domain, I gratefully appreciate the permission issued by Dr. S.M. Mandaville. For permissions issued to reproduce original figures from dissertations/protocols of his students, I remain thankful to Dr. S. Sudhakar, who has also provided me his consent to reproduce unpublished figures. Dr. P. Murugesan has kindly provided me a hand drawn figure. For advancing our knowledge in this area by their rich contributions, I thank all my fellow scientists, whose publications are cited in this book.

July, 2018   **T. J. Pandian**
Madurai-625 014

Contents

1

Introduction

1.1 Annelidan Science*

Annelids are bilaterally symmetrical, triploblastic, schizocoelomatic, metamerically segmented vermiform invertebrates. Their segmented body has a well-developed ladder-like central nervous system with a bi-lobed cereberal ganglion and sense organs, a closed blood-vascular system, coelom, an excretory system and a fairly well developed endocrine system. Among aquatic fauna, they are a fascinating taxon displaying (i) epitoky, a spectacular unique phenomenon involving transformation from benthic to meroplanktonic reproductive morphism, (ii) osmotrophism displayed by the gutless tubificids acquiring cent percent nutrients across the body surface from ambient sea water, (iii) partial (tubificids) and complete (vestimentiferans) chaemoautotrophism by engaging symbiotic microbes to draw energy, (iv) poecilogony, another unique feature shared only by some gastropods (Pandian, 2017), (v) metamerism, the most distinguishing feature of annelids and (vi) regeneration, which may be followed by bidirectional (a genet divided into two ramets) and multidirectional (a genet divided into multiple ramets) asexual reproduction. They are classified into Archiannelida, Polychaeta, Oligochaeta and Hirudinea (Table 1.1). In polychaetes, a pair of bilaterally located parapodia facilitates burrowing and locomotion. It is, however, reduced to setae in oligochaetes and totally missing in archiannelids and hirudineans. The marine interstitial archiannelids are phylogenetically enigmatic annelids and include five families. The exclusively marine polychaetes include burrowing, crawling, digging, drifting (*Tomopteris* spp), and tubiculous forms (Fig. 1.1) commonly known as sea mice (*Aphrodita*), scale- (*Polynoe*), paddle- (*Notophyllum*), pile/rag/clam- (*Nereis virens*),

---

* Names of most annelids species are listed following Worms—World Register of Marine Species; however, some are named, according to author's citation.

TABLE 1.1

Systematic resume of Annelida (compiled from Barnes [1974] and others; terrestrial and amphibious taxa are indicated by bold and italic letters; representative family names alone are indicated)

Phylum: Annelida (15,000 [Wildlife J Junior, 2017], 16,763 [Chapman, 2009], 22,000 [Aguado et al., 2014])

Class: Archiannelida, *Polygordius, Nerilla* (60 species [Westheide, 1984])

Class: Polychaeta (8,000 [IASzoology.com], 10,000 species [Minelli, 1993], 13,000 [Australian Museum, 2015])

Sub class: Erranta

Families: Aphroditidae, Polynoidae, Siboglinidae, Phyllodocidae, Amphinomidae, Nereididae, Pisionidae, Alciopidae (pelagic), Tomopteridae (pelagic), Hesionidae, Syllidae, Nephtyidae, Glyceridae, Eunicidae, Lysaretidae, Arabellidae, Lumbrineridae, Histriobdellidae, (Ectoparasites), Ichthyotomidae (Ichthyoparasites), Myzostomidae

Sub class: Sedentaria

Families: Orbiniidae, Spionidae, Siboglinidae, Magelonidae, Chaetopteridae, Cirratulidae, Flabelligeridae, Opheliidae, Capitellidae, Echiuridae, Cossuridae, Arenicolidae, Maldanidae, Oweniidae, Sabellariidae, Pectinariidae, Ampharetidae, Terebellidae, Sabellidae, Serpulidae, Fabriciidae

Class: Oligocheta (1,700 species [Martin et al., 2008])

Orders: Lumbriculida (145 species, Ferraguti et al., 1999 [*Lumbriculus*]), Moniligastridae

Order: Haplotaxida

Sub order: Haplotaxina

Sub order: Tubificina

Families: Enchytraeidae (670 species, Schmelz and Collado, 2015), Tubificidae (1,000 species, Martin et al., 2008), Naididae (175 species, Ferraguti et al., 1999), Phreodrilidae, Opistocystidae, Dorydrilidae

Sub order: *Lumbricina* (33–670 species and subspecies in ~ 48 genera [Wikipedia])

Families: Alluroididae, **Glossoscolecidae**, **Lumbricidae**, **Megascolecidae** (***Pheretima***), **Eudrilidae**

Class: Hirudinea (700–1,000 species [Govedich and Bain, 2005])

Order: Acanthobdelida, Rhynchobdellida

Families: Glossiphoniidae (*Placobdella*), Piscicolidae (*Pontobdella*)

Order: *Gnathobdellida*

Families: *Hirudinidae*, **Haemadipsidae** (50+10 species [Won et al., 2014]) (***Hirudo, Haemopis***)

Order: Pharyngobdellida: Primarily aquatic with some semi-terrestrial species

---

nuclear- (*Namalycastis*), fire- (*Amphinome*), blood- (*Glycera*), tube- (*Onuphis*), lug- (*Arenicola*), bamboo- (*Maldane*), trumpet- (*Pectinaria*), spagehetti- (*Terebella*), fan- (*Sabella*) and feather duster- (*Hydroides*) worms. The drifting holoplanktonic polychaetes are included in two families Alciopidae and

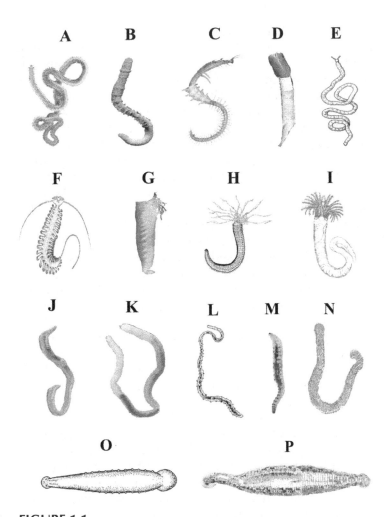

FIGURE 1.1

A. *Nereis*, B. *Arenicola*, C. *Chaetopterus*, D. *Riftia* (Public domain, Wikipedia), E. *Polygordius* (freehand drawing from Fraipont), F. *Tomopteris kils* (redrawn), G. *Ampharete*, H. *Amphitrite*, I. *Hydroides* (H and I freehand drawings from Dales, 1963), J. Earthworm, K. *Amphichaeta* (permission by limnes@chebucto. ns.ca), L. *Tubifex*, M. *Aeolosoma* (copyright free projects.ncsu.edu), N. *A. hembrichi*, O. *Piscicola* (freehand drawing from Mann, 1962), P. Human leech (A, B, C, F, G, J and O are all copyright free images from dreamstime.com).

Tomopteridae and are being expanded to Lopadorhynchidae, Pontodoridae, Iospilidae, Poeobiidae and Typhloscolecidae (Bonifazi et al., 2016). The fire worms (Amphinomidae) are known for their brittle poisonous setae and some blood worms can be venomous (Aguado et al., 2014).

Though the swarming palolo worms *Eunice viridis* are collected and consumed by Samoan Islanders (Caspers, 1984), annelids do not directly contribute to the global fisheries. However, they serve as key bait in recreational fisheries. More than 65 million anglers from Europe, USA, Canada and Australia catch annually 2 million ton (mt) high value fishes (Pandian, 2015). Between 1966 and 1980, 27–38 million pileworms *Nereis virens* were harvested annually (Creaser and Clifford, 1982). The estimated market value of the bait worms for Europe alone is ~ US$ 250 million (Olive, 1999). van der Have et al. (2015) have listed some 25 polychaete (pile-, lug-, blood-, sand- and nuclear-worms) species used as bait by anglers in Europe, Korea and China. The demand for these bait worms is so high that *N. virens* is cultured in the UK and Netherlands; the bloodworm *Diopatra bilobata* and Korean lug worm *Perinereis* spp are also cultured in Vietnam, Korea, China and Japan (for export). Among freshwater oligochaetes, *Tubifex* and the like serve as live feed for domestic ornamental fishes. Many publications (e.g. Marian and Pandian, 1984, 1985, Marian et al., 1989) and privately circulated books (e.g. Pandian and Marian, 1985a) are available describing the procedures for rearing and harvesting the worm are available.

Being ecosystem engineers, services rendered by the earthworms needs no emphasis. These worms accelerate degradation of organic matter and molecules produced by plants and other organisms, and render nutrients, especially nitrogen reusable by plants. Total production of mineral nitrogen by the worms ranges between 30 and 50 kg/hectare (ha)/year (y). By altering porosity, these worms contribute to soil structure and thereby water absorption and retention; for example, water infiltration rate through soil can be increased by the worms from 15 to 27 mm/hour (h), resulting in reduced runoff. In soil formation, they breakdown the primary minerals and incorporate them with organic matter. Their aquatic counterparts, the polychaetes and tubificids serve also as ecosystem engineers to turbulate (see Hutchings, 1998) and make the organic matter from sediments and deposits available for benthic productivity. For example, *Tubifex benedii, Amphichaeta sannio, Paranais litoralis* (naidid) and *Manayunkia aestuarina* (polychaete) jointly contribute 50–90% of the total invertebrate production in the Forth estuary, Scotland (see Giere, 2006). Inhabiting 1.6 km long, 3 m width and 30 cm depth of intertidal zone of the Pacific coast, USA, the small (25 mm long) opheliid *Euzonus mucronata* annually turbulates 14,600 ton (t) sediments. Hence, the indirect contribution by the annelids to the global benthic fisheries must be of high order. However, not much information is yet available on quantitative contribution by the polychaetes to the trophodynamics of many ecosystems (cf Pandian, 2016). Time memorial, leeches, especially *Hirudo medicinalis* have been used in hirudotherapy. *H. medicinalis* secretes hirudin, a 65 amino acid peptide that inhibits thrombin-catalyzed conversion of fibrinogen to fibrin and prevents host blood from clotting. By inflicting the deepest bite and the most-prolonged post-bite extravasation, a leech can engorge maximally with 50–100 ml human blood (Govedich and Bain, 2005).

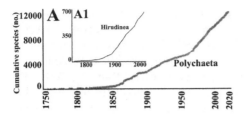 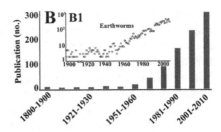

FIGURE 1.2

A. Described polychaete species during the period from 1750–2020 (from Polychaeta Statistics, modified). A1 in window shows the same relationship for hirudineans (modified from Sket and Trontelj, 2008). B. Number of publications relevant to polychaetes during the period from 1800–2010 (from Faulwetter et al., 2014, modified). B1 in window shows the same relationship for earthworms (modified from Sturzenbaum et al., 2009).

For polychaetes, relatively more information is available on the number of species and publications for the past two centuries. Syllidae, Nereididae, Spionidae and Serpulidae are speciose families each comprising > 500 species (Faulwetter et al., 2014). Described polychaete species number has remained < 2,000 until the 19th century and is expected to increase to ~ 13,000 species by 2020 (Fig. 1.2A). This trend also holds true to hirudineans (Fig. 2.1A1) and perhaps to oligochaetes. Publications relevant to polychaetes, which have remained less than a dozen per decade until 1960, have increased rapidly to ~ 300/decade during 2000–2010 (Fig. 1.2B). This type of spurt may hold true for earthworms (Fig. 1.2B1) and other clitellates also. Hence, this book provides only a 'snap-shot' of annelid reproduction and development rather than an in depth or exhaustive description of each item listed in the 'Contents'.

1.2 Species and Structural Diversity

A vast majority of annelids are gonochores and reproduce sexually. Not surprisingly, this feature is reflected in their species diversity. For example, any macrofaunal sample from the Australian soft sediments is reported to hold from 24% (Bass Strait) to 36% (Port Phillip Bay) polychaete species (see Hutchings, 1998). Indicating the sustained contribution to annelidan taxonomy, the described species number has progressively increased from 8,700 (Barnes, 1974) to 14,000 in polychaetes, 4,000 + in oligochaetes and 800 in hirudineans (Rouse and Pleijel, 2006), 16,763 (Chapman, 2009, see also Westheide and Purschke, 2013) and to 22,000 (Aguado et al., 2014). More recently, Bleidorn et al. (2015) have considered that annelids comprise >

17,000 species (Table 1.2, see also p 5). This number may further increase, as the described polychaete species number alone, which was around 10,000, is estimated to shoot to 25,000 (Hutchings, 1998). Similarly, the number of hirudinean species has also increased from 500 (Barnes, 1974) to 700–1,000 (Govedich and Bain, 2005). Martin et al. (2008) reported that of 1,700 valid species of aquatic oligochaetes, of which the most speciose Tubificidea holds over 1,100 species. However, it must be stated that the annelidan taxonomy remains fluid but dynamic. Erected new species by mis/wrong identification is being continuously corrected. For example, a check-list of polychaete species from the Black Sea reveals that 51 species reported between 1868 and

TABLE 1.2

Distribution and number of annelid species in marine, freshwater and terrestrial habitats

Taxon	Habitat (no.)			Total (no.)	Reference
	Marine	**Freshwater**	**Terrestrial**		
Archiannelida	60	–	–	60	Westheide (1985)
Polychaeta	> 13,000	~ 12†	2**	13,002	Australian Museum (2015), Erseus (1994)**
Hirudinea	102	482	92	684	Sket and Trontelj (2008)
Oligochaeta	+	+			Martin et al. (2008)
Moniligastrida			+		
Lumbriculida		145 + 24*		169	Feragutti et al. (1999)
Haplotaxina		++	+		
Tubificina	16†			16	Giere (2006)
Enchytraeidae			~ 670	670	Schmelz and Collado (2012)
Tubificidae	600	1,100 + 13*	+	1,713	Martin et al. (2008)
Naididae	–	175	–	175	Ferraguti et al. (1999)
Phrocodriliae	+	+			
Ophistocystidae					
Dorydrilidae		++			
Lumbricina		+	+		
Alluroidae		++			
Lumbricidae			432	432	Rhoden (2015)
Megascolecidae			+		
Eudrilidae			+		
	13,776	1,939	1,202	16,911	Bleidorn et al. (2015)

* subterranean species recognized by Chatelliers et al. (2009), + freshwater, ++ mostly freshwater, – absent, † riverine polychaetes

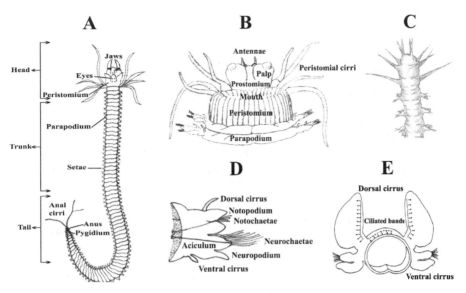

FIGURE 1.3

A. An errant *Nereis* showing head/peristomium, trunk and tail regions. B. Ventral view of *Nereis* head. (modified and drawn from Snodgrass). C. Anterior portion of *Hesionides*. Note the ventrally directed parapodia adapted for crawling. D. Ventral view of parapodium of *Nereis*. E. Surface ciliation in filter feeding *Phyllodoce* (modified and drawn from Segrove). Arrows indicate direction of water current (A. all are free hand drawings; A and D, courtesy Dr. P. Murugesan).

2011 has been synonymized with other species (Sahin and Cinar, 2012). In fact, many reviewers refer only generic names, as species-level revisions are an ongoing research.

The annelidan body is distinctly divided into a series of similar segments. These segments are arranged in a linear series along the antero-posterior axis. The prostomium and pygidium, which are not true segments, are located at the anterior and posterior ends of the body. The formation of new segments occurs just anterior to the pygidium. The oldest segments are therefore anterior and the youngest are posterior (Fig. 1.3). Considering polychaetes as representative taxon, the number of segments ranges from a few to as many as 300 in the orbiniid *Nainereis dendritica*, the body length from 2 mm in the cirratulid *Monticellina serratiseta* to 3 m in an *Eunice* sp and the width from 0.4 mm in the cirratulid *Caulleriella lafolla* to 6 mm in the orbiniid *Phylo nudus*. Obviously, thinning of the body facilitates burrowing to deeper depths. Table 1.3 and Fig. 1.3 briefly summarize the remarkable structural diversity in prostomium as well as structures like the parapodia associated with each segment of the body. Sensory organs attached to the prostomium and bilaterally located parapodia are greatly modified to suit the burrowing, crawling and tubiculous mode of life and to collect food, as well.

TABLE 1.3

Diversity of segmental structures in annelids

Polychaeta

Crawlers

Well developed preoral prostomium bears antennae, palps, eyes, nuchal organ. Fleshy biramous parapodia consisting of an upper notopodium and lower neuropodium terminate by an invaginated series of setae. Cirri form processes arise from the bases of these podia. The segment is supported by one or more chitinous rods (Fig. 1.3A, D)

Drifters

Structures like the crawlers. Transparent. Enormously large eyes in alciopids and membranous parapodial pockets in tomopterids (Fig. 1.1). Setae are absent

Burrowers

Reduced prostomium. Antennae, palps and eyes absent but carry food collecting structures (e.g. tentacles of *Amphitrite*, Fig. 1.1A). Reduced uniramous parapodia represented by transverse ridges, which may be modified into hooks

Borers

Spionid *Polydora* burrows into the shell aided by a viscous fluid, which dissolves the exposed calcite crystals of the shell (Pandian, 2017)

Diggers

Heads of the pectinariid worms bear rows of large conspicuous seta used for digging

Tubiculous worms

Carnivores: Segmental structures do not greatly differ from those of crawlers. Inhabit in vertically or horizontally straight hyaline/membranous tubes

Others: Prostomium and sensory structures reduced or absent. Specialized food collecting structures present. Sabellids build membranous sand grain tubes. Serpulids secrete calcareous tubes. Oweniids carry their tubes

Oligochaeta

No parapodia. Setae can be long or short, straight or curved, heavy or needle-like and blunt, pointed, forked, pinnate or plumose. The setal shaft is S-shaped with middle swelling or nodule. Longer setae are characteristics of aquatic species. Setal number is fixed at 8 in aquatic species and *Lumbricus*. In mature oligochaetes, certain anterior adjacent segments thickened and swollen by glands that secrete mucus for copulation and for formation of cocoon. Their glandular area is called clitellum, which forms a girdle around the body

Hirudinea

No head appendages but eyes are present. No parapodia and setae except in *Acanthobdella*. Dorso-ventrally flattened body with suckers at the anterior and posterior ends. Segment number is fixed at 34. Septa absent

1.3 Geographic Distribution

Geographic range is the horizontal and vertical areas, in which populations of a species are distributed. It is a species specific trait with evolutionary consequences inclusive of determination of life span and exposure to different environmental and biological factors. In freshwaters, the oligochaetes *Tubifex*

tubifex is cosmopolitan in distribution; so is the polychaete *Nereis virens* in the intertidal zone of marine environment. The archiannelids and polychaetes are exclusively marine, although a couple of polychaetes *Peregodrilus heideri* and *Hrabeiella periglandulata* are almost terrestrial inhabiting moist soil (see Rota et al., 2001), and a dozen of them inhabit rivers and estuaries. On the other hand, oligochaetes and hirudineans are found in marine, freshwater and terrestrial habitats. Experts have provided counts on the number of species for groups within, oligochaetes but a total of which does not tally with that reported for the class (Table 1.2). As a result, the sum of subtotals does not tally with recently reported 22,000 species of annelids (Aguado et al., 2014). For example, the estimate of 5,000 clitellate (Oligochaetea + Hirudinea) species by Martin et al. (2008) does agree with the subtotal estimates reported by different experts for the suborders and families. Despite this constraint, and considering 17,000 species number for annelids (Bleidorn et al., 2015), ~ 11.5 and 7% annelids are distributed in freshwater and terrestrial habitats, respectively; the remaining 81% are marine inhabitants. Reports on geographic distribution of a specific taxon in some of these habitats are available; for example, polychaetes in the Black Sea (Sahin and Cinar, 2012), naidids in African freshwaters (Grimm, 1987) and terrestrial enchytraeids in Latin America (Schmelz et al., 2013). A report on distribution within the major geographical zones is also available but only for freshwater hirudineans (Table 1.4). In general, the geographic range is increasingly limited in the following descending order: planktotrophy < lecithotrophy < brooding/viviparity < exclusive asexual reproduction. Within Cirratulidae, the range is extended from the Washington coast of North America to Puget Sound, Washington (DC) for *Chaetozone acuta*, which broadcasts small (50–60 μm) planktotrophic egg; but it is limited between California and Mexico in the west Pacific coast for *C. corona* spawning 75 μm egg, to arctics alone for *C. setosa* releasing 120–160 μm egg, and to the east and west coasts of North America around the equator for *Aphelochaeta monilaris* spawning 275–300 μm egg (see Blake, 1996).

TABLE 1.4

Number of freshwater hirudinean genera and species present in major geographical zones (compiled from Sket and Trontelj, 2008)

Zone	Genera (no.)	Species (no.)
Palaearchtic	45	187
Nearchtic	24	79
Neotropric	27	107
Afrotropic	22	50
Indotropic	27	64
Australisian	15	34

Polychaetes dominate the benthic macrofauna (Gremare et al., 1998). From their long- (20 years) term study on polychaetes at two sites in the western English Channel, Quiroz-Martinez et al. (2012) have analyzed the dynamics of abundance as function of species diversity. With increasing abundance up to 8,000 individuals, species diversity is also increased from ~ 40 to ~ 60 in a site but ~ 25 to ~ 35 species in another site.

Understandably, vertical expansion especially into abysmal depth encounters more environmental challenges than horizontal distribution. The maximum depth, at which the soft-bodied annelids have been collected, is in the range of ~ 4,000 m (Table 1.5). This depth is comparable with the external shell-less molluscs like that aplacophorans (3,200–4,000 m) and cephalopods (2,430–4,850 m) (Pandian, 2017). However, the asteroid and ophiuroid echinoderms with internal skeleton penetrate to 6,035 m (Pandian, 2018). Strikingly, the shelled molluscs are capable of expanding up to 6,370 m (bivalves) and 9,050 m (prosobranchs) (Pandian, 2016). The presence of hard shell(s) enables these shelled molluscs to expand up to the greatest depths.

The abysmal depths, from which annelids have been collected, ranges from 1,500 m for the orbiniid *Phylo nudus* to 4,016 m for the cirratulid *Chaetozone gracilis* for polychaetes and 700 m for the naidid *Nais abissalis* to 1,680 m for four taxa of Tubicinae in Lake Baikal and to 4,900 m depth in the Indian Ocean for the tubificid *Abyssidrilus stilus* (Table 1.5). Describing the pattern and scale of biogeographic variability of abyss in the North Atlantic Ocean, Smith et al. (2006) have reported that the diversity of polychaete species declines perceptibly below 3,000 m depth. Bathymetric distribution of polychaetes seems to be determined by the availability of dissolved oxygen. For example, the Black Sea becomes anoxic and is contaminated with hydrogen sulfide below 150–200 m depth. The number of polychaetes species present in the sea progressively decreases from ~ 100 between 0 and 10 m to 0 below 150 m depth (Sahin and Cinar, 2012). Obviously, the polychaetes in the Black Sea are unable to switch over to anaerobiosis and colonize the anoxic depths. However, Hartman (1966, 1967) has indicated that Antarctic collections have included 23 polychaete species belonging to 14 genera from the depths of 4,930–4,963 in the South Sandwich trench. In freshwater, the naidids occur up to 50 m depth but rarely *Nais abissalis* has been collected from 700 m depth in Lake Baikal. Typically, they feed on phytoplankton (Brinkhurst and Gelder, 1991). With increasing depth, they do not find adequate phytoplankton below 100 m depth (Bondarenko et al., 1996). However, the other oligochaetes, feeding on detritus from sediments, represented by enchytraeids (1,600 m), haplotaxids, lumbriculids and tubificids have been collected from the depth of 1,680 m in Lake Baikal but with perceptible decline from ~ 1,700 no./m^2 at ~ 50 m depth to < 10 no./m^2 at 1,680 m depth (Martin et al., 1999). In contrast to the Black Sea, both water column and sediments of Lake Baikal is well oxygenated at all depths (Martin et al., 1999). Considering, hydrostatic pressure, light, temperature and food

TABLE 1.5

Vertical distribution of annelids (from Erseus, 1994*, Blake, 1996, Dean, 1995†, Martin et al., 1999**)

Species	Depth (m)	Location
Polychaeta		
Orbinidae (14–15 species)		
Califa calida	470–2,000	Off California
Phylo nudus	400–1,500	California
Paraonidae (85 species)		
Arcidea monicae	200–300	Mediterranean
	590–1,745	California
A. wassi	80–1,480	California
A. ramosa	10–2,000	California
	44–2,400	Western Pacific, Japan
A. simplex	100–3,000	California
Apisthobranchidae (5 species)		
Apisthobranchus	3,000	
Spionidae (90 species)		
Spiophanes anoculata	463–2,400	California
S. kroeyeri	3,500	Australia-Antartic
Poecilochaetidae		
Poecilochaetus johnsoni	90–189	California-Mexico
Chaetopteridae (30 species)		
Phyllochaetopterus limicolus	119–3,000	California
Cirratulidae (46 species)		
Chaetozone spinosa	280	Japan
	2,623–2,955	California
C. gracilis	4,016	Catalina Island
Tharyx kirkegaardi	1,260–2,400	California
	255–3,000	Atlantic
Cossuridae (16–17 species)		
Cossura pygodactylata	1–2,720	Western France
C. candida	11–2,400	Mexico, Baja California
C. modica	985–2,955	Oregon-California
C. brunnea	1,600–2,200	North Carolina-Mexico
C. rostrata	6–3,348	Oregon-Western Mexico
Oligochaeta		
Nais abissalis (Naichidae)	700	Lake Baikal**
Propappus glandulosus (Enchytraeidae)	1,600	Lake Baikal**
Haplotaxis sp (Haplotaxidae)	1,680	Lake Baikal**
Stylodrilus asiaticus (Lumbriculidae)	1,680	Lake Baikal**
Balkaiodrilus maievici (Tubificidae)	1,680	Lake Baikal**
Abyssidrilus stilus (Tubificidae)	4,900	Indian Ocean*
Ctenodrilidae		
Raricirrus variabilis	4,000	Virgin Islands†

availability, Martin et al. (1999) have found that other than oxygen level, food availability is the most dominant factor that determines the bathymetric distribution of the haplotaxids. Incidentally, the cocoons of *Tubifex tubifex* and *Potamothrix hammoniensis* deposited at the sediment surface are all eaten by the fish *Abramis brama* but > 99% of them survive, when deposited at 20 mm depth (Newrkla and Mutayoba, 1987). There are adequate indications that in the deep anoxic sediments, tubificids may become anaerobic (e.g. Narita, 2006). With abundance of food at 4,900 m depth, *A. stilus* may have switched to anaerobiosis. Interestingly, the highest altitude, at which the enchytraeid *Buchholzia appendiculata* has been collected, is ~ 2,000 m above sea level (asl) in the montane regions of South America (Schmelz et al., 2013). Amazingly, the oligochaetes have a range of vertical distribution of 6,900 m, i.e. from the depth of 4,900 m to an altitude of 2,000 m.

In the absence of moisture/water, the oligochaetan earthworms are unable to penetrate below 80 cm depth (e.g. *Glossodrilus*, Jimenez and Decaens, 2000). Soil moisture potentials, measured in –kPa unit, reduce the optimum for growth and reproduction at –2 kPa to almost 0 at –50 kPa (Johnston et al., 2014). In the presence of water at the peculiar subterranean habitats, the stigobiont oligochaetes are known to flourish (Chatelliers et al., 2009). With their elongated, segmented and flexible body shape, these stigobiont oligochaetes are pre-adapted to the subterranean habitats and do not exhibit troglomorphic features like the absence of body pigments and eyes, elongated appendages and increased sensory structures. The stigobiont oligochaetes belonging to 42 genera in 17 families are reported from the subterranean aquatic habitats. Of them, the number of species belonging to the lumbriculid *Trichodrilus* and tubificid *Rhyacodrilus* accounts for 23 and 11%, respectively (see Chatteliers et al., 2009).

In his informative taxonomic description of Californian polychaetes, Blake (1996) has provided useful data on body length and depth, at which orbiniids, paranoids, spionids, cirratulids and others have been collected. When data on body length are plotted against depth, different trends become apparent (Fig. 1.4), clearly indicating that body size may not be a factor for polychaetes to penetrate into greater depths. Unlike oligochaetes, not all polychaetes switch over to anaerobiosis at hypoxic/anoxic depths (e.g. Black Sea, Sahin and Cinar, 2012). Hence, it is likely that the oxygen levels of water and sediments as well as respiratory structures may prove to be important factors in bathymetric distribution of polychaetes.

Endemism: Depending on limited powers of motility and larval dispersal being at the mercy of waves and currents, many annelids are endemic. Not surprisingly, of 155 oligochaete species, 114 are endemic to the truly ancient and long-lived Lake Baikal (Martin et al., 1998). Data on the distribution of stigobiont oligochaetes in the subterranean habitat suggest pronounced endemism. More than 60% species are known only from the type locality (see Chatelliers et al., 2009). Likewise, hydrothermal vent-inhabiting

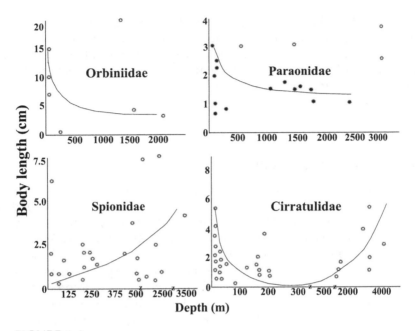

FIGURE 1.4

Vertical distribution of polychaetes belonging to four selected sedentarian families (drawn from data reported by Blake, 1996).

vestimentiferan polychaetes are also endemic. Though polychaetes are continuously occupying a wide vertical depth range, for example *Cossura candida* from 11 to 2,400 m depth off California, Mexican waters (Table 1.5), many of them are reported as endemic. From 19 expeditions carried out during the last 124 years for polychaete taxonomic research around the southernmost tip of the South American continental shelf, Martin et al. (2005) have recorded 431 species belonging to 108 genera and 41 families. Subregions on the Pacific and Atlantic sides are characterized by the presence of 10% endemic species. Investigation on the geographic distribution of the endemic 178 polychaete species between 18° and 56° S of South America has revealed the marked peak endemic hotspot between 36° and 41° S, corresponding to a peak in species richness (Moreno et al., 2006).

Symbiosis: Many annelids symbiotically inhabit as (e.g. *Dipolydora commensalis* on hermit crab within the shell, Lindsay and Woodin, 1993) or as endosymbiont in another polychaete, e.g. *Veneriserva pygoclava meridionalis* in aphroditid host *Laetmonica product* (Micaletto et al., 2002) and within the canals of aquiferous systems of sponges. For example, 33 syllid species constituting > 9% of all the known symbiotic polychaetes are reported to inhabit the canal system of sponges. As many as 600 *Haplosyllis spongicola* happily inhabit within a small (16.5 cm³) *Ciona* sp but without disturbance to the host. *H. spongicola* can inhabit in 36 species of host sponges (Lopez et al., 2001). The earthworm *Eudrilus eugeniae* symbiotically engage *Bacillus endophyticus* to draw riboflavin, an essential nutrient for regeneration (Samuel et al., 2012, Subramanian et al., 2017).

1.4 Gutless Oligochaetes

Since the discovery of a couple of gutless tubificids in coralline sands of Bermuda (Giere, 1979), over hundred species belonging to two phallodrilan genera *Olavius* and *Inanidrilus* (= *Phallodrilus*) with no digestive system and excretory organs have been described (Erseus, 2003). In *I. leukodermatus*, the integument surface of the long, slender worm is much annulated and highly folded into numerous tiny irregular ridges. These expansions increase the worm body surface nearly 10 times (Giere, 1981). Unlike in other annelids, the epidermal layer is unusually thick with extensions crossing the cuticle (microvilli) and ending in epicuticular projections. Consequently, a wide cuticle-epidermal interface is present. All the studied phallodrilan 30 species are reported to incorporate a fairly thick layer of extracellular bacteria beneath the cuticle. Only in two species *Olavius algarvensis* and *O. ilvae* from the Island Elba, Italy, the microbes are enclosed by the epidermal cells and thus attain an intra-cellular position (Giere and Erseus, 2002). In these gutless tubificids, the symbiotic microbes belonging to the following phytotypes are present: 1. Large, oval-shaped γ *Proteobacteria*, which are shown to be sulfide oxidizers (Dubilier et al., 1995). 2. Small rod shaped α *Proteobacteria*, which reduce sulfate into sulfide. This unique 'cyclotrophism' by these symbiotic microbes enables the gutless tubificids to thrive not only in sulfide/oxic interfaces but also in oxic layers (Giere et al., 1991). 3. Some phytotypes, including *Spirochaeta* also occur; however, their symbiotic function is not known (Giere, 2006). Yet, the quantum of symbiotic microbes harbored in relatively smaller area of these gutless tubificid oligochaetes is too small, in comparison to the massive trophosome filled with prokaryotic symbionts in the hydrothermal vent-inhabiting vestimentiferan siboglinid polychaetes. Hence, the overall contribution by the symbiotic microbes is ranked low (Giere et al., 1984).

Interestingly, these gutless tubificids draw nutrition through (i) osmotrophism and (ii) symbiosis. With relatively larger body surface area to volume ratio and thin cuticle, the oligochaetes are well adapted to uptake nutrients from ambient water across the body wall. Absorption of many amino acids and glucose by osmotrophic fauna may proceed from extremely low concentration against concentration gradients of 4–6 orders magnitude; on accumulation, these organic substances are metabolically used (see Pandian,

1975). Southward and Southward (1980) have estimated that the gutless worms can absorb glucose to meet 30% of the metabolic needs. The pore water within the sediments, from which the gutless tubificid *I. leukodermatus* have been collected, contains an extremely high concentration of glucose and fairly high levels of amino acids. *I. leukodermatus* is able to absorb mainly hexoses (but not pentoses) at the rate of ~ 150 µg glucose/h. Among amino acids, aspartate is preferably absorbed in comparison to glutamate (Giere et al., 1984). As in hydrothermal vent inhabiting vestimentiferans, the tubificids display substantial activity of ribulose-1,5-biphosphate-carboxylase, a marker enzyme of the Calvin-Bensen cycle, known to be present only in bacteria and plants. The high levels of ATP-sulfurylase and sulfide oxidases indicate that enzymatic sulfur metabolism is carried by the symbiotic bacteria. These enzyme studies suggest that the gutless tubificids are able to draw ATP through the symbiotic microbes and oxidize sulfide for the fixation of inorganic CO_2 from ambient water. Hence, the gutless tubificids like *I. leukodermatus* can also thrive in sediments containing high concentrations of hydrogen sulfide. The preferred zone of *I. leukodermatus* lies between 5 and 7 cm depth, which correspond to +50 and –50 mV, i.e. in and around the redox discontinuity (RPD) layer (Fig. 1.5), where extremely high concentrations of sugars and amino acids are present in the pore water within the sediments. The worm keeps migrating between the upper oxic and lower anoxic depths. At the lower anoxic depth, it acquires and accumulates reduced sulfur but the necessary binding of oxygen occurs at the upper micro-oxic sediment layers (Giere et al., 1991). Strikingly, these gutless tubificids have successfully conquered and colonized an ecological niche, so far unoccupied by any other interstitial fauna (Giere et al., 1984).

In these gutless tubificids, the location of the microbial symbionts and complicated mode of reproduction suggests vertical transmission. In

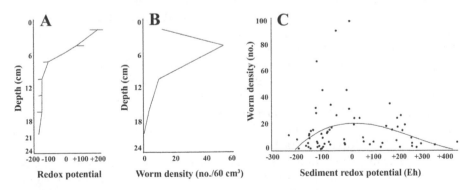

FIGURE 1.5

Vertical distribution of the gutless oligochaete *Inanidrilus leukodermatus* as functions of A. redox potential and B. worm density, and C. worm density as function of sediment redox potential (compiled, modified and redrawn from Giere et al., 1991).

fact, studies on *I. leukodermatus* from Bermuda have revealed the vertical transformation of oval sulfide oxidizing γ-*Proteobacteria*, infecting the sticky eggs laid singly on the sediment grains and not deposited in cocoons, as in other oligochaetes. However, the differences in the bacterial association among the populations of *O. algarvensis* indicate the horizontal transmission, i.e. environmental acquisition of the symbionts. Among the vent-inhabiting bivalves, vertical transovarian transmission occurs in the vesicomyid clams but horizontal environmental acquisition in mytilid mussels (see Pandian, 2017). It is likely the vertical transmission occurs in *Inanidrilus* species but horizontal one in species belonging to *Olavius*.

1.5 Hydrothermal Vents and Cold Seeps

Deep sea hydrothermal vent habitats are characterized by temperatures higher than ambient for deep sea, lower oxygen concentration, higher levels of toxic sulfide compounds, iron, and other metals and gases (Chevaldonne et al., 1997). Mixing of the hot water arising from the volcanic spring with cold ambient water creates biologically habitable regimes (Nyholm et al., 2008). Typically, the hydrothermal vents are linear, discontinuous and shift along the ridges and thereby generate short-lived habitations. In the Pacific, the vents are distributed between 48° N and 22° S (Lutz, 1988, however see also Van Dover, 1994). The startling discovery of hydrothermal vents in 1977 heralded the description of dense benthic fauna in the vents. In these benthic communities, the sole base of the food chain is the chemoautotrophic production by microbial symbionts using hydrogen sulfide as geothermal energy source (see Cavanaugh et al., 1981). The chemoautotrophic symbionts harness geothermal energy through oxidation of reduced chemical substances, fix inorganic compounds, and use them for biosynthesis and growth, although it is not yet known how the vestimentiferans acquire nitrogen for protein synthesis. Indeed, hydrothermal vents are now known to support extensive but endemic biological communities (Nyholm et al., 2008) that are found world over at marine hot springs distributed along the mid-ocean system (Hurtado et al., 2004). Expectedly, island-like patchy endemic populations are distributed intermittently along tectonically or volcanically active spreading vent segments, which are separated by tens to hundreds of kilometers (Hurtado et al., 2004).

In these vents, the faunal assemblage includes a few crustacean species (e.g. galatheids: *Munidopsis subsquamosa*, *M. lentigo*; brachyuran *Bythograea thermydron*; caridean shrimp *Alvinocaris lusca*), the bivalve molluscs belonging to Vesticomyidae, and Bathymodiolinae, a single clade within the family Mytilidae (see Pandian, 2017) and ~ 32 polychaete species (Lutz, 1988). Perhaps, the vestimentiferan tubeworms are the first to be discovered. Though initially grouped with Pogonophora, phylogenetic and embryological evidences have convincingly shown that these gutless deep sea tubeworms are siboglinid polychaetes (McHugh, 1997, Marsh et al., 2001, Rouse et al., 2008). Within the family Siboglinidae, three subfamilies are included: Vestimentifera, Monilifera (a small group of gutless worms living on submerged wood) and Osedacinae (with *Osedax* as the only genus). The subfamiliy Vestimentifera is again subdivided into three infra-families: Lamellibrachiinae, Escarpiinae and Tevniinae (Karaseva et al., 2016). The tevniinids inhabit exclusively on the rocky substrates of hydrothermal vents of the Pacific Ocean. The lamellibrachiinids and escarpiinids occupy both soft and rocky substrates of cold seeps (see Kobayashi et al., 2015) and the periphery of hydrothermal vents (Karaseva et al., 2016). Descriptions for 19 valid vestimentiferan species and their bathymetric distribution (Fig. 1.6) are available. Notably, their distribution ranges from > 100 m to 3,500 m depth,

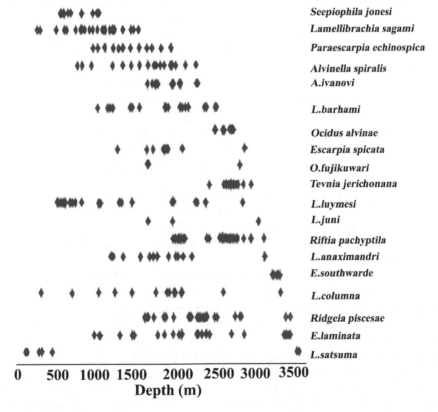

FIGURE 1.6

Bathymetric distribution of vestimentiferan tubeworms in hydrothermal vents (from Karaseva et al., 2016, modified).

limited between 600 m and 1,000 m depth for *Seephiophila jonesi* but widely scattered for *Lamellibrachia* spp.

Like the moniliferans, the vestimentiferan tubeworms also lack a functional digestive system. They derive nutrition from chemoautotrophic microbial symbionts harbored in the trophosome located on their elongated trunk. Not surprisingly, > 50% of the total body length (1.5 m) of the giant tubeworm *Riftia pachyptila* is occupied by the most extensive trophosome and the gonad (Cavanaugh et al., 1981). Scanning and transmission microscopic studies have confirmed the presence of prokaryote in the trophosome. Direct count has shown the presence of > 3.7 × 10⁹ microbes/g live trophosome tissue. In 21 of 31 specimens of *Riftia* examined, crystals (100 μm) of elemental sulfur have been found within the trophosomal tissue. The presence of sulfur crystals within the tissue suggests that the prokaryotic microbes are able to chemotrophically oxidize sulfide (Cavanaugh et al., 1981). Felbeck (1981) has reported that high levels of the enzymes thiosulfate transferase, APS reductase and ATP sulfurylase are involved in generation of ATP through sulfide oxidation. Further, the extended vascular system within the trophosome (Jones, 1981) and the relative insensitivity of the blood oxygen-carrying capacity to changes in levels of temperature and CO_2 suggests the presence of special adaptation in this worm to ensure sustained supply of O_2 and CO_2 to the trophosomal symbionts (Arp and Childress, 1981). Structures similar to the symbionts have also been described within the trophosome of other vestimentiferans like *Lamellibrachia luymesi* and *L. barhami*. Hence, it is likely that the vestimentiferan tubeworms draw energy through the same chemoautotrophic pathway. Nyholm et al. (2008) have reported the metabolite flux and transcriptomic data for *Redgeia piscesae* and indicated that the host sustains substrate availability, which potentially regulates the host's transcriptome or symbiont's cell cycle. The growth rate of *R. piscesae* ranges widely from 0 to 25 cm/y and is highly variable between individuals within the same aggregate and between different vent sites. Consequently, the minimum age of the aggregate also ranges from 10 to 30 years (Urcuyo et al., 2007).

Considering the similarity of faunal assemblages in these geographically widely separated hydrothermal vents, and sessile/sedentary nature of the adult worms, dispersal can occur only through the larval stage. In this context, the following observations are relevant: (i) *R. pachyptila* produces neutrally buoyant small (~ 100 μm) but lecithotrophic eggs. Its larvae can persist in the water column for ~ 34 days (Marsh et al., 2001), during which they may disperse over 100 km. (ii) Adults of *Alvinella prompejana* can exit from their tubes and swim vigorously. They release larger (~ 200 μm) negatively buoyant lecithotrophic eggs. The eggs may drift with bottom currents and are characterized by intermittently arrested embryonic development, which extends the dispersal duration (Pradillon et al., 2001). (iii) Megaplumes generated by volcanic eruption rises as much as 1,000 m above the axial walls and can potentially transport the buoyant larvae and juveniles across vast distances (Mullineaux et al., 1995). However, the juveniles may have to withstand the non-vent conditions for a certain period (see Chevaldonne et al., 1997). Hurtado et al. (2004) have studied the dispersal on a few polychaete vent species across ~ 7,000 km of the East Pacific Rise (EPR) and Galapagos Rift (GAR). The dispersing vent larvae have to overcome the following filters/barriers: (i) East Microplate Region (EMR), and (ii) that separating EPR and GAR populations, (iii) equator separating northern and southern EPR populations and (iv) Rivera Fracture Zone (RFZ). These filters are shown to work on different time scale and to different degrees among the examined vent taxons. The equatorial region exhibits combinations of deep oceanic currents and topographical features that limit faunal exchange between EPR and GAR and across the equator along the EPR.

In the vestimentiferan, the mode of development, dispersal and genetic exchange have been repeatedly discussed. Of ~ 19 vestimentiferans, spermiogenesis has been described only for five species: *Riftia pachyptila*, *L. luymesi*, *L. barhami*, *Paraescarpia echinospica* and *Ridgeia piscesae* (see Marotta et al., 2005). Sperms are released as bundles (e.g. *Tevnia jerichonana*) and spermatozeugma (e.g. *R. pachyptila*). The sperm bundles eventually disintegrate in sea water and the freed spermatozoa are capable of swimming. Sperm masses adhering to female's body (*R. piscesae*) or in the spermatheca (*T. jerichonana*) have been reported. Direct sperm transfer from males to females has been observed in *R. piscesae*. However, brooding of embryos has never been observed in any vestimentiferan. Although apparent spawning has been observed in *R. pachytptila* (Van Dover, 1994) and *L. luymesi* (Hilario et al., 2005), it is not known whether the 'spawn' consists of unfertilized eggs/ zygotes or developing embryo. Presumably, the eggs are fertilized either in the ovisac just before spawning or externally, as eggs are released.

According to Hurtado et al. (2004), the annelids in the Pacific hydrothermal vents adopt different strategies for long distance dispersal. For example, *Alvinella prompejana* spawns large negatively buoyant eggs (~ 200 μm) that may be drifted by demersal currents. *R. pachytptila* releases neutrally buoyant (~ 100 μm) with adequate resource to survive in the water column for ~ 38 days. *Branchipolynoe symmytilida* produces very heavy large eggs (~ 500 μm) and shows no genetic differentiation over long distance. In fact, low mitochondrial diversity is reported in many vent polychaete species (e.g. *T. jerichonana*) is indeed great.

1.6 Energy Budget

Resource allocation for reproduction may ultimately depend on one or more of the energy transformation steps, namely, food acquisition, digestion and absorption, assimilation, respiration, somatic growth and reproduction (Fig. 1.7). For example, blood ingestion shows a direct relation to reproductive output in *Hirudo medicinalis* (Davies and McLoughlin, 1996). In animals, energy budget is assessed by estimation of $C = F + U + R + P$, where C is the food energy consumed, F, U and R, the energy lost on feces, urine and metabolism, respectively and P, the energy gained due to growth (e.g. Pandian, 1987). In ecological energetics, one or more energetics components have been estimated at population level in many polychaetes (e.g. Dixon, 1976, Banse, 1979) and oligochaetes (e.g. Dash and Patra, 1977, Senapati and Dash, 1983). The physiological energetics is associated with relatively more precise estimates of these components (e.g. Ivleva, 1970, Goerke, 1971). Obviously, energetics becomes important for resource allocation for reproduction and development. For allocation to regeneration, the earthworm *Eudrilus eugeniae* engages symbiotic microbes like *Bacillus endophyticus* to draw riboflavin (Subramanian et al., 2017). However, estimation of each one of the energetics components in annelids is cumbersome and unwieldy for the following reasons: (i) osmotrophism in gutless annelids, (ii) contribution to C by symbiotic microbes, (iii) consumption of egested feces, (iv) anaerobiosis, (v) aestivation/hibernation, (vi) regeneration and (vii) clonal reproduction.

Food (C) and acquisition: Besides the conventional heterotrophic acquisition of food by a vast majority of annelids, some gutless annelids are osmotrophic and/or chemoautotrophic. 1. A vast majority of polychaetes and oligochaetes are microphages and feed on bacteria, fungi and protozoa adhering to sediments, sands and other substrates. 2. Terrestrial oligochaetes especially, the earthworms undergo inactive periods of seasonal aestivation (for definition, see Pandian, 2017) for a few months (e.g. see Dash, 1987). In deep sediments of Lake Baiwa (Japan) with no oxygen during four summer months from July to September, the tubificid *Rhyacordrilus hiemalis* undergoes anaerobic aestivation (Narita, 2006). Polychaetes inhabiting intertidal zone undergo a short or longer period of intermittent tidal aestivation. During these periods, facultative anaerobes suppress aerobic respiration but subsequently switch over to energetically costlier anaerobiosis. Apart from the inaccessibility to food (and thereby reducing C), energy expended on R becomes complicated in these aestivating annelids. 3. Many annelids have the ability to regenerate the lost tissue(s)/organ(s). Some of them can regenerate the fraction of lost anterior or posterior body, i.e. unidirectional cloning. Others can undergo architomic or paratomic asexual reproduction namely multi-directional cloning. Apart from switching the mode of food acquisition, the loss of both feeding palps reduces the feeding duration in a spionid *Pseudopolydora kempi japonica* but not in another spionid *Rhynchospio glutaeus* (Lindsay and Woodin, 1995). In yet another spionid *Pygospio elegans*, C limits asexual reproduction (Wilson, 1985). Hence, regeneration may considerably alter C but reduced C can inhibit cloning.

Consumption (C): Feeding behavior ranges from herbivory (e.g. naidids feeding on phytoplankton) to carnivory (e.g. leeches) and the feed ranges from suspended to deposited food particles. Hence, estimation on C is more difficult, cumbersome and unwieldy. However, new methods to estimate the microbial feed of oligochaetes (Jones and Mollison, 1948) and ingestion rate of polychaetes (Cammen, 1980a) have been developed. In general, ingestion rate is inversely related to body size and nutritional value of the substrate. In oligochaetes, the rate increases from 80 mg/g worm/d fed on dung in *Allolobophora caliginosa* to 7,000 mg/g/d in soil feeding *Millsonia anomala* (see Dash, 1987).

Hirudineans are carnivores and feed on animal preys or suck their blood. Antiserum and precipitation test of the gut contents of *Glossiphonia complanata* consist mostly (85%) molluscs, oligochaetes and chironomids (see Dash, 1987). The gnathobdellid leech *Limnatus nilotica* requires seven–nine or six feedings of blood from frogs or dogs to attain sexual maturity at the age of 17–20 months and body size of 0.5–2.0 g (Negm-Eldin et al., 2013). On a saturated feeding, a leech may consume 4- (e.g. *Limnatus, Johanssonia*), 5- (e.g. *Hirudo*) and 10-times (e.g. *Haemodipsa*) of its own normal body weight. The duration of digestion in *L. nilotica* lasts from 10–20 days to > 8 months, depending on leech body size and frog's or dog's blood. Briefly, feeding in gnathobdellids may be regular once a day or once a few days/months, as in glossiphoniids. Apart from this unwieldy procedure for estimation of C, osmotrophism in polychaetes deter precise estimation of C. An account on chemoautotrophism was provided earlier.

1.6.1 Osmotrophism

In surface waters of the ocean, total Dissolved Organic Matter (DOM) amounts to ~ 1 mg/l. Free amino acids, comprising 5% of DOM, occur at concentrations of 5×10^{-7} M/l in free water and 1.1×10^{-4} M/l in interstitial water of sediments (see Pandian, 1975), in which most of the soft-bodied annelids thrive. Indeed, sea water is an organic 'soup'. Not surprisingly, primary productivity of the osmotrophic, heterotrophic bacteria (0.2–0.5 g C/m²/d) exceeds that (0.2–0.3 g C/m²/d) of autotrophic phytoplankton in the Pacific waters (see Pandian, 1975). As early as in 1909, Putter rightly made the first claim that DOM may also be absorbed across the body surface and used as energy source by animals, as well. In animals, the uptake rates of DOM follow Michael-Menten kinetics, with rates increasing to a threshold as a substrate concentration increases. An argument against Putter's claim is that if DOM can be absorbed through body surface against concentration gradients, the DOM from body fluids can also be leaked through body surface into the ambient sea water (see Pandian, 2017). Ferguson (1971, 1982) has measured both influx and efflux of free amino acids in many invertebrates including polychaetes and found that the net influx of amino acid is overwhelmingly inward, with a single exception of glycine. Siebers (1976) has made a more detailed study on absorption of neutral and basic amino acids across the body surface by *Enchytraeus albidus* and *Nereis diversicolor*. He has found that neutral amino acids (glycine, L-valine) are transported against the gradient in a saturable process but not the basic amino acids like L-arginine and L-lysine. The uptake of amino acids alone contributes up to 25% of the nutrient requirements in *Nais elinguis* (Petersen et al., 1998). Not only sugars (cf Southward and Southward, 1980) and amino acids but also fatty acids are also absorbed by the polychaete *Cirriformia spirobranchia* (Testerman, 1972). Notably, the uptake of DOM by polychaetes is reduced under anaerobic conditions (Jorgensen and Kristensen, 1980b). The uptake of DOM from ambient sea waters by the polychaetes has remained a hot area of research. Between 1963 and 1982 alone, as many as 17 groups of researchers have examined the ability of 22 polychaete species to assess the net influx of DOM into the worm's body. Most of these authors have concluded that a smaller or large fraction of metabolic requirement is accounted by the net uptake of DOM (Table 1.6). An objection raised by Jorgensen (1976) is that the interstitial amino acid concentrations of 50 µM may facilitate substantial uptake but a negligible one in the overlying surface water with a concentration of 0.5 µM. Secondly, a few polychaetes that have been investigated are burrowers (e.g. *Glycera americana*, Ferguson, 1982) or tube-dwellers (e.g. *Lanice conchilega*). The DOM content of water in the burrow and tube may vastly differ from that of interstitial water, as water in the burrow/tube is being continually exchanged by ventilation. The third but a more serious objection is the failure of some authors to establish relevant to environmental concentration for the target substrates (e.g. Siebers, 1982). However, these objections may not hold water for following reasons: 1. Of ~ 21 group of authors, 11 groups have confirmed the net uptake. 2. There are > 100 gutless tubificids, which mostly depend on osmotrophism alone.

TABLE 1.6

Polychaete species that are reported to uptake DOM

Net uptake of DOM measured

1. *Capitella capitata* (Stephens, 1975), 2. *Cirratulus hedgepethi*, 3. *Diopatra cuprea* and 4. *Glycera americana* (Ferguson, 1982), 5. *G. dibranchiata* (Stevens and Preston, 1980, Preston and Stevens, 1982), 6. *Marphysa sanguinea* and 7. *Pareurythoe californica* (Costopulos et al., 1979), 8. *Nereis diversicolor* (Ahearn and Gomme, 1975, Stephens, 1975), 9. *N. succinea* (Jorgensen and Kristensen, 1980a, b), 10. *N. virens* (Jorgensen, 1979, Jorgensen and Kristensen, 1980a, b), 11. *Stauronereis rudolphi* (Testerman, 1972)

Uptake of DOM alone measured

1. *Clymenella torquata* (Stephens, 1963), 2. *Eunereis longissima*, 3. *Goniada* sp and 4. *Nephtys* sp (Southward and Southward, 1972), 5. *Lanice conchilega* (Ernst and Goerke, 1969), 6. *Lumbrinereis* spp, 7. *Nainereis dendritica* and 8. *Podarke pugettensis* (Testerman, 1972), 9. *Neanthes arenaceodentata* (Reish and Stephens, 1969), 10. *Nereis limnicola* (Stephens, 1975)

Feces (F): Polychaetes have the ability to digest a wide variety of organic substances (Cammen, 1987). They extract the required carbon equally from both the carbon-rich microbes and carbon-poor detritus by increasing the ingestion rate of the latter by ~ 30 times (e.g. *Nereis succinea*, Cammen, 1980b). Similarly, the earthworms also increase the ingestion rate to acquire adequate nutrients. The Gut Load (GL), the amount of (dry) substrates ingested as a percentage of body weight, varies with nutrient quality; for example, the GL increases from 100% of moist loam soil to 600% of soil with cellulose and to 700–800% of soil. The feces of oligochaetes are voided as worm casts containing semi-digested organic material of organismal origin/urine and soil and thereby enrich the agriculture fields (Dash, 1987). The egestion rate ranges from 100 to 460% of body weight/d for three tropical earthworm species (Dash et al., 1984, Dash et al., 1986). The worm cast production also ranges from 2 to 247 t/ha in different sites and thereby brings up layers of soil between 1 mm and 5 cm (see Dash, 1987). In aquatic oligochaetes, the quantification of F is difficult and methods adopted for the estimation of F have yielded widely different values. For example, in the inverted method developed by Appleby and Brinkhurst (1970), the worms are kept in an

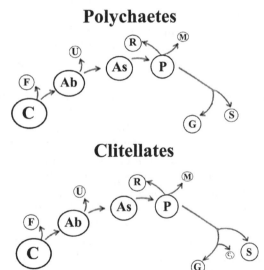

Polychaetes

Clitellates

FIGURE 1.7

Energetic components in annelids. C = Consumption, F = Feces, U = Urine, R = Respiration, M = Mucus, P = Production, S = Somatic growth, G = Gonad growth, C_1 = cocoon, Ab = Absorption, As = Assimilation. Due to dual sexual systems and cocoon production, clitellates invest relatively less on gamete production than the gonochoric polychaetes.

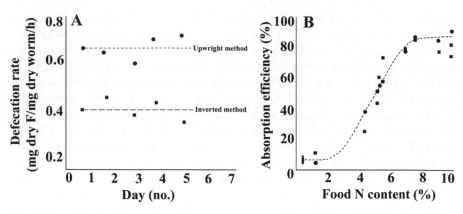

FIGURE 1.8

A. Defecation rate of *Limnodrilus hoffmeisteri* as estimated by upright and inverted methods (drawn using data reported by Kaster et al., 1984). B. Absorption efficiency of polychaetes as function of food nitrogen content. ■ Observed values, ● Predicted values (drawn from protocol record of T.J. Pandian).

inverted (posterior end down, anterior end up) position but in that described by Kaster et al. (1984), they are kept in normal oriented upright position. Figure 1.8A shows that F estimates for the tubificid *Limnodrilus hoffmeisteri* averages to 0.4 mg (dry) F/mg (dry) worm/h in the inverted method but 0.6 mg/g/h in the upright method. The latter value on F is 50% more than that of the former.

In general, the recovery of F of aquatic animals is not only cumbersome but also incomplete for following reasons: (i) loss of soluble material, (ii) loss owing to decomposition, (iii) difficulty in distinguishing food from F especially in deposit-feeding polychaetes (e.g. *Pectinaria gouldii*, Gordon, 1966) and (iv) chances of F being reingested by polychaetes, which switch from suspension- to deposit-feeding (e.g. *Nereis diversicolor*, Harley, 1950). Hence, the need for an easier indirect method to estimate absorption efficiency, Ae (A = C – F = absorption, Ae = A/C × 100) is obvious. Pandian and Marian (1985b) have considered nitrogen (N) content of C as an indicator of Ae. Considering 11 relevant publications, which have reported adequate information of N content of C ranging from 0.4% (sediment, Pollack, 1979) to 9.8% (95% *Mytilus edulis* + 5% agar, Neuhoff, 1979), they have found an almost linear relationship between N content of food and Ae (Fig. 1.8B). The obtained trend includes Ae values reported for *Arenicola marina* feeding on sediment containing 0.4% N to *Nereis virens* fed on mussel tissues containing 9% N. The values are also related to polychaetes of different weight classes (1 to 1,156 mg) and exposed to different temperatures (12 to 29°C). The contributions made by variations in Ae by temperature and body size on N-Ae relationship are negligible. Earlier, Tenore (1983) had considered two types of *Spartina alterniflora* detritus containing 0.9 and 1.0% N but

4.45 and 2.61 Kcal/g dry weight, respectively and found that energy content rather than N content of food determines the incorporation of *S. alterniflora* detritus by *Capitella capitata*. However, he has not estimated Ae of *C. capitata* fed on these detritus. Incidentally, the indirect method of N content of food as a marker of Ae holds good for fishes, aquatic insects and reptiles (Pandian and Marian, 1985c, d, e).

Urine (U): In polychaetes, the main excretory products are ammonia, urea and uric acid (Pandian, 1975). The terrestrial oligochaetes excrete ammonia and urea and their urine contains some mucus protein (Bahl, 1947). However, quantitative value for the substance or energy content of U for annelids is not yet available. The only available value seems to be that reported for blood sucking *Haementeria ghilianii*. In this leech, Sawyer et al. (1981) have estimated the ingested blood retained after urine elimination. Considering their consecutive three values, the substance lost on U by the leech accounts for 23.2% of C.

Respiration (R): As in other animals, oxygen uptake ultimately liberates biologically useful energy for body functions in annelids too. The fraction $(C - [F + U] = As)$ of assimilated energy ranges from 55 to 75% in annelids. On an average, ~ 65% of the As is used for metabolism (R) in aquatic polychaetes (Cammen, 1987) and terrestrial oligochaetes (Dash, 1987), leaving ~ 35% of As for body growth. A reason for the high (75%) R value is that some annelids switch over to energetically less efficient anaerobiosis under oligoxic/anoxic conditions.

1.6.2 Anaerobiosis

Many polychaetes utilize mostly glycogen and various amino acids as a major energy source (Hochachka et al., 1973). In many polychaetes, the simultaneous catabolism of glycogen and amino acids maintains the redox balance in the cell during anaerobiosis. The first phase of glucose oxidation in anaerobic glycolysis is through the Embden-Myerhoff-Parnas (EMP)-pathway. In this pathway, each molecule of the 6'-C glucose is degraded to two pyruvate molecules, which are then converted into lactate. In some animals, the duration of anaerobiosis is too short to permit accumulation of lactate without excess reduction in pH of tissue fluids. When a favorable condition returns, a considerable fraction of the accumulated lactate is converted back into pyruvate and oxidized to water and CO_2. This oxidation results in an increased 'oxygen demand' by tissues over normal metabolic requirement and constitutes the so called 'oxygen debt' (*Neoamphitrite figulus*, Theede, 1973). To prevent lethal reduction of the pH of the body fluids, pyruvate may, however, be excreted (e.g. *Nereis* spp, Schottler, 1979) and/or reutilized by gluconeogenesis (e.g. *Nereis pelagica*, Theede, 1973). Not only pyruvate but also other end products are excreted as such. For example, propionate and acetate excreted by the anaerobic *Arenicola marina* is in the range of 18–27 μmol/g live weight/d (Surholt, 1977).

When anaerobic duration is prolonged in the facultative anaerobic and obligate anaerobic worms, the carboxylation of pyruvate yields oxalo-acetate (OXA), which is converted to fumerate and finally to succinate through the malate route. In this new pathway of glucose degradation and phosphoenolpyruvate (PEP) to OXA, succinate (e.g. swamp worm *Alma emini*, Mangum et al., 1975) and alanine are produced as major end products of prolonged anaerobiosis (see Pandian, 1975) and finally to volatile fatty acids (propionate being the major product) and acetate (Schottler et al., 1983). For example, the opheliid *Euzonus mucronata* can survive the periodic exposure to anoxic condition for 18–20 days, producing succinate and propionate as major end products of glycolysis (Ruby and Fox, 1976). Metabolic pathways terminating with succinate and propionate have two advantages over the EMP glycolytic pathway: 1. They produce twice as much ATP/mol substrate and reduce by 50% the accumulation of potential toxins. 2. On restoration from anaerobiosis, succinate and propionate can readily be converted to OXA, which directly leads into the Krebs cycle, while lactate must first be converted to pyruvate and then to OXA before entering the Krebs cycle (Cammen, 1987).

Growth (P): Of ~ 35% of As energy stored in the body for growth, small or larger fractions are used for body growth (S), reproductive output (G) and mucus (M). The serpulids construct a calcareous tube using organic matrix and divert as much as 65% of As for tube construction (e.g. *Mercierella enigmatica*, Dixon, 1980). The sabellids secrete mucus for construction of a sand tube. Filter feeding polychaetes also secrete mucus to capture food particles. Of the total mucus secreted by *M. enigmatica*, 94% is secreted by the food capturing structure, the 'crown' and the remaining 6% used by the body surface and gut, i.e. each 3%. The mucus of *M. enigmatica* consists of mostly mucopolysaccharides (Dixon, 1976). Yet, no estimate has been made on the fraction of As secreted as M in the energy budget of *M. enigmatica*. However, the nitrogenous excretion of earthworms is through mucus. Their urea consists of 64% mucus, 16% urea and 20% ammonia. Dash and Patra (1979) have estimated that the quantum of excretion by the earthworms *Lampito mauritii* and *Ocnerodrilus* is in the order of 142.3 g mucus/m^2/y, amounting to 6.6 g N/m^2/y; evidently, the chemical nature of mucus of the earthworms is different from that of polychaetes.

Reproduction (G): For the terrestrial oligochaetes, assimilated energy allocated for reproduction averages to < 2% (Dash, 1987). Corresponding values are not available for the aquatic oligochaetes and polychaetes. Values on the fraction of P energy allocated for R ranges from 12% in *Harmothoe imbricata* (Gremare and Olive, 1986) to 30–40% in many other polychaetes (Cammen, 1987). A single value reported for the serpulids *M. enigmatica* indicates that 12% of the assimilated energy is allocated for G, the gamete production (Dixon, 1976, 1980). Understandably, the broadcast spawning, gonochoric polychaetes allocate a greater fraction of the assimilated energy for reproduction than the simultaneous hermaphroditic oligochaetes, which ensure a relatively higher survival of their progenies through internal fertilization and safer deposition of developing eggs within the protective cocoon. In fact, the increased cost of developing and maintenance of dual reproductive systems reduces energy allocated for G by 50% in hermaphroditic polychaetes too. For example, the number of eggs spawned by seven polychaete species of *Ophryotrocha* averages to 9.4 eggs/d and 4.45 eggs/d for gonochorics and hermaphrodites, respectively (calculated data from Premoli and Sella, 1995). Apparently, the oligochaetes have chosen to invest a greater fraction of assimilated energy to meet the costs of dual reproductive system and cocoon.

1.7 Life Span and Generation Time

Generation Time (GT) is the duration of time required by an animal from its egg stage to egg releasing stage. The investment on GT as a percentage of Life Span (LS) indicates the remaining fraction of LS for investment on reproduction. Hitherto, annelidan reproductive biologists seem to have not paid adequate attention on GT/LS. As a result, even the available bits and pieces of information are widely scattered. Incidentally, there is time lag between sexual maturity and oviposition/spawning. For example, the lag between the appearance of clitellum, a morphological marker of sexual maturity in clitellates, and oviposition ranges from 6 days in *Perionyx excavatus* to 19 days in *Metaphire houlleti* (Joshi and Dabral, 2008).

Timm (1984) is perhaps the first to consolidate the relevant information on potential age for 36 aquatic oligochaete species. From Table 1.7, the following may be noted: 1. The 20 criodrilid, lumbriculid, naidid and tubificid species cultured at prevailing laboratory temperature underwent sexual reproduction only (cf Table 4.1). The estimated maximum age was 7.0, 9.5, > 11 and 14 years for the lumbriculid *Stylodrilus heringianus*, tubificid *Tubifex tubifex*, naidid *Spirosperma ferox* and criodrilid *Criodrilus lacuum*, respectively. 2. In them, the maximum age was decreased in the following order: criodrilid < naidid < tubificids < lumbriculid. 3. The age was consistently shorter for these worms (except in *Potamothrix moldaviensis*) from the fields, where they underwent sexual and/or asexual reproduction. However, no description was given as to how the age of these worms, especially in the fragmenting ones was determined. 4. The maximum age was also decreased in all the species, when reared at 20–25°C. In asexually reproducing naidids *Stylaria lacustris*, the doubling duration decreased from 11.1 days at 10°C to 5 d at 20–25°C (Schierwater and Hauenschild, 1990). Further, the annual potential

TABLE 1.7

Potential life span (y) of representative aquatic oligochaetes. In laboratory, reproduction was exclusively sexual. In the fields, seasonal temperature ranged from 5°C during winter to 15°C during summer (from Timm, 1984, modified)

Species	Field	Laboratory	At 20–25°C
Criodrilidae			
Criodrilus lacuum	–	> 14	–
Naididae			
Spirosperma ferox	> 12	> 11	8
Psammoryctides barbatus	> 12	> 8	~ 4
Ilyodrilus templetoni	8	5	3
Potamothrix hammoniensis	~ 4.5	~ 3.5	3.5
P. moldaviensis	~ 3.5	~ 4.5	2.5
Lumbriculidae			
Stylodrilus heringianus	> 12	7	3.5
Rhynchelmis limosella	> 6.5	3	–
Tubificidae			
*Tubifex tubifex**	> 11	9.5	3.8
*T. tubifex***	8.5	6.3	4.3
*T. tubifex****	6.5	5.5	4.2

* from limnophilus, Roosna Alliku spring, **from reophilus, *** from limnophilus, Lake Peipsi-Pihva

number of generations of *S. lacutris* decreased from 30 with unlimited food supply to 20, 14 and 12, when food supply was limited to 50, 20 and 10%, respectively. These values were 42, 24, 13 and 9 for *Nais* sp and 66, 39, 23 and 18 in *Chaetogaster diastrophus* (Lohlein, 1999). Consequent to the differences in temperature and food supply in the fields, the mean maximum age of these worms was ~ 4 years only.

Yet the LS of annelids in the natural fields, where they are subjected to an intense predation, ranges between a few days and months. Table 1.8 summarizes available information on LS and GT of representative hirudineans, earthworms, terrestrial and aquatic oligochaetes and polychaetes. It includes information for gonochoric (*Dinophilus gyrociliatus*) and hermaphroditic species as well as sexually reproducing (e.g. leeches) and/ or asexually reproducing oligochaetes and polychaetes. The following may be noted: 1. Adequate information on gonochoric polychaete is not yet available. 2. There are fast and slow growing enchytraeids (e.g. *Marionina clevata*, Springett, 1970) and tubificids characterized by short (semelparous) and long (iteroparous) (e.g. *Tubifex tubifex, Limnodrilus hoffmeisteri*, Poddubnaya, 1984) LS. Despite these constraints, the data assembled in Table 1.8 permit for the first time the following new findings: The percentage investment on GT is decreased from 40–46% in large (10–23 g) sanguivorous leeches to ~ 30% in

TABLE 1.8

Realized life span (LS) (d) and generation time (GT) of some annelids

Species	Size	LS	GT	GT/LS (%)	Remarks	Reference
Hirudineans						
Johanssonia arctica	10 g	955	446	46	♀, Boreal 1–2°C	Khan (1982)
Hirudo medicinalis	23 g	713	280	40	♂	Davies and McLoughlin (1996)
Terrestrial oligochaetes						
Aporrectodea trapezoides	2.8 g	490	142	29	Singleton, earthworm, Spain	Fernandez et al. (2010)
		490	165	34	Twins, earthworm, Spain	
Hyperiodrilus africanus		196	56	22	♂	Tondoh (1998)
Enchytraeus variatus		254	~ 26	10		Bouguenec and Giani (1989)
Polychaetes						
Ophyrotrocha socialis	3 mm	> 1098	> 60	5	Boreal at 10°C, ♂	Ocklemann and Akesson (1990)
O. adherens	5 mm	196	30	15	♂	Paavo et al. (2000)
O. diadema YY, YW		350	~ 35	10	♂	Akesson (1982)
WW		245	~ 35	14	♂	
Dinophilus gyrociliatus	< 1 cm	70	11	15	Gonochoric	Akesson and Costlow (1991)
Streblospio benedicti		38	9.5	25	♀, ♂, planktotrophic	Levin and Bridges (1994)
		43	13.5	32	♀, ♂, lecithotrophic	Levin and Bridges (1994)
Capitella capitata	~ 8 mm	420	195	47	♀, ♂, sexual, Argentina	Martin and Bastida (2006)
Neanthes limnicola		548	171	31	♀, ♂, viviparous, California	Fong and Pearse (1992)

Table 1.8 contd. ....

...*Table 1.8 contd.*

Species	Size	LS	GT	GT/LS (%)	Remarks	Reference
				Aquatic oligochaetes		
Lumbricillus rivalis		60	20	15	♂♀	(see Lindegaard et al., 1994)
Branchiura sowerbyi	< 2 cm	364	35	10	♀, Naidid	Lobo and Alves (2011)
Enchytraeus variatus		62	12	19	♂♀	Bouguenec and Giani (1989)
Tubifex costatus	~ 2.3 mg	200	60	20	♂, Marine, sexual + asexual	Brinkhurst (1964)
T. tubifex (Italy)	< 25 mg	200	60	30	♀, Sexual/Parthenogenic	Pasteris et al. (1996)
T. tubifex (Boreal)		95	52	54	♀, Sexual/Parthenogenic	Poddubnaya (1984)
		380	72	19	♀, Sexual/Parthenogenic	
Limnodrilus hoffmeisteri		150	93	62	♀, Sexual/Parthenogenic	
		349	90	26	♀, Sexual/Parthenogenic	

medium sized (2.8 g) sediment/detritus-feeding earthworm, to ~ 19–30% in tubificids and to 5–15% in small (2–5 mm body length) terrestrial oligochaetes and hermaphroditic polychaetes. However, it is not clear whether the GT is determined by other factors like (i) LS (from a few days in aquatic oligochaetes to a few years in leeches), (ii) feeding habits (detritivore, herbivore [naidids] and sanguivore) or (iii) egg size (ranging from a few μm to mm, p 73). One or more of the above cited factors may determine GT. In this area, researches are required. Of course, GT may be regulated by juvenile hormone (JH) and the like in crustaceans (Pandian, 2016).

1.8 Gametogenesis

In polychaetes, for example in temperate *Kefersteinia cirrata*, oogenesis commences with the accumulation of primary oocytes and proceeds through prophase to deplotene. At this stage, the progress of meiosis is arrested and oocyte nucleus expands to form the germinal vesicle. After a period, vitellogenesis begins usually during autumn-winter months— as in many temperate echinoids (see Pandian, 2018). Oogenesis as well as spermatogenesis and spermiogenesis are completed between December and February and the worm is ready for spawning by spring (Olive and Pillai, 1983).

Oogenesis: In polychaetes, two patterns of oogenesis have been described (Eckelbarger, 1983, 1986). In the intra-ovarian pattern, almost the entire process of oogenesis including vitellogenesis occurs within the fairly large and structurally complicated ovary. The maturing oocytes receive yolk precursors from (i) closely associated nurse cells (abortive or sibling oocytes), (ii) follicle cells, (iii) closely associated circulatory system or (iv) a combination of these sources. In about eight families (e.g. Capitellidae, Sabellariidae, Orbiniidae, Eckelbarger, 2005), in which the intra-ovarian oogenesis is reported to occur. There are significant differences regarding the type of precursors utilized during vitellogenesis, their metabolic pathways and chemical nature of yolk bodies themselves. Besides, there is no apparent relationship between the ovarian type and oogenic mode, egg size or larval development mode.

In the extra-ovarian pattern, the ovary is small, structurally simple and transient in nature. The small pre-vitellogenic oocytes are released either solitarily (e.g. Sabellidae, Serpulidae, Oweniidae, Glyceridae) or in clusters into the fluid-filled coelom; they may also be released as follicles into the coelom (Alciopidae, Nereididae, Phyllodocidae, Terebellidae, Cirratulidae, Ampharetidae, Pectinariidae). The clusters may subsequently be separated, as in Sphaerodoridae and Pholoidea or rarely remain intact in some syllids, tomopterids and onuphids until vitellogenesis is completed. The oocytes may (i) develop without an association with accessory cells, (ii) be accompanied by [a] coelomocytes specialized for triglyceride synthesis or [b] eleocytes involved in yolk precursor production, or (iii) be closely associated with nurse cells playing nutritive role (Eckelbarger, 2006). In *Nereis virens*, for example, the eleocytes extra-ovarially synthesize yolk protein precursors, the vitellogenins and subsequently transport into the oocytes by receptor-mediated endocytosis (Hafer et al., 1992).

Spermato- and Spermio-geneses: Typically, gonial cells, produced from the testes, are dropped into the coelomic cavities (seminal vesicles), where they undergo a series of variable but species specific number of mitosis prior to entry into meiosis. In oligochaetes, spermiogenesis involves (i) nuclear shaping, (ii) condensation of chromatin, (iii) formation of acrosome, (iv) reduction in the number of mitochondria and (v) development of a long flagellum (Ferraguti, 1984). A typical sperm consists of (a) an apical acrosome with an acrosomal tube in clitellates but not in other annelids, (b) a condensed thin nucleus, (c) two (e.g. *Tubifex*) to eight (e.g. *Spargnophilus*) tightly packed straight (e.g. *Lumbricus*) or coiled (e.g. *Phreodrilus*) mitochondria (d) a mid-piece (Erseus, 1999) (absent in polychaetes, however see Blake and Arnofsky, 1999) (Fig. 1.9) and (e) a central long, thin flagellum. Shaping of the sperm head differs between broadcasting polychaetes and brooding sabellids/serpulids. Of 10 species investigated, seven brooding species belonging to Sabellidae, Serpulidae and Fabricinae have an elongate head and the remaining three broadcasting serpulids spherical head. Among 23 sabellinaeids, not only the 11 broadcasters but also five brooding species have spherical head. Briefly, broadcasting polychaete species consistently have spherical head

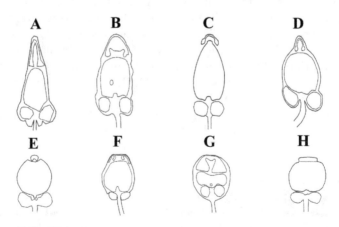

FIGURE 1.9

Representative sperms of some polychaete families. A. Eunicidae, B. Sabellidae, C. Phyllodocidae, D. Terebelliformiae, E. Spionidae, F. Scolecidae, G. Cirratuliformia and H. Amphinomidae (rough sketches made from Rouse, 2006).

but the brooders may have elongate or spherical head (Rouse, 1999). The vestimentiferan polychaetes have attracted studies on sperm ultrastructure. In *Riftia pachyptila*, *Ridgeia piscesae* and *Siboglinum ekmani*, the sperm structure is essentially similar. Their mature sperm are filiform with an elongate nucleus and flagellum. Two to three mitochondria wrap around the nucleus. The acrosome is helical and lies at the apex of the nucleus in *S. ekmani* but on the side of nucleus in *R. pachyptila* (see Rouse, 1999). Notably, the sperm are aflagellate in members of Alvinellidae, Histriobdellidae and Pisionidae (Westheide, 1988). Sperm of *Ophryotrocha* are immotile and resemble spermatids (Pfannenstiel and Grunig, 1990).

Parasperm: In a few eccentric species of aquatic of tubicine oligochaetes, two types of spermatozoa are produced (Ferraguti, 1984): (i) fertilizing eusperm with regular haploid DNA content and (ii) protective and transporting parasperm with a much reduced DNA content. The spermatogonia are released from the testis into the coelomic cavity of the tenth segment (seminal vesicles) as cysts, each consisting of a group of four cells interconnected by cytoplasmic bridges. The cysts undergo three spermatogonial divisions and form 8, 16 or 32 cells. In the euspermic line, primary spermatocytes and euspermatids are produced through regular meiosis, giving rise to 128-cell cysts. On the other hand, paraspermatid cysts are formed by a much greater number (550–3,000) of cells. Consequently, they are larger (100–150 μm) than euspermatid cysts (70–80 μm). A mitotic spindle is never formed in these fragmenting cysts of paraspermatocytes. As a result, they undergo a peculiar nuclear fragmentation. During this fragmentation, the DNA is distributed unevenly among spermatids, giving rise to a great and variable number of parasperm with variable DNA contents (Boi et al., 2001). The eusperms are characterized by the presence of acrosome, smaller mitochondria (half the size of parasperms), long (30 μm) body (3 μm in parasperm) and shorter tail. Following the gametic exchange between hermaphroditic partners, sperms are bundled as spermatozeugmata in spermatheca. The bundles are made up of parallel core of eusperm surrounded by parasperms. The observations of Boi and Ferraguti (2001) indicate the commencement of spermatogenesis on the third week and appearance of spermatids on the sixth week. The peaks of spermatid appearance are in successive waves and sperm production is a cyclic event.

In members of Arenicolidae, Maldanidae, Syllidae and Terebellidae, sperms are bundled into spermatozegumata. In many other annelids, the sperms may be tightly packed into a spermatophore, which is a capsule containing sperms with their head toward the center. In polychaetes, spermatophores have been reported from Spionidae (e.g. *Spio filicornis*, Greve, 1974), Capitellidae, Arenicolidae, Hesionidae, Syllidae and Histriobdellidae, Nerillilidae, Protodrilidae and Myzostomidae (see Schroeder, 1989). Though a spermatophore can be a device to ensure fertilization of eggs during cocoon deposition, its function is doubtful. In fact, the experimental

TABLE 1.9

Effect of spermatophore removal on reproduction in the earthworm *Eisenia foetida* (from Monroy et al., 2002, modified and added)

Traits	Spermatophore	
	Intact	Removed
Cocoon (no./worm)	15.7	13.9
Cocoon size (mg)	166	146
Cocoon viability (%)	68	81
Cocoon survival (no./worm)	21.7	25.9
Hatchling (no./worm)	64	67
Hatchling size (mg)	10.7	11.3
Hatchling (no./cocoon)	2.0	2.3

removal of spermatophore in the lumbricid *Eisenia foetida* slightly increases the reproductive success (Table 1.9). Calculations of cocoon survival and hatchling per cocoon in the worm indicate that in each cocoon 12% of additional eggs have been fertilized after the removal of spermatophore.

Sperm types: In polychaetes, the sperms are grouped into three types: (i) ect-aquasperm, (ii) ent-aquasperm and (iii) introsperm (Jamieson, 1986). The ect-aquasperms are liberated freely into the ambient water, in which fertilization with con-specific eggs occurs. In about 30 species, it is characterized by (a) small cylindrical acrosome resting in a depression on the anterior end of the nucleus, (b) spherical or ovoid nucleus, (c) a few rounded cristate mitochondria and (d) a free axoneme with the 9 + 2 arrangement of microtubules (Blake and Arnofsky, 1999) (e.g. Amphinomidae: *Eurythoe complanata*, Sabellidae: *Idanthyrsus pennatus*, Serpulidae: *Galeolaria caespitosa*, Chaetopteridae: *Chaetopterus variopedatus*). The ent-aquasperm is also shed into the ambient water but it is drawn in by the inhalant or feeding current of the female/hermaphrodite. In them, internal fertilization invariably occurs (e.g. Sabellidae: *Fabricia*, *Oriopsis*, Maldanidae: *Micromaldane*). In contrast, the introsperm never enters the water in aquatic species and occurs in all terrestrial annelids. Of > 28 polychaete species, Blake and Arnofsky (1999) have found the confirmed or probable coexistence of spermatophore and introsperm in 20 species and lack of spermatophore in the remaining eight ecto-aquaspermic species. Introsperm may be transferred from male to female by copulation in some hesionid and saccocurid polychaetes (see Schroeder, 1989) but by pseudocopulation, in which the apposed male and female shed gametes directly into a 'cocoon' in dorvillid *Ophryotrocha* with aflagellate sperms; however, sperms transferred into the spermatheca of a mating partner are subsequently shed into the eggs contained within a cocoon in a few polychaetes and most oligochaetes (Jamieson, 1986). Within the interstitial polychaete genus *Microphthalmus*, Westheide (1967, 1979) describes a gradation of simple sperm transfer by spermatophore in *M. aberrans* to hypodermic impregnation into the female opening of the receptacular tissue in *M. listensis*. In still others, sperms, as in earthworm *Lumbricus terrestris* (Koene et al., 2005) or spermatophore, as in some hesionids (e.g. Westheide, 1967) and leeches (see Jamieson, 1986) are transferred directly into the body surface of the mating partner and the sperm pierce through the body wall to reach the eggs. In *L. terrestris*, 40–44 copulatory setae pierce into the partner's skin causing damage, while injecting the semen drawn from its setal glands.

Spermatheca: Named as seminal receptacles in polychaetes and uterus in hirudineans, the spermatheca serves to receive and store sperms/spermatophores. For detailed description of its structure, Adiyodi (1988) may be consulted. In oligochaetes, the spermatheca is generally paired saccular organs with their ducts opening to the outside. Their number ranges from zero in parthenogenic lumbricid *Bimastos* to one pair (Naididae, Enchytraeidae, most Tubificidae and Moniligastridae) to seven pairs in megascolecid *Perionyx polytheca*. In hirudineans, they are a bilobed but unpaired structure. In some polychaetes (e.g. spionid *Polydora ligni*), the sperms are stored in pockets of female nephridium (see also Rice, 1980). The stored spermatozoa remain motionless for varying periods and may be compacted into a cylindrical bundle called spermatozegumata. In polychaetes, the stored sperms may undergo some morphological changes, suggestive of a capacitation-like process (e.g. *Pisione alikunhi*, Alikunhi, 1951) and *P. remota* (Strecher, 1968).

1.9 Reproductive Modes and Dispersal

With a soft body and low motility, annelids are prone to predation. Unlike molluscs with external protective shell(s), they do not have structural and/or chemical (however see p 48–49) defense mechanism to avoid predation. As a consequence, a key driving force in their evolution and speciation seems to have been an extraordinary diversification of reproductive modes. Interestingly, gonochorism in consort with multiplication of reproductive modes has facilitated a greater speciation in polychaetes with ~ 13,000 species. Conversely, the consistent presence of hermaphroditism and/or parthenogenesis along with internal fertilization and direct development within the cocoon in both aquatic (e.g. tubificids) and terrestrial (e.g. earthworms, see Fig. 8.10) 'herbivorous oligochaetes' has reduced species diversity to ~ 2,000 species (see Table 1.2). Carnivory/sanguivory has further reduced species diversity to ~ 800. It is in this context, reproductive modes in annelids become important and interesting.

Polychaetes display fascinating and incredibly diverse reproductive modes (Fischer and Fischer, 1995). Broadly, the eggs may be freely spawned or brooded. For the first time, Wilson (1991) made an extensive survey of taxonomic distribution of these reproductive modes across 307 species belonging to 36 families and 10 orders. He brought them under the following six groups: 1. Free spawned (fs) eggs, 2. Freely released embryos encapsulated in gelatinous mass (gel), 3. Brooding eggs outside the body, say, on substratum (br–ext), 4. Eggs brooded inside the tube (br–tube), 5. Brooding encapsulated embryos inside the tube (br–enc–tube) and 6. Eggs brooded inside the body (br–int). The first group was further divided into three sub-groups namely 1. fs into (i) with feeding and dispersing planktotrophic (PLK) larvae, (ii) with dispersing lecithotrophic (LEC) larvae and (iii) direct developers (DIR). Each of the remaining five groups were divided into 2. gel into (i) PLK, (ii) LEC and (iii) DIR, 3. br–ext into (i) PLK, (ii) LEC and (iii) DIR, 4. br–tube into (i) PLK, (ii) LEC and (iii) DIR, 5. br–enc–tube (i) PLK, (ii) LEC and (iii) DIR as well as 6. br–int into (i) PLK, (ii) LEC and (iii) DIR. In all, as many as 18 reproductive modes were recognized. Expectedly, Wilson's report was more of taxonomic. Analysis of his data on a holistic polychaete level, a different picture emerged (Table 1.10), from which the following may be inferred: 1. Surprisingly, > 47% of polychaetes brood their eggs/embryos; the remaining 53% freely spawn their eggs (40%) or release embryos in encapsulated jelly mass (13%), 2. Of 307 species surveyed, 126 species, i.e. 41% of them develop through a feeding and dispersing PLK stage during the period of indirect development, 3. Another 23.5% (72 species) of them are LEC; hence they pass through the short larval period of dispersal in the pelagic realm, 4. Of the remaining 35.5% undergo direct development. Incidentally, Blake and Arnofsky (1999) listed the number of spionids that (i) broadcast thick eggs and (ii) brood thin eggs. An estimate of these two

TABLE 1.10

Distribution of reproductive modes in polychaetes (from Wilson, 1991, modified). PLK = Planktotrophic, LEC = Lecithotrophic, DIR = Direct development; all values in parantheses are in %

Mode	PLK	LEC	DIR	Total
Freely spawned eggs				
Freely spawned (fs)	79	34	10	123 (40.0)
Gel encapsulated (gel)	10	16	13	39 (12.7)
Subtotal	89 (28.9)	50 (16.3)	23 (7.5)	162 (52.7)
Brooded eggs				
External (br-ext)	6	7	15	28 (9.1)
Tubular (br-tube)	7	11	39	57 (18.6)
Encapsulated tube (br-enc-tube)	24	3	15	42 (13.7)
Internal (br-int)	0	1	17	18 (5.9)
Subtotal	37 (12.0)	22 (7.2)	86 (59.3)	145 (47.3)
Total	126 (41.0)	72 (23.5)	109 (35.5)	307

TABLE 1.11

Development and dispersal of polychaete larva. A resurvey of information presented in Appendix of Carson and Hentschel (2006). E = Epitokous, MA = Mobile adults, AS = Asexual

Item	Development Mode	Subtotal (no.)	Total (no.)	Dispersal
	1. Indirect development		173	**High/Medium**
	1.1 Free spawner (fs), planktotrophic (PLK)		109	**High**
1.1.1	Pelagic, PLK	35		High
1.1.2	PLK, Trochophore	20		High
1.1.3	E, Benthic spawning, pelagic PLK > 3 w- mo	12		High
1.1.4	fs, PLK, mitraria	6		High
1.1.5	fs, PLK, rostraria	2		High
1.1.6	fs, PLK, aulophore	1		High
1.1.7	Pelagic, PLK, MA	24		High
1.1.8	Pelagic, PLK, AS	5		High
1.1.9	fs, LEC, larval duration > 10 d	1		High
1.1.10	Pelagic, gel. Encapsulated	2		High
1.1.11	Pelagic, gel. tube	1		High
	1.2 Free spawner (fs), lecithotrophic (LEC)		56	**Medium**
1.2.1	E, LEC	8		High
1.2.2	fs, LEC, pelagic for < 7 d	43		Medium
1.2.3	fs, LEC, pelagic for < 7 d + rafter	3		Medium
1.2.4	pelagic, LEC, MA	2		Medium
	1.3 Benthic spawner (bs)		8	**Medium**
1.3.1	bs	3		Medium
1.3.2	bs, MA	2		Medium
1.3.3	E, benthic egg mass	3		Low
	2. Brooded development		82	**Medium/Low**
	2.1 Brooded + pelagic PLK		33	**Medium**
2.1.1	Internal pelagic, PLK	4		High
2.1.2	Brooded PLK for > 13 d	1		High
2.1.3	Brooded but pelagic PLK larva	20		Medium
2.1.4	Externally brooded PLK ~ 8 d	1		Medium
2.1.5	Brooded but released at 3 setiger	3		Medium
2.1.6	Brooded, pelagic LEC	4		Medium
	2.2 Brooded but no pelagic PLK		49	**Low**
2.2.1	Tube brooded egg mass	8		Low
2.2.2	Brooded, pelagic for 1 d	1		Low
2.2.3	Brooded, direct development	2		Low
2.2.4	Tube brooded, direct development, MA	12		Low
2.2.5	E, brooded, benthic, AS	10		Low
2.2.6	Internal/External brooding	15		Low
2.2.7	Brooded benthic larva	1		Low

Table 1.11 contd. ...

...Table 1.11 contd.

Item	Development Mode	Subtotal (no.)	Total (no.)	Dispersal
	3. Direct developers (Dd)		99	Low
3.1	Egg mass, short swimming, neochaete larva	5		Medium
3.2	Eggs in jelly mass, MA	1		Medium
3.3	Egg mass	33		Low
3.4	Direct developers	7		Low
3.5	Dd, presumed no pelagic stage	14		Low
3.6	Dd, MA	5		Low
3.7	Dd, MA, AS	25		Low
3.8	Brooded up to direct development	2		Low
3.9	LEC	3		Low
3.10	Viviparous	4		Low

groups suggests that 47% of them brood their eggs and 53% are broadcast spawners. Understandably, the sedentary tubiculous spionids opt more for brooding rather than broadcast spawning.

Unlike brooding limited to embryos and release of the larvae in crustaceans, some brooding annelids (e.g. 82 species Table 1.11) continue to brood the larvae too. The larvae are released from three up to 15–24 chaetigers stage. In polychaetes, the planktotrophic larval duration may last to 50 days or longer. In Fig. 1.10A, relevant data reported by Blake and Arnofsky (1999) on egg size of some spionids are plotted against larval duration. Unusually, the duration increases with increasing egg size, although individual values are scattered. Understandably, the larvae arising from larger eggs tend to postpone the departure from brood at 3rd to 15th–24th chaetiger stage (Fig. 1.10B). On plotting the values reported by Prevedelli and Simonini (2003) for body size (length) of three nereidids, a dorvilleid and a dinophylid against egg size, a direct relationship becomes apparent (Fig. 1.10C). At interspecies level, the same holds true for the relationship between body size and fecundity for free spawners and brooders (Fig. 1.11). Rouse and Fitzhugh (1994) have listed values for body size, egg size and fecundity of some sabellids that broadcast spawn and brood eggs intra-tubularly or extra-tubularly. On plotting mean values for the spawners and brooders, the emerging trends indicate that egg size increases with increasing body size but fecundity decreases with increasing body size (Fig. 1.10C). Hence, the limited available information in sabellids seems not to fall in line with that of nereidids and others. Incidentally, more information is available on the effects of body size on fecundity and egg size of brooders and spawners belonging to sabellinids and fabricinids (Rouse and Fitzhugh, 1994). To establish phylogenetic relation among brooders and spawners in the subgroups,

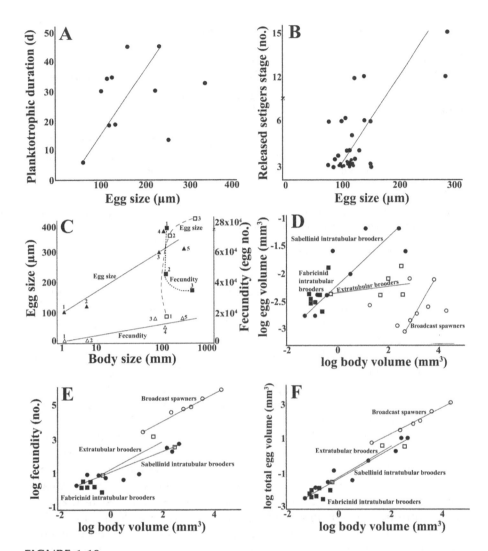

FIGURE 1.10

Effects of egg size on A. planktotrophic duration and B. release of chaetiger stage in polychaetes (drawn using data compiled by Blake and Arnofsky, 1999). C. Effects of body size on egg size (▲—▲) and fecundity (△—△) in 3 nereid species (1 = *Marphysa sanguinea*, 2 = *Perinereis rullieri*, 3 = *P. cultrifera*), a dorvilleidid (4 = *Ophryotrocha labronica*) and a dinophilid (5 = *Dinophilus gyrociliatus*) (drawn using data reported by Prevedelli and Simonini, 2003). Effect of body size (length) on egg size (■—-—■) and fecundity (□—-—□) of broadcast spawning (x, mean of 6 species), extra-tubular brooding (x, 3 species) and intra-tubular brooding (x, 3 species), terebellids (drawn using data compiled by McHugh, 1993). Effects of body volume on D. egg volume, E. fecundity and F. total egg volume in broadcast spawning, extra-tubular brooding, intra-tubular brooding fabricinid and intra-tubular brooding sabellinid polychaetes (modified from Rouse and Fitzhugh, 1994).

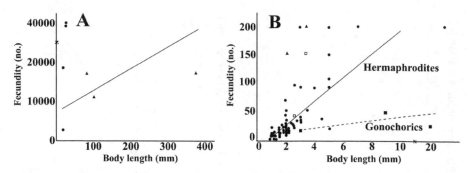

FIGURE 1.11

Fecundity as function of body length in A. free spawning and B. brooding polychaetes. Note the values shown by ● are simultaneous hermaphrodites, ■ gonochorics and □ with feeding larvae (drawn using data reported by Kupriyanova et al., 2001 and also data (▲) reported by Prevedelli and Simonini (2003) are also included).

an attempt has been made by Rouse and Fitzhugh, who have found that body size (volume) holds linear relations at different levels with egg size (Fig. 1.10D). However, linear trends for the relation *per se* become apparent (Fig. 1.10D), when values reported for species with unknown life history are eliminated and statistical analysis is not applied. Accordingly, the level and degree of the slopes of these trends differ. Notably, the slope decreases in the following order: broadcast spawners < sabellinid intra-tubular brooders < extra-tubular brooders < fabricinid intra-tubular brooders. This is also true for the relations between body size vs fecundity (Fig. 1.10E) as well as body size vs egg size (Fig. 1.10F). It must be noted that all these linear relationships are based on log values. Different trends may become apparent, when the normal values are considered (see p 75). Remarkably, sabellinids larger than log 1.33 mm³ alone are broadcast spawners. Fabricinids smaller than log − 1.67 mm³ are all intra-tubular brooders. Both extra-tubular as well as intra-tubular sabellinids and fabricinids have a wide size range from log − 1.33 to log 3.0 mm³. It is not known whether the presence of nurse eggs/larvae in the intra-tubular brooding is responsible for the widening of the size range in these polychaetes. Interestingly, Garraffoni et al. (2014) have also listed available values for egg size, fecundity and adult size of terebellids characterized by indirect (seven species) or direct (five species) life cycle. The mean values for the direct developers are 0.1 mm³ for egg size, 242 eggs for fecundity and 8,608 mm³ for adult size, in comparison to 0.006 mm³, 408 eggs and 10,488 mm³ for egg size, fecundity and adult size of indirect developers, respectively. Although, the differences between these mean values are justifiable for terebellids with direct and indirect life cycles, the individual values range very widely; for example, the adult size of direct brooders ranges from 85 to 49,480 mm³.

For polychaetes, relevant information on dispersal is widely scattered. Thankfully Carson and Hentschel (2006) have summarized the relevant information from ~ 200 samples holding 501 polychaete species that inhabit 10–120 m depth of the Santa Catalina Island, off Los Angeles, California and other surrounding six island shelves with an area of 2,739 km². This publication includes appended data, which have been subjected to more analyses in this account to draw additional information. Of these 501 species belonging to 226 genera and 49 families, the life history is known only for 354 species and for the remaining 152 species belonging to 78 genera and 24 families, it is not yet described, especially for Parasonidae (27 species), Sabellidae (18 species), Ampharetidae (14 species) and Lumbrinereidae (12 species). In 1991, when Wilson has made the first survey, the history has already been known for 307 species. After a passage of 16 years, the history is described at the rate of three polychaete species/y (see also Giangrande, 1997). Clearly, there is an urgent need for more research input into this area.

Carson and Hentschel (2006) have divided the 354 polychaete species into (i) high-dispersal 'open marine' category comprising of species that are capable of dispersing and exchanging larvae with different populations over tens of kilometer areas, (ii) medium-dispersal category denoting species capable of dispersal on the scale of one kilometer and (iii) low-dispersal 'closed' category that includes species dispersing in the range of < 1 km, as their larvae are cued to settle close to their parents (Table 1.11). In this categorization, they have considered not only the established PLK, LEC and DIR but also other complicating life history factors like epitoky (E) (vertical distribution), Mobile Adults (MA) (horizontal distribution) and asexual reproduction (AS) (local distribution). From the present analysis, the following are inferred: 1. Not surprisingly, the diversification of reproductive modes have gone to 41 recognizable types, 2a. Despite the incidence of epitoky (benthic spawning + PLK), mobile adults and/or asexual reproduction, the PLK lasting from > 10 days to months ensures high dispersal of polychaetes that freely spawn eggs/embryos in encapsulated jelly mass. 2b. Levin (1984) has also reported that *Polydora ligni* (7–42 days larval duration) and *Pseudopolydora paucibranchiata* (> 7 days larval duration) have dispersal capability over longer distance. *Streblospio benedicti* and *Capitella* with < 3 days larval duration have the ability for small scale dispersal only. However, LEC followed by pelagic larval duration of < 7 days ensures only medium level of dispersal. 2c. Benthic spawning, even with MA also limits the dispersal to medium level. 3a. Brooding + pelagic PLK for more than 8 days and larvae released at three chaetigers stage ensures medium dispersal. 3b. However, brooding resulting in direct development with zero or 1 day pelagic larval duration limits the dispersal to low level. 4. Despite the presence of neochaete larva, MA, MA + AS, DIR results in low dispersal. 5. Of 354 species, 28.5% (109 species) polychaete species are broadcast spawners followed by

PLK larvae with potential for high dispersal. The remaining (18.1%) free spawners (LEC 56 + bs 8) are characterized by medium dispersal, as their LEC larval duration lasts for 0–3 days only. Of 82 brooder species, 33 of them pass through a period of PLK larval duration lasting for ~ 8 days. They are also characterized by medium dispersal. Brooding with no pelagic larval duration (49 species) and DIR (99 species) limit the dispersal to low level. Briefly, 49% of polychaetes are free spawners and are characterized by high or medium dispersal potential; another 9.3% (33 species) brood their eggs/embryos/larvae but release their larvae for (medium level) dispersal. Of the remaining 42%, some are brooders with no pelagic larval duration (14%) and others (28%) are characterized by DIR with low dispersal capacity. From Wilson's (1991) survey, it is shown that 47 and 53% of polychaete species are brooders and broadcast spawners, respectively. The survey by Carson and Henstchel indicates that only 42% are brooders. Notably, the annelid larvae with high potency for dispersal are capable of dispersing on the scale of tens of kilometers only.

1.10 Fertilization Site and Success

The broadcast spawning polychaetes reproduce by releasing gametes into water, where fertilization and subsequent development occur. In them, the ambient sperm concentration, into which spawned eggs are released, is a key factor in deciding Fertilization Success (FS). Motile polychaetes may form mating aggregation and often display a high degree of synchronized spawning, as in palolo worm *Eunice viridis* (Caspers, 1984). In eggs of gregarious species, too many sperm may lead to polyspermy. Hence, laboratory studies are required on gamete longevity and optimal egg-sperm ratio that ensures the highest FS. The experimental study on a representative gregarious serpulid *Galeolaria caespitosa* provides some basic information on FS. From a well designed laboratory study, Kupriyanova (2006) has found that 1. despite variations between 58 and 66 µm, the egg size remains equal at 61.3 µm in worms weighing 10 to 50 mg (Fig. 1.12A). 2. Likewise, swimming velocity of sperm averages to 107.3 µm/second (s) in males weighing up to 25 g (Fig. 1.12B). 3a. Expectedly, FS increases sigmoidally with increasing sperm concentration up to 10^8/ml and 3b. Achieves the highest FS of 75% and > 80% at 10^8/ml sperm concentration, when given 1 and 10 minutes duration for fertilization, respectively (Fig. 1.12C). 4. Sperm velocity progressively decreases from ~ 145 µm/s and perceptibly from > 100 µm/s 2 hours after activation and finally too < 5 µm/s 8 hours after activation (Fig. 1.12D). More or less similar decreasing trends are noted for the decrease in FS (Fig. 1.12E, F). Surprisingly, the longevity of eggs and sperms is almost

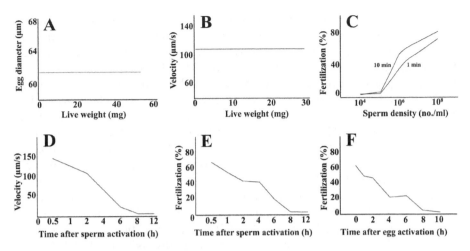

FIGURE 1.12

Experimental fertilization in the serpulid *Galeolaria caespitosa*. A. Effect of body weight on egg size, B. Effect of body weight on sperm velocity, C. Effect of sperm density on fertilization success, D. Sperm swimming velocity after activation, E. Fertilization success after sperm activation and F. Fertilization success after egg activation (compiled and modified from Kupriyanova, 2006).

equal. In general eggs remain fertilizable for longer durations than the sperms.

Williams and Bentley (2002) have undertaken a comparative study on the gametic longevity of an errant *Nereis virens* and sedentary *Arenicola marina*. In them, fertilizable egg age lasts for 72 days to ensure 100% FS. However, the age of sperm is just 30 hours and 60 hours to ensure 100 and 60% FS in sea water, respectively. In the siboglinids inhabiting thermal vents, cleavage commences 24 hours after fertilization. On *in situ* incubation of eggs, cleaved embryos average to 65% (range: 40–84%) in the laboratory and 90% (74–98%) in *Riftia pachyptila* and *Lamellibrachia luymesi*, respectively. But, they average to 86% (76–91%) in the field for *R. pachyptila* (Hilario et al., 2005). Understandably, FS in these *in situ* and *ex situ* incubated eggs ranges between 40 and 98%. Apart from these gonochorics, experimental FS values are also available for hermaphroditic annelids, in which 100% FS is expected. As already indicated, the removal of spermatophore in the earthworm *Eisenia foetida* increases FS in 11% of more eggs within a cocoon (see Table 1.9). In selfing sabellid polychaete *Laonome albicingillum*, FS is 83%, in comparison to 88% in outcrossing individuals (Hsieh, 1997). In another polychaete *Ophryotrocha diadema*, grouping of dozen individuals doubles the percentage of non-egg layers (20%) against just 10% of non-egg layers in isolated individuals (Henshaw et al., 2015). Evidently, FS remains at 80 and 90% in grouped and isolated individuals, respectively.

From field observation, only very few FS values are reported for polychaetes. In epitokers, aggregation of spawners synchronizes spawning and chemical cue-guided swimming sperm may lead to > 90% FS. Behavioral strategies also bring the mating partners close together and ensure high FS in ambient waters. The 'nuptial dance', during which gametes are released, may bring the mating partner close together to ensure high FS, as in *Autolytus prolifer*. In others, pseudocopulation, during which the gonopores of the mating partners are brought together, may bring the partners still closer to achieve higher FS. However, no information is yet available on this aspect. For benthic spawners, FS is low (40–60%). A series of publications (Hardege et al., 1996, Watson et al., 1998 and Williams and Bentley, 2002) have reported that the lugworm *A. marina* female lays eggs within its burrow, where the laid eggs remain fertilizable for 72 hours. Male lugworm releases sperm bundles on the sediment surface. Carried by incoming tides over the sediment surface and guided by volatile organic pheromone, the sperms are drawn into the burrow at the density of 10^6/ml in the water column. Fertilization is successful but varies widely from 0 to 90% with most values falling within 40–60%.

In 45% of polychaetes (165/354 species, see Table 1.11), fertilization is external. It is also external in another 9% brooding polychaetes, in which larvae are released after brooding embryonic and/or a short or longer larval stages. In these 9% brooders, the sperms are either delivered on the female's body or the female draws the sperms using chemical cues, as in *Arenicola marina* or filtering the ambient water by the branchiae and other structures for food acquisition. For example, sperms are gathered in the short oral tentacles of the female before mature eggs are passed over the tentacles for fertilization in *Nicolea zostericola*, which broods its embryo in an extra-tubular cocoon (Eckelbarger, 1984). This is also the case in the intratubular brooder *Neolepoa septochaeta* (see McEuen et al., 1983). In all, fertilization is external in 54% of polychaetes. In other brooders characterized by direct development/viviparity, fertilization seems to be internal. Polychaetes engage an array of organs that can be termed as 'penis'. The penis is inserted into the female's genital pores (e.g. *Saccocirrus eroticus*) or female's body (e.g. *Stratiodrilus novaehollandiae*). In females with no sperm receptacle, the sperms are transferred by hypodermic injection. Sella and Ramella (1999) list five dinophylids, which engage hypothermic injection for sperm transfer.

Fertilization is also internal and occurs in the ovisac or cocoon. In oligochaetes pseudocopulation is not uncommon, during which mutual transfer of sperm/spermatophore occurs. In *Eisenia foetida*, for example, the spermatophores are implanted between the 21st and 24th segments, adjacent to the spermathecal anlage. In 86% of the mating partners, only one spermatophore is implanted (Monroy et al., 2003). The preferred segments for spermatophore implantation in the leech *Haementeria parasitica* are also close to those segments, in which eggs are borne and thereby the distance to be travelled by the sperms to reach the eggs is minimized (see Sawyer et al., 1981). In most glossiphoniid, piscicolid and erpobdellid leeches, which lack penis, sperm are transferred by a hypodermic injection. In other earthworms like *Lumbricus terrestris*, the copulatory setae pierce into partner's skin and inject sperm as well as an allohormone that inhibit sperm digestion (Koene et al., 2005). At concentrations of 7×10^4/ml and 1.5×10^6/ml, sperm from more than one male *G. caespitosa* increases FS from ~ 25–30% to ~ 90–95% (Marshall and Evans, 2005).

Whereas information on FS for gonochores (e.g. *Arenicola marina*) and protandric hermaphrodites (e.g. *G. caespitosa*) is limited, available information on FS of simultaneous hermaphroditic annelids is also limited. Available publications elucidate the adjusted resource allocation in hermaphrodites (e.g. polychaetes: *Ophrytrocha diadema*, Schleicherova et al., 2006, Cannarsa et al., 2015). For example, on enforced isolation, the number of cocoons generated is 12/♀ and 2/♀ in the oligochaetes *Eisenia andrei* and *E. foetida*, respectively, clearly indicating that the latter is able to reduce allocation with enforced isolation and self-fertilization. However, hatching is limited to 2.5 cocoons per female in *E. andrei* and 1.5 cocoons per female in *E. foetida*. Interestingly, Hsieh (1997) is perhaps the only author, who has reported FS (in unit of egg production) as a function of number of eggs used by self- and cross-fertilization in simultaneous sabellid polychaete *Laonome albicingillum*. From Fig. 1.13, the following may be noted: 1. With increasing number of eggs made available/used for fertilization, FS increases in both self- and cross-fertilization. 2. But outcrossing worms are able to achieve 50 to 95% FS, while selfers achieve 50 to 75% FS only. Competition for mating in *E. andrei* also increases FS. For example, FS is increased from ~ 60% with a single

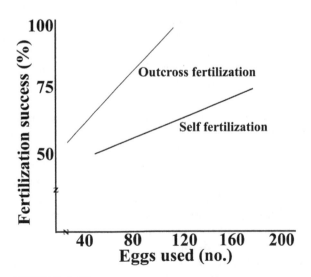

FIGURE 1.13

Fertilization success in self- and outcross-fertilization in the hermaphroditic sabellid polychaete *Laonome albicingillum* (drawn from data reported by Hsieh, 1997).

mating partner but > 80%, when the number of mating partners is increased to two–six (Porto et al., 2012).

1.11 Annelidan Larvae

As indicated, a majority of polychaetes display indirect development involving typically the trochophore larva, or one or other modified form of trochophore. Many genes, whose expression is known to regulate one or other events of embryonic and larval development, have been identified; for more details, Irvine and Seaver (2006) may be consulted. In his narrative account, Rouse (2006) has listed 10 annelid larval forms, and described a range of ciliary bands and tufts that are used for feeding and locomotion. Accordingly, the ciliary bands are designated as (i) *Prototroch* with a single equatorial compounded ciliary band dividing the larva into an anterior episphere and posterior hyposphere, (ii) *Akrotroch* with complete ciliary ring around the episphere, as in eunicids and other six families, (iii) *Meniscotroch* with a crescent shaped area of stout cilia on the episphere, as in glycerids and 14 other families, (iv) *Metatroch* characterized by the post-oral ciliary band, beating from posterior to anterior direction, e.g. Amphinomidae and 10 other families, (v) *Neurotroch* with a longitudinal ciliary band running from behind the mouth to near the anus, as in siboglinids and 30 other families, (vi) *Mesotroch*, in which the transverse ciliary ring adorns the larval midbody, as in *Chaetopterus*, (vii) ventral *Gastrotrochal* and dorsal *Nototrochal* ciliary rings on segment, as in spioniform larvae and (viii) *Telotroch* with locomotory ciliary ring located at the posterior end of larva, as in hesionids and 11 other families. Table 1.12 briefly summarizes the morphology and occurrence of 10 different trochophore and trochophore-like larvae reported from polychaetes.

During development, some spionids, draw nutrients from the so called nurse eggs and larvae within a brood. Nurse eggs serve as source of extra embryonic nutrients for developing embryos. On activation, these eggs too release polar bodies but subsequently are compartmentalized with a loss of nuclear DNA (e.g. *Boccardia proboscidae*, Gibson et al., 1999, Smith and Gibson, 1999). As a climax, larvae (e.g. *Streblospio benedicti*, see Blake and Arnofsky, 1999) ingest fellow larvae prior to hatching and release from brood. As much as 95% of the eggs brooded by *Polydora cornuta* are nurse eggs and are ingested by adelphophagic larvae hatched at 3rd chaetiger stage (Mackay and Gibson, 1999). Some *P. cornuta* females switch between adelphophagic and planktotrophic larvae arising from a single brood or in successive broods. Both nurse eggs and adelphophagic larvae are an adaptive mechanism to accelerate early cleavages in smaller eggs with less yolk that

TABLE 1.12

Description of polychaete larvae (condensed from Rouse, 2006, all figures are sketches from Rouse, 2006 and others)

Larva	Description	Figure
Proto-trochopore	Completely or nearly completely ciliated trochophore, e.g. oweniids	
Trochophore	Typical larva with opposed-band method of feeding by involving ciliary bands of prototroch and metatroch	
Meta-trochophore	Normal trochophore with clear signs of segmentation. Non-functional parapodia, if present, e.g. *Lanice conchilega*	
Nectochaeta	Trochophore with functional parapodia, e.g. phyllodocids	
Aulophore	Metatrochophore is long living in a tube, e.g. terebellids, pectinariids	
Chaetosphaera	Swims by undulation as well as cilia. Can roll up into a ball, e.g. spioniformids, sabellarids	
Nectosoma	More like chaetosphaera larva but cannot roll up into a ball, e.g. *Poecilochaetus*	
Rostraria	A pair of distinctive tentacle is used for feeding via ciliary bands, e.g. amphinomid	
Mitraria	Normal trochophore but undergoes 'catastrophic metamorphosis' by casting off much of its body prior to settling as juvenile, e.g. Oweniidae, *Myricola*	
Erpochaete	More a 'juvenile larva' that creeps over the sediment using its chaetae	

facilitates the embryo to grow faster (see Pandian, 2016). Interestingly, Schneider et al. (1992) have shown that with ~ 90% yolk content, egg size of *Platynereis massiliensis* is 10 times that (with 64% yolk content) of the sibling species *P. dumerilii*. As a consequence, the cell cycles upto 5th cleavage are 3.7 times slower in *P. massiliensis* than in *P. dumerilii*. Hence, it is advantageous to have smaller eggs, whose embryos/larvae draw extra-embryonic nutrients from nurse eggs and larvae.

1.12 Defense and Parental Care

The soft-bodied, slow motile/sedentary annelids are not readily prone to predation, as it has been considered earlier. Many annelids provide protection to their eggs/embryos by (i) encapsulating them in jelly matrix or gelatinous envelope, (ii) synthesis and accumulation of deterrents in their larvae and adults, (iii) providing internal or external structures for brooding eggs/embryos/larvae as well as provision of hard cocoon or soft cocoon followed by parental care.

Jelly matrix: In general, polychaetes are broadcast spawners. However, some of them shed pear-shaped or spherical egg masses, in which eggs are embedded in a jelly matrix and attach them to suitable substrate or the parent's tube. In echinoderm eggs, the jelly facilitates floatation of pelagic eggs and increases the chances for sperm-egg collision and hence fertilization success (see Pandian, 2018). In polychaetes, the jelly matrix is considered to protect the developing embryos, supply nutrients but limit dispersal of the offspring to settle in and around the 'suitable' habitat (Sato et al., 1982). From an experimental study, Sato and Osani (1996) have found that (i) de-jellied unfertilized eggs of the polychaete *Lumbrinereis latreilli* lose the capacity for sperm binding and hence become unfertilizable, (ii) fertilizability can be restored by addition of jelly to the de-jellied eggs. An electron microscopic study has revealed that the sperm-egg binding occurs only in the presence of jelly and an unknown interaction between the jelly matrix and spermatozoa may be a pre-requisite to induce acrosome reaction. Incidentally, there are others, in which the jelly layer is formed shortly after fertilization (e.g. *Nereis falcaria*, Read, 1974). In *Platynereis dumerilii*, a huge jelly coat is formed immediately after fertilization (Kluge et al., 1995).

Chemical defense: The recent past has witnessed increasing number of publications on chemical defense of polychaete worms. For example, the LEC spionid *Streblospio benedicti* contains at least 11 chlorinated and brominated hydrocarbons (alkyl halides) and the brooding capitellid *Capitella* sp I contains three brominated aromatic compounds. Quantifying these alkyl halides, Cowart et al. (2000) have found that *S. benedicti* contains 1.8, 8.3, 4.7 and 28.9 ng/mm^3 in the pelagic LEC larva, post-release, new recruit and adult, respectively. This increasing trend suggests that the halides are synthesized during post-embryonic developmental stages. With contrasting life history of *Capitella* sp I, the halogenate compound contains 1150 and

126 ng/mm^3 in the LEC larva and adult, respectively. At these concentrations, the haloaromatics are known to deter predation.

Investigating the chemical and structural defense by external strategies in six tubiculous sabellids, Giangrande et al. (2014a) have reported toughness of the branchial crown, structure and strength of the tube as well as anti-bacterial lysozyme activity in the mucus. The tube strength, as measured by tearing weight, decreased from 800 g in hard substratum-inhabiting *Sabella spallanzanii* to 200 g in *S. spectabilis*. However, the anti-bacterial lysozyme activity in the mucus of the soft substratum burrowing *S. spectabilis* remains high. In another interesting study, Iori et al. (2014) have observed that on being attacked by the fish *Oryzias melanostigma*, the oenonid *Halla parthenopeia*, after emitting purple mucus, quickly moves away (similar to inking and clouding by sea hares and cephalopods, Pandian, 2017) and subsequently release transparent mucus. The purple mucus is toxic, due to the presence of halochrome a 1, 2 anthraquinone.

Brooding: In aquatic animals, affording protection of eggs, embryos and/ or larvae is not uncommon. The afforded protection ranges from simple embryo and/or larval guarding to brooding them until a certain larval stage or complete development. Brooding may occur in burrows (e.g. *Aphelochaeta*, Petersen, 1999). In *Ophrytrocha puerilis puerilis*, female spends more time in brooding than male (Berglund, 1986). *Nereis acuminata* reproduces monogamously and exhibits male parental care, a rare reproductive mode in marine invertebrates (Weinberg et al., 1990). In others like *Neanthes arenaceodentata* and *Platynereis massiliensis*, females lay eggs and die, males fertilize the eggs, and protect and ventilate the fertilized eggs (see Premoli and Sella, 1995). Brooding occurs within a flexible cocoon in *O. adherens*; *O. socialis* produces a system of branching mucous tubes, in which eggs are laid. In *O. hartmanni*, mucous tubes are ventilated by parents (Paavo et al., 2000). However, quantification of guarders and brooders has not been made. There are hints for other taxonomic groups. For example, brooding occurs in 3 and 27% of ophiuroids and gastropods, respectively. Notably, brooding occurs in one or more species in all the listed phyla (Table 1.13) and within each phylum in all classes (e.g. echinoderms). About 42–47% of polychaetes are brooders, in which ~ 36% brooding lasts until the directly developed young ones are released. Brooding ranges from ~ 1% in echinoderms to 96% in crustaceans. However, it is terminated with a release of young ones in ~ 9% of annelids but ~ 19% of crustaceans. It seems that aquatic invertebrates can ill-afford viviparity; for example, true viviparity occurs in a single ophiuroid *Amphipholis squamata* in echinoderms and six species of annelids. On the other hand, it occurs in 577 species of teleostean fishes (Pandian, 2013).

Cocoon: A distinguishing feature of clitellate annelids is the presence of specialized segments comprising the clitellum (Sayers et al., 2009). The clitellum secretes proteinaceous egg case, the cocoon, into which eggs

TABLE 1.13

Brooding and direct development in some aquatic vertebrates and invertebrates (from Pandian, 2011, 2013, 2016, 2017, 2018)

Taxon	Species (no.)	Brooded (%)	Direct Development (%)
Teleostean fishes	32,000	22	1.8
Crustacea	52,000	96	~ 10 +
Mollusca	~ 1,00,000	~ 34 +	~ 25 +
Annelida	17,000	~ 42	~ 28 +
Echinodermata	7,000	~ 1 +	1 species only

are deposited. In clitellates, the cocoon provides a microenvironment for embryonic development and prevents desiccation/imbibition, predation (however, see Young, 1988) and microbial invasion. It is highly resistant to physical and chemical decay and renders the preservation of fossilized spermatozoa within the 50-million year old cocoon (Bomfleur et al., 2017). In the glossiphoniid leech *Theromyzon tessulatum*, the cocoon formation is initiated with secretion of a thin, external mucous layer, into which fibrous proteinaceous matrix is deposited, forming a sheath surrounding the clitellum (see also Westheide and Muller, 1996). After the eggs are shed through the female pore, the sheath and its contents are passed over the worm's head and sealed at either end with glue-like plugs, called opercula. In leeches internal fertilization occurs either in the ovisac or cocoon (see Sayers et al., 2009). Clitellates secrete three types of cocoons: (i) mechanically strong, (ii) hard-shelled cocoons that are abandoned, leaving the embryos to develop independently on nutritive cocoon fluid (albuminotrophy), thermally and chemically resilient (iii) membranous and (iv) gelatinous cocoons. The large yolky (oviparous) eggs, deposited in the membranous cocoons, are brooded by the parent (Rossi et al., 2016). In the aquatic leech *T. tessulatum*, cocoon formation is a simple but dynamic series of coordinated events. Just a week prior to egg laying, cell type I proliferates and differentiates into cell types II and III, depending on the position in cell type I within the clitellum. Type II cells secrete alcian blue-staining granules that form strong, malleable and bioadhesive opercula (see also Rossi et al., 2013). Type III cells secrete azocarmine-staining granules that build the cocoon wall. Type I and V cells make minor contributions and type IV plays supporting/signaling role (Sayers et al., 2009). The jawless leeches (Erpobdellidae) produce relatively more (~ 1,000) but smaller (~ 50 μm) eggs, enclosed in a cocoon, which is cemented on a substratum and left totally uncared (Table 1.14). In others like glossophoniids, the cocoons contain < 60 large (~ 600 μm) eggs, supplied with albumin and protected. Besides supplying albumin, *Helobdella stagnalis* protects and feeds the young one. With the evolution of parental care in the glossophoniids leeches, mortality of eggs, hatchlings and juveniles have been almost totally reduced to zero.

TABLE 1.14

Increasing level of parental care from erpobdellid to glossophoniid leeches (from Kutschera and Wirtz, 2001, modified)

Features	*Erpobdella octoculata*	*Glossiphonia complanata*	*Helodella stagnalis*
Cocoon (no./season)	~ 120	~ 3–4	~ 5–6
Fecundity (no./season)	~ 1,000	~ 60	~ 50
Egg size (μm)	50	600	500
Albumin	No	Yes	Yes
Ventilation	~ 10 min	30 d	50 d
Egg mortality	High	No	~ 0
Care for hatchling	No	Yes	Yes
Hatchling mortality	High	Low	~ 0
Care for juvenile	No	Yes	Yes + Fed
Juvenile mortality	High	Low	~ 0

1.13 Metamorphosis and Settlement

It must first be recognized that although metamorphosis and settlement are closely related, they are different temporally separate processes. The former is defined as the process, by which a larva undergoes a series of changes to terminate the larval phase. But the settlement is the process, by which a planktonic larva explores and selects a suitable substratum, toward which it moves to finally settle (see Qian, 1999). Larval settlement on large spatial scale is primarily determined by hydrodynamics; however, the successful settlement by competent larvae (starved and aged larvae may not be competent) on smaller spatial scale is mediated by abiotic and biotic cues. These cues may originate from host plants/animals, bacterial microfilms or habitats. The ease with which the microscopic settling larvae can visually be recognized by the formation of milky white calcareous tube in serpulids has facilitated many publications. The segmented larvae of some terebellids and many polynoids undergo metamorphosis as plankton and remain in the pelagic realm for days and weeks, and eventually settle and start benthic life (see Qian and Uwe-Dahms, 2006). In some spionids, sabellarids and oweniids, the metamorphosing larvae possess enlarged erectile anterior setae that serve as floating device (Bhaud and Cazaux, 1990). Settlement of pelagic larvae is of both academic and economic importance. As foulers, the settling larvae create problems in harbors, ships and on coolant screens in power stations. Sabellarids and serpulids often settle gregariously and form colonies. In this regard, *Phragmatopoma californica*, *Sabella alveolata*, *Spirobranchus polycerus*, *Hydroides dianthus* and *H. ezoensis* have received much attention (Qian and Uwe-Dahms, 2006).

TABLE 1.15

Factors that induce or inhibits settlement of polychaete larvae

Species	Factor	Reference
Polydora ligni	Starvation reduces settling ability	Qian and Chia (1993)
Hydroides elegans	Aged larva loses settling ability	Qian and Pechenik (1998)
Spiorbis borealis	Dark substratum attracts settlement	James and Underwood (1994)
Capitella sp	Organic rich sediments (but not hydrogen sulfide, Cuomo, 1985) attracts settlement	Dubilier (1988)
Capitella sp	Juvenile hormone arising from sediment attracts settlement	Biggers and Laufer (1992)
S. borealis	*Fucus serratus* attracts settlement	Williams (1964)
S. rupestris	*Lithothammoni* attracts settlement	O'Connor and Lamont (1978)
Spirobranchus giganteus	*Diploria strigosa* attracts settlement	Marsden et al. (1990)
H. elegans	*Bugula neritina* attracts settlement	Bryan et al. (1998)
H. elegans	Levels of glycojuvenate secreted by rod bacteria attract settlement	Lau and Qian (1997)
Janua brasiliensis (experimental)	Extracellular polysaccharides and glycoprotein attract settlement	Kirchman et al. (1982)
Phragmatopoma californica (experimental)	Fattly acids: cis eicosapentaenoic acid, palmitic acid, palmitoletic acid attract settlement	Pawlik (1986)
H. elegans (experimental)	Amino acids: glycine, glutamine, aspartic acid arising from leechate of *B. neritina* attract settlement	Harder and Qian (1999)

Available information on settlement is summarized in Table 1.15. Earlier, sea grass and coral species that attract the serpulid settlement had been identified. Bryan et al. (1998) have recognized that alcohol extract of dried aqueous *Bugula neritina* leechate carries the compound responsible for attraction of *H. elegans* larvae to settle. Subsequently, experimental investigations have shown that one or other polysaccharides, fatty acids and amino acids can also attract the settlement. Recent studies have shown that the microbial film (Beckmann et al., 1999), and chemical signals like the polar aliphatic amino acids emanating from the biofilm (Hadfield et al., 2014) are not individually responsible for final settlement but the sorbent-like substratum acting as a co-factor is also responsible to induce final settlement of *H. elegans* larvae (Harder et al., 2002). In view of the fouling problems on the ships, harbors, pipelines and screens in power stations, this aspect merits more attention.

2

Sexual Reproduction

Introduction

Annelids display divergent expressions of sex. In them, sexuality ranges from parthenogenesis to gonochorism and self-fertilizing to sequential and serial hermaphroditism. Sexual reproduction in gonochoric and hermaphroditic annelids is characterized by gametogenesis, meiosis and fertilization of gametes arising from one (self-fertilizing) or more than one (multiple matings, polyandry) individuals. In parthenogenic polychaetes, females are always present and males may appear sporadically. In parthenogenic earthworms, male reproductive system is eliminated either partially or totally. Barring hirudineans, many polychaetes and oligochaetes undergo agametic reproduction. Sexual reproduction may succeed or co-exist with agametic cloning. With occurrence of parthenogens, hermaphrodites and agametic cloners, sexual reproduction in annelids is a fascinating subject for research.

2.1 Reproductive Systems

Polychaetes: In majority of polychaetes, the gonads are dispersed and not discrete organs. Masses of eggs and sperm are generated as projections or swellings from undifferentiated peritoneal lining of the coelom. Most segments generate gametes, as in errant nereidids and eunicids. With specialization of the thorax and abdomen in sedentary polychaetes, the gamete generation is, however, limited to the abdominal segments. Increasingly, the number of gametogenic segments is limited (*Amphitrite*) to only six segments in *Arenicola* and further to only two ovarian segments in *Spirorbis spirorbis* (Daly, 1978a). Table 2.1 lists some examples for the anlage of gametogenic segments in gonochoric, protandric and simultaneous hermaphroditic polychaetes. In the intra-ovarian polychaetes, oogenesis is completed in the projections/swellings of the coelomic lining. But gametogonia are shed into the coelom, where further genesis and maturation take place in extra-ovarian polychaetes (Eckelbarger, 1983, 1986). In gametogenic segments, the nephridia serve as exits for the genital and excretory products. However, the ruptured body serves as an exit in epitokous polychaetes (Barnes, 1974).

Oligochaetes: Strikingly, the number of gametogenic segments is very much limited to reduce the reproductive cost in the simultaneous hermaphroditic oligochaetes. In some of them, the nephridia serve as exits for the sperms.

TABLE 2.1

Examples for the presence of ovary and testis in the same or separate gametogenic segments in some hermaphroditic polychaetes. SH = simultaneous hermaphrodites

Species, Reference	Features
Gonochorics	
Nereidids, Eunicids	Almost all segments are gametogenic
Arenicola	Only 6 gametogenic segments
Spirorbis spirorbis, Daly (1978a)	Only 2 ovarian segments
Protandrics	
Ophrytrocha diadema Schleicherova et al. (2010)	Until the age of 21st day, testes are present in the 4th and 5th segment. Becomes SH, when 14–17 segments are added. Functions as SH from 30th–40th d
O. gracilis Sella et al. (1997)	Four testicular segments from 3rd to 6th. Followed by 14 ovarian segments from 15/16 to 30th. External fertilization. Generates 11.2 cocoons/w for 13 weeks
SH with gonads in different segments	
Aracia sinaloae Tovar-Hernandez and Dean (2014)	First five abdominal ovarian segments followed by last eight testicular segments
Laonome albicingillum Hsieh (1997)	Anterior most 10 abdominal ovarian segments and subsequent the 33 abdominal testicular segments. Extra-ovarian maturation. No seminal receptacle
Neanthes limnicola Fong and Pearse (1992)	Viviparous. Oocytes collected from posterior one third of the body
SH with gonads in the same segment	
Branchiomma bairdi Tovar-Hernandez et al. (2009)	Gametogeneic segments are present from 7th thoracic to entire abdomen. Ovary and testis co-exist in the same segment but separated topographically
Diopatra marocensis Arias et al. (2013)	Oogenesis and spermatogenesis occur in temporally and topographically separated regions of the same segment. Spermatogenesis occurs close to the ventro-lateral branches of the blood vessel from February to April. Oocytes mature in the coelom during March–June. Self-fertilization. SH at 2 no./m^2, protandric at 4 no./m^2 and gonochoric at > 10 no./m^2. Female ratio ranges from 0.67 to 0.80

In other oligochaetes, the gametogenic segments are located anteriorly (Fig. 2.1A). The female gametogenic segments are located posterior to the male segments. Notably, the paired gonads are developed as discrete organs with the respective ducts so that a discrete reproductive system is present. The female components include a pair of ovaries, ovisacs and oviducts as well as female pores in the 13th and 14th segments. The male component

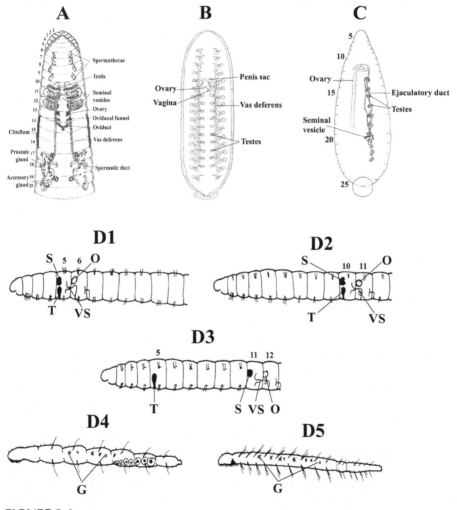

FIGURE 2.1

Reproductive system in A. an oligochaete earthworm (freehand drawing), B. hirudinean *Hirudo medicinalis* (modified and redrawn from Mann), C. *Glossiphonia complanata* (modified and redrawn from Harding and Moore) and D1. naidid, D2. tubificid, D3. enchytraeid, D4. potomadrillid and D5. aeolosomatid. Segments number are indicated by Arabic numerals, T = testis, VS = vas deferens, S = spermatheca, O = ovary, G = gonad. Note the changes in distribution of reproductive components in different segments (drawn from Lassarre, 1971).

comprises testes, male funnels in the 10th and 11th segments and variable number of seminal vesicles in 9–12 segments as well as different ducts and male pores (see Diaz-Cosin et al., 2011). The number of gametogenic segments along the body varies in different families (Barnes, 1974). Figure 2.1D$_1$–D$_5$ shows the locations of ovary, spermatheca and vas deferens in naidid, tubificid, enchytraeid, potamodrillid and aeolosomatid. In the exceptional aeolosomatids, sexual reproduction is a very rare phenomenon. Genital organs are known only in some species. Paired gonads differentiate at the ventral coelomic epithelium with ovaries in the mid-body and testis in front of them. Usually, only one ovary is fully mature. Gonoducts are lacking. Sperms are discharged through metanephridia and eggs are shed through a simple porous body wall. Copulation has never been observed. Spawning and normal sexual development are reported only from *Aeolosoma quaternarum* (Bunke, 1986). For more details, Jamieson and Ferraguti (2006) may be consulted.

Hirudineans are also hermaphrodites, in which the number of ovary is limited to a single pair, as in *Hirudo medicinalis* (Fig. 2.1B) but the pair is extended to a few segments, as in *Glossiphonia complanata* (Fig. 2.1C). However, the number of testicular segments is 10. Notably the anlage of the ovaries is anterior to those of testes. The oocytes leave the ovaries to ovisac, where they mature (Barnes, 1974).

Regrettably, not many authors have described the exact locations of the ovarian and testicular segments in polychaetes. Available information including that of a passing remark for *Neanthes limnicola* is listed in Table 2.1. In *Ophryotrocha* spp, two–four anterior testicular segments appear first; subsequently, posterior segments are added, of which 4–16 segments harbor ovaries. In them, the anlage for the testicular and ovarian segments is clearly separated. This is also true of some SH like *Aracia sinaloae* and *Laonome albicingillum*. Notably, the ovarian segments appear first in the anterior abdominal segments. In other SH, both ovaries and testes co-exist in the same segment, although topographically separated, as in *Branchiomma bairdi* and temporally also in *Diopatra marocensis*. Strikingly, not all SH commence as SH. Manifestation of hermaphroditism is dependent on population density. For example, *D. marocensis* functions as SH, protandric and gonochoric at densities of 2, ~ 4 and > 10 no./m^2, respectively (see Arias et al., 2013).

Differences in structural organization of the reproductive system among annelids have the following implications: 1. The Primordial Germ Cells (PGCs) are responsible for the manifestation of sex. Derived from PGCs, the Oogonial Stem Cells (OSCs) manifest female sex in some individuals and that of Spermatogonial Stem Cells (SSCs) male sex in others. Understandably, mutations involving one or more genes have retained both OSCs and SSCs in hermaphroditic annelids. 2. In polychaetes, the transition from crawling errant mode of life to sedentary/sessile mode has involved reduction in the number of gametogenic segments to facilitate the sharing of resource allocation between gametogenesis and brooding/parental care. 3. In oligochaetes and hirudineans, the cost of manifesting and maintaining dual reproductive systems is facilitated by reduction in the number of the 'costlier' ovary to a single pair, in comparison to that of gonochoric polychaetes, in which the minimum is six pairs, as in *Arenicola*. Internal fertilization and protection of eggs within the cocoon increase survival and recruitment of progenies. To escape from 'inbreeding depression' and to increase genetic diversity, every effort is made by hermaphroditic annelids to ensure cross fertilization (see later). On enforced isolation, some of them reduce allocation for cocoon production (e.g. *Eisenia foetida*).

2.2 Gonochorism

Sex is a luxury, and costs time and resources but ensures recombination and genetic diversity, the raw material for evolution and speciation. Not surprisingly, the polychaetes are more diverse and speciose than oligochaetes and hirudineans. Both male-(XX-XO) and female-(ZZ-ZW) heterogametic sex determination mechanisms are reported to operate in gonochoric polychaetes. Whereas the mechanisms are stable in most gonochoric polychaetes, labile mechanisms have been demonstrated in a few polychaetes. Besides, Premoli et al. (1996) have reported heritable variation in sex ratio of gonochoric *Ophryotrocha labronica* and advanced a hypothesis that sex is determined by multilocus (polygenic) genetic systems in polychaetes.

2.2.1 Sex Ratio

The ratio represents the cumulative end products of sex determination and differentiation processes. It may serve as a key index to assess potential natality in a population. Incidentally, no female is yet known in *Parenterodrilus taeniodes* (Purscke, 2006). However, sex cannot easily be distinguished morphologically in gonochoric polychaetes. Not surprisingly, many authors have refrained from reporting the ratio. However, sex can be distinguished in 42–47% of brooding polychaetes. It can also be distinguished in species, in which the color of the testis differs distinctly from that of the ovary. For example, the abdomen of a ripe *Pomatoceros* male appears white and that of a female bright pink or orange colored due to the difference in color of the sperms and eggs (see Barnes, 1974). Encountering immature and sporadic occurrence of hermaphroditic individuals, some authors have assigned an additional ratio for them (Table 2.2). In a few polychaetes, sex ratio remains stable at or around 0.5 ♀ : 0.5 ♂, indicating the chromosomal mechanism

TABLE 2.2

Sex ratio of some gonochoric polychaetes

Species	Sex Ratio ♀ : ♂ : ⚥	Reference
Chromosomal mechanism of sex determination		
Amphisamytha galapagensis	0.50 : 0.50	McHugh and Tunicliffe (1994)
Hediste japonica	0.50 : 0.50	Sato (1999)
Sabella spallanzanii	0.50 : 0.50	Currie et al. (2000)
Marphysa sanguinea	0.50 : 0.50	Prevedelli and Simonini (2003)
Perinereis cultrifera	0.50 : 0.50	Prevedelli and Simonini (2003)
P. rullieri	0.50 : 0.50	Prevedelli and Simonini (2003)
Glycera dibranchiata	0.55 : 0.45	Creaser (1973)
Polygenic mechanism of sex determination		
Bispira brunnea	0.33 : 0.37 : 0.22*	Davila-Jimenez et al. (2017)
Branchipolynoe seepensis	0.19 : 0.28 : 0.29*	Jollivet et al. (2000)
Nicolea upsiana	0.19 : 0.46 : 0.35†	Garraffoni et al. (2014)
N. zostericola	0.45 : 0.38 : 0.13†	Eckelbarger (1975)
Nereis virens > 30 cm	0.50 : 0.50	Creaser and Clifford (1982)
< 30 cm	0.67 : 0.33	
Capitella capitata		Petraitis (1985b)
♀ × ♂	0.50 : 0.50	
⚥ × ♂	0.03 : 0.97	
Dinophilus gyrociliatus	0.75 : 0.25	Prevedelli and Simonini (2003)
D. gyrociliatus		Prevedelli and Vandini (1999)
cereals	0.50 : 0.50	
tetramin	0.67 : 0.33	
Ficopomatus enigmaticus		Obenat and Pezzani (1994)
spring cohort	0.39 : 0.61	
autumn cohort	0.26 : 0.74	
Pomatoleios kraussi	0.67 : 0.33	Nishi (1996)
Polydora ligni	0.71 : 0.29	Zajac (1991)
Halla parthenopeia	0.28 : 0.72	Osman et al. (2010)
Lumbrinereis funchalensis	0.19 : 0.81	Osman et al. (2010)

* hermaphrodites, † immature, hence sex not known

of sex determination. In others like *Capitella* sp I, the homogametic ZZ males change sex to hermaphrodites and a cross between ZZ male and ZZ hermaphrodite skew the ratio in favor of male. In male heterogametic (XO) *Dinophilus gyrociliatus*, selective fertilization of larger eggs by sperms bearing X chromosome and small eggs by sperm carrying no sex chromosome nullifies the chromosomal mechanism of sex determination and sex ratio, as well. In all others, sex seems to be determined by polygenic system, i.e. one or more genes harbored on autosome(s) have an overriding effect in sex differentiation, following sex determination by sex chromosome. As a result, female ratio, for example, varies widely from as low as 0.19 in *Lumbrinereis funchalensis* to as high as 0.71 in *Polydora ligni*. Apparently, the autosomic genes in them express in response to one or other environmental factors like temperature in *Ficopomatus enigmaticus* and algal food *Halopteris scoporia*. In *Typosyllis prolifera*, homo/heterogametic sex chromosome determines irrevocably the primary sex of male, as its primary sex ratio is 0.5 ♀ : 0.5 ♂. However, female differentiation is labile, whereas differentiation in primary males remains absolutely stable throughout the life. Hence, some females switch to males depending on density. The number of females switching to males increases from 62% in Porec population to 77% in Pula population in Yugoslavia. Further, laboratory experiments have shown that the social condition is a more important factor in females switching to males in this 'genetically unbalanced protogynous/serial hermaphrodite'. Control of sex ratio by environmental factors is explained in more details elsewhere (Chapter 7).

2.2.2 Ovary Somatic Index

Gonado Somatic Index (GSI) relates to gonad weight to body weight. The nearest equivalent value is reported for the terebellid polychaete *Eupolymnia nebulosa* (Gremare, 1986). In view of difference in resource allocation, values for the Ovary Somatic Index (OSI) and Testis Somatic Index (TSI) are reported during recent years (e.g. fishes, Pandian, 2015). In *E. nebulosa*, the OSI value increases from 9 at age 2 to 21 at 4 years (Fig. 2.2A). However, TSI remains

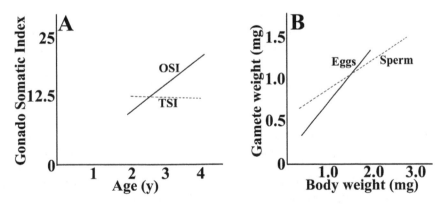

FIGURE 2.2

A. 'Gonado Somatic Index' as function of age in *Eupolymnia nebulosa* (modified and redrawn from Gremare, 1986). B. Gamete weight as function of body weight in *Mercierella enigmatica* (modified and redrawn from Dixon, 1976).

TABLE 2.3

Ovary Somatic Index (OSI) of some polychaetes (compiled from McHugh, 1993, Rouse and Fitzhugh, 1994). G = gonochoric, fs = free spawner, ds = discrete spawner, cs = continuous spawner, D = direct development, S = semelparous, Sqb = sequential brooder

Species	Reproductive Mode	Fecundity	OSI
Terebellids (McHugh, 1993)			
Eupolymnia nebulosa†	G	–	15 ♀, 12 ♂
E. crescentis	G, fs, ds	128, 500	6.7
Neoamphirite robusta	G, fs, ds	829, 833	8.6
Thelepus crispus	G, D, cs	51, 555	5.9
T. crispus	G, D, cs	51, 555	0.32*
Ramex californiensis	G, cs, Sqb	410	2.4
Sabellids (Rouse and Fitzhugh, 1994)			
Caobangia abbotti	Intra brooder	1	0.03
Fabricinuda trilobata	G, fs, Intra brooder	2	0.02
Manayunkia aestuarina	G, fs, riverine, Intra brooder	2	0.05
M. caspica	G, fs, riverine, Intra brooder	2	0.08
M. speciosa	G, fs, riverine, Intra brooder	4	0.10
Augeneriella alata	Intra brooder	2	0.11
Fabricia stellaris	Intra brooder	4	0.14
Amphiglena marita	Maternal nutrients drawn	10	0.05
A. terebro	♂, Intra brooder	7	0.29
A. mediterranea	♀, Intra brooder	9	0.01
A. nathae	♂, Intra brooder	9	0.16
Amphicorina brevicollaris	Intra brooder	10	0.14
Demonax medius	Intra brooder	1200	0.11
Potamilla torelli	Intra brooder	202	0.06
Potamilla sp	Intra brooder	518	0.63
Perkinsiana sp	♂, asynchronous extra brooder	8666	0.07
Perkinsiana antarctica	♂, asynchronous extra brooder	300	0.01

† Gremare (1986), * data reported by Garraffoni et al. (2014)

constant at ~ 12 in the same age classes. Dixon (1975) has also estimated the dry body weight and gametes of the serpulid *Mercierella enigmatica*. Contrastingly, his calculated values indicate that (i) with increasing body weight, both TSI and OSI values increase, (ii) are in the range of > 67 at the body size of 1.5 mg (dry) body weight and (iii) the TSI values begin to exceed those of OSI beyond 1.5 mg size (Fig. 2.2B). The loss of ash-free dry weight of a spawning *Nephtys caeca* is 45 mg and amounts to OSI of 24.3; the values are 53 mg and 30.3 for *N. hombergii* (Olive et al., 1985). For the hermaphroditic oligochaetes and hirudineans, the OSI values are not available. However, it may be difficult to estimate it in a leech, for example, which engorges by sucking blood and increases its body weight. But its body weight decreases due to progressive cocoon depositions and with advancing time/age. In the hermaphroditic polychaetes, ~ 80% of the gonad consists of the ovarian component (e.g. *Ophryotrocha* spp, see p 89).

Direct measurements on the OSI have been made for four terebellid species by McHugh (1993). These values range from 2.4 for the continuous breeder and sequential brooder *Ramex californiensis* to 8.6 for the broadcast spawner *Neoamphitrite robusta*. Rouse and Fitzhugh (1994) and Garraffoni et al. (2014) have also reported data for the total egg volume and body volume of a few sabellids and terebellids, respectively. Hence, it is possible to calculate OSI on volume basis for these polychaetes (Table 2.3). These values are several orders lower than those reported from the direct estimates made on the weight basis by Dixon (1976) and Gremare (1986). As fecundity of some sabellid species is > 10 eggs, the calculated OSI values suggests that in such species characterized by the intra-ovarian oogenesis may have low OSI values. Notably, the OSI values for *Thelepus crispus* directly estimated and those calculated are 5.9 (McHugh, 1993) and 0.32 (Garraffoni et al., 2014), respectively. It is also difficult to comprehend that the semelparous terebellid *Amaena occidentalis* has an OSI of 0.036 (Garraffoni et al., 2014). Hence, it is recommended that OSI and TSI values are directly estimated on the basis of weight to enable the comparison of reproductive performance between annelid species and others, as well.

2.3 Hermaphroditism

It is defined as the expression of both female and male functions in a single individual either simultaneously or sequentially. Three patterns of functional hermaphroditism have been recognized: (i) simultaneous hermaphrodites functioning as a male or female at a time (unilateral mating partners, e.g. *Ophryotrocha gracilis*, Sella et al., 1997) or as both male and female within a short span of time (reciprocal mating partners, e.g. *O. gracilis*). They may not usually undergo natural sex change, although *O. gracilis* can do it. The sequentials do it once in a life time in a single direction but the serials do it more than once in either direction (e.g. *Syllis amica*, Table 2.4). The sequentials are further divided into (a) male to female sex changing protandrics and (b) female to male sex changing protogynics. The former is more common among aquatic invertebrates (e.g. molluscs, Pandian, 2017, echinoderms, Pandian, 2018) but the latter among teleostean fishes (Pandian, 2011). Of

TABLE 2.4

Hermaphroditic polychaetes (updated and compiled from many sources including those indicated by bold letters by Schroeder and Hermans, 1975)

Simultaneous hermaphrodites

Archiannelidae: *Mesonerilla armoricana, M. fagei, M. roscovita*

Arenicolidae: *Branchiomaldane vincenti*

Capitellidae: *Capitella hermaphrodita, Capitomastus minimus*

Cirratulidae: *Aphelochaeta* sp, *Caulleriella parva, Chaetozone vivipara, Ctenodrilus serratus*

Ctenodrilidae: *Raphidrilus nemasoma, Trilobodrilus*

Hesionidae: *Hesione sicula, H. mazima(?), H. pantherina, Microphthalmus fragilis, M. listensis, M. oberrans, M. sczelkowii, M. similis, M. tyrrhenicus, M. urofimbriatus*

Nereididae: *Neanthes lighti, N. limnicola* (viviparous) (Fong and Pearse, 1992), ***Namalycastis indica, Namanereis quadraticeps, Platynereis massiliensis***

Onuphidae: *Diopatra marocensis* (Arias et al., 2013)

Orbiniidae: *Nainereis laevigata*

Pectinariidae: *Lagis koreni*

Polygordiidae: *Polygordius triestinus*

Polynoidae: *Macellicephala violacea*

Sabellidae: *Amphiglena marita, A. mediterranea, A. nathae, A. terebro, Aracia sinaloae, Branchiomma luctuosum, B. bairdi, **B. cingulata, Caobangia billeti**, Demonax pallidus, **Euratella salmacidis**, Laonome albicingillum, **Manayunkia aestuarina**, Ophryotrocha bacci, O. diadema, O. gracilis, O. hartmanni, **O. labronica**, O. maculata, O. socialis, Perkinsiana antarctica, Perkinsiana* sp

Serpulidae: *Apistobranchus glacierae, Filograna / Salamacina complex, Pomatoceros lamarckii, Serpula polycerus, Spirobranchus polycerus**

Sphaerodoridae: *Ephesiella mizla*

Spionidae: ***Polydora gigardi**, P. hermaphroditica, P. ligni, P. commensalis?, Spio filicornis**

Spirorbidae: *Bushiella abnormis, B. atlantica, B. granulata, B. kofiadii, B. quadrangularis, Bushiella* sp, *B. similis, Circeis armoricana, C. oshurkovi, C. paguri, Circeis* sp, *Eulaeospira convexis, Helicosiphon platyspira, H. biscoensis, Janua pagenstecheri, Metalaeospira clasmani, M. pixelli, M. tenuis, Neodexiospira alveolata, N. brasiliensis, N. foraminosa, N. formosa, N. kayi, N. lamellosa, N. pseudocorrugata, N. steueri, Nidificarisa nidica, N. palliata, Paradexiospira vitrea, Paralaeospira levinseni, P. malardi, P. parallela, Pileolaria* sp1, *P.* sp2, *P. berkeleyana, P. daijonesi, P. darkarensis, P. lateralis, P. marginata, P. militaris, P. pseudoclavus, P. spinifer, P. tiarata, Pillaiospira trifurcata, Protolaeospira(?) eximia, P. pedalis, P. striata, P. tricostalis, P. triflabellis, P. stalagmia, Romanchella pustulata, R. quadricostalis, R. scoresbyi, R. solea, Simplaria postwaldi, S. pseudomilitaris, Spirorbis bifurcanis, S. corallinae, S. cuneatus, S. gesae, S. inornatus, S. infundibulum, S. rothilisbergi, S. rupestris, S. spatulatus, S. spirorbis, S. strigatus, S. tridentatus, Spirobranchus cariniferus, Vinearia zibrowii*

Syllidae: *Bollandia anthipathicola?, **Brania protandrica**, Brania pusilloides, Exogone naidina, Grubeosyllis neapolitana, Haplosyllis spongicola, **Myrianida pinnigera**, Pionosyllis neapolitana, Sphaerosyllis hermaphrodita, **S. corruscans, Typosyllis variegata**, T. vittala*

Terebellidae: *Alkmaria romijni*

Tomopteridae: *Enapleris euchaeta*

Typhloscolecidae: *Sagitella kowalewskii* (?)

Table 2.4 contd. ...

...Table 2.4 contd.

Protandric hermaphrodites

Archiannelidae: *Nerillidium gracile, N. macropharyngeum, N. mediterraneum, N. renaudae, N. simplex, N. troglochaetoides, Nerillidopsis hyaline, Troglochaetus beranecki*

Dorvilleidae: *Sabellastarte spectabilis*

Myzostomida: *Asteromyzostomum grygieri, Contramyzostoma, Cyclocirra, Cystimyzostomum, Endomyzostoma scotia, E. neridae, Hypomyzostoma jasoni, H. jonathoni, Mesomyzostoma botulus, M. katoi, M. lanterbecqae, M. leukos, M. lobus, M. okadai, M. reichenspergeri, Myzostoma cirriferum, M. debiae, M. deformator, M. eeckhauti, M. fuscomaculatum, M. glabrum, M. hollandi, M. indocuniculus, M. josefinae, M. kymae, M. laurenae, M. miki, M. pipkini, M. pulvinar, M. susanae, M. tertiusi, Notopharyngoides, Protomyzostomum lingua, P. roseus, Pulvinomyzostomum inaki, P. messingi*

Serpulidae: *Ficopomatus enigmaticus, Galeolaria caespitosa, G. hystrix, Hydroides elegans, H. norvegica, Mercierella enigmatica, Paradexiospira vitrea, Pomatoceros lamarckii, P. triqueter, Protolaeospira translucens, Sabellastarte magnifica, Salmacina australis, S. incrustans*

Syllidae: *Brania clavata, Exogone gemmifera, E. verugera, Janus knightjohnsi*

Spionidae: *Dipolydora commensalis*

Protogynic hermaphrodites

Serpulidae: *Sabella microphthalma*

Syllidae: *Syllis variegata*

Serial hermaphrodites

Dorvillidae: *Ophryotrocha puerilis puerilis*

Syllidae: *Syllis amica, S. prolifera* (Policansky, 1982), *Trypanosyllis zebra, Typosyllis prolifera*

Leech: *Helobdella striata* (Kutschera and Wirtz, 1986)

*of two morphs, one is hermaphroditic

~ 115,000 molluscan species, 23 and 2% are simultaneous and sequential hermaphrodites, respectively. The simultaneous hermaphroditic pulmonates (24,000 species) invariably carry an ovotestis (see Pandian, 2017) but none of the annelids bear an ovotestis. Incidentally, Davison (2006) considers ovotestis as an underdeveloped organ of evolution.

From a survey, Schroeder and Hermans (1975) have reported that 67 polychaetes belonging to 23 families are hermaphrodites; of them, 31, 32 and 4 species are recognized as simultaneous, protandric (*Ophryotrocha* spp) and protogynic (e.g. *Sabella microphthalma*) hermaphrodites, respectively. Among the 23 families, the incidence of hermaphroditism is more common in Hesionidae, Syllidae, Sabellidae, Serpulidae and Nereididae. After a period of > 25 years, the reviews of Premoli and Sella (1995), Giangrande (1997) have virtually repeated the same conclusions arrived by Schroeder and Hermans. For some reasons, no attempt has been made to validate and update the conclusions of Schroeder and Hermans. From an intensive survey of relevant literature up to 2017 and computer search, the incidences of hermaphroditism in 207 species belonging to 23 families are listed in

Table 2.4. This updating has revealed that the number of hermaphrodites in polychaetes is more than doubled and implies that > 1.35% (207 out of 13,002 species) of polychaetes are hermaphrodite. This is the first updated estimate of hermaphroditism in polychates and the number is likely to increase. Notably, all the 70 species belonging to the family Spirorbidae are SH (Kupriyanova et al., 2001). A computer search has revealed the presence of ~ 26 myzostomid species, all of them have been considered as protandrics. Though broadly classified into (i) simultaneous, (ii) protandric, (iii) protogynic and (iv) serial hermaphrodites, Schroeder and Hermans have grouped them into seven types but assigned ~ 20 species to one or other type with the question mark. Hence, the classification of hermaphroditic types remains very fluid. At least, a few species may be shifted from one to other type.

A vast majority of polychaetes are gonochores. However, of ~ 13,000 polychaete species (Table 1.2), > 207 species (see Table 2.4) (1.35% of polychaetes, 0.6% of annelids) are hermaphrodites. Both oligochaetes (3,155 species, 18.7% of annelid species) and hirudineans (684 species, 4.0%) are hermaphrodites. This is perhaps the first estimate to indicate that the annelid comprises of ~ 76% gonochorics and 24% hermaphrodites. This estimate of 76% gonochorism in annelids may be compared with 75% in molluscs (Pandian, 2017), 92% in crustaceans (Pandian 2016) and 99% in echinoderms (Pandian, 2018).

Notably, a few observations on the polychaete hermaphroditism may be mentioned. In the serpulid *Spirobranchus polycerus*, Marsden (1992) has reported the presence of seven opercular horned SH morph inhabiting singly or in a small group on the live hydrozoan coral *Milliporea complanata* and two horned gonochoric morph living in the low tidal zone. Lacking seminal receptacle, the serpulid *Laonome albicingillum* is a self-fertilizing SH (Hsieh, 1997). With sequential sex change, the protandrics exhibit different sex ratios. Figure 2.3 summarizes the ontogenetic pathways, through which protandric to female, protogynic to male and SH are generated by sex change in polychaetes. Notably, the female gonad is simply added in the protandric to SH. Hence, these protandrics do not undergo sex change but only adds female sex. The ratio of ♀ : ♂ in the protandric hermaphrodites is 0.61 : 0.39 in *Ophryotrocha diadema* and 0.70 : 0.30 in *O. gracilis* (Sella and Ramella, 1999). An interesting protandric is the sabellid *Sabellastarte spectabilis*. The expression of sex ratio has been assessed as ♂ : ♀ : ⚥ : 0 (unknown) in small (6–8 mm), medium (9–10 mm) and large (11–13 mm) worms. The ratios increase from 0.13 to 0.30 and 0.07 to 0.31 in females and hermaphrodites, respectively. In the same sizes, it decreases from 0.56 to 0.25 and 0.24 to 0.14 in males and unknowns, respectively (Bybee et al., 2006). Notable is the persistence of males even in the largest size of protandric polychaete hermaphrodites at ratios ranging from 0.14 in *S. spectabilis* to 0.39 in *O. diadema*.

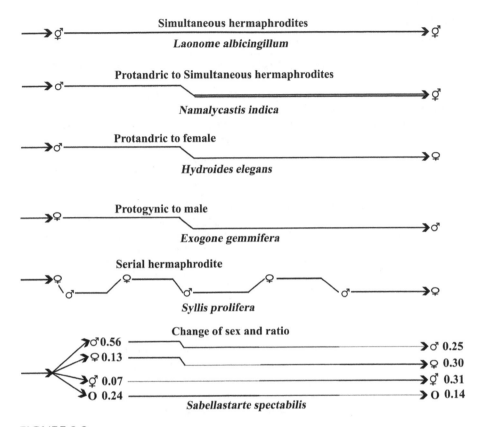

FIGURE 2.3

Ontogenetic pathways through which sex is changed or sex ratios are changed in some hermaphrodites. O represents unknown sex.

2.4 Parthenogenesis

A number of authors have defined parthenogenesis (e.g. Suomalainen, 1950, Beatty, 1967). In view of polyploid automictic parthenogens, in which triploid (3n) parthenogenic eggs require activation by a haploid sperm from diploid sexual morph, parthenogenesis is defined as the generation of embryo from an egg without any genetic contribution by a sperm. For the first time, the incidences of parthenogenic annelids are assembled in Table 2.5. Of > 75 parthenogenic annelid species, 75% are earthworms. The ability of earthworms for asexual reproduction is limited (e.g. Xiao et al., 2011), in comparison to that of enchytraeids and naidids. Understandably, earthworms have opted for parthenogenesis to reduce the cost of developing and maintaining a dual reproductive system. The incidence of parthenogenesis

TABLE 2.5

Parthenogenic annelids (compiled from many sources)

Polychaetes: *Dodecaceria concharum, D. pulchra* (Gibson, 1977), *Brania pusilla* (Hauenschild, 1955)

Tubificids: *Tubifex tubifex, Limnodrilus hoffmeisteri, L. mastix, Lumbriculus variegatus* (Poddubnaya, 1984, Brinkhurst, 1986)

Enchytraeids: *Cognettia glandulosa, Fridericia galba, F. ratzeli, F. striata, Lumbricillus lineatus, Mesenchytraeus glandulosus, Achaeta bilobisa?, F. bisetosa?, F. connata?* (Christensen, 1961)

Earthworms: *Allolobophora trapezoides, Amynthas bileatus, A. catenus, A. chilanensis, A. corticus, A. cruxus, A. diffringens, A. gracilis, A. hohuanmontis, A. hupiensis, A. shinammontis, A. tokioensis, Aporrectdodea caliginosa, A. rosea, A. trapezoides, Bimastos beddardi, B. gieseleri, B. heimburgeri, B. longicinctus, B. palustris, B. parvus, B. tumidus, B. welchi, B. zeteki, Drawida hattamimiju, D. nepalensis, Dendrobaena octaedra, D. rubida, Dendrodrilus rubidus, Eisenia rosea, Eiseniella tetraedra, Eukerria saltensis, Heliodrilus hachiojii, Lumbricus eiseni, L. terrestris, Microscolex dubius, M. phosphoreus, Metaphire hilgendorfi, Ocnerodrilus occidentalis, Octolasion cyaneum, O. occidentalis, O. tyrtaeum, Onychochaeta windlei, Pheretima agrestis, P. bicinta, P. diffringens, P. hilgendorfi, P. hupiensis, P. levis, P. loveridgei, Piutellus papillifer, P. umbellulariae, Pontoscolex corethrurus, Dichogaster bolaui?, Allolobophora muldali* (PF) (Reynolds, 1974, Jaenike and Selander, 1979, Blakemore, 2003, Tsai et al., 2007).

---

is limited to a couple of polychaetes and to a few species in tubificids and enchytraeids. At present, there is no report on the incidence of parthenogenic hirudinean. Incidentally, Gibson (1977) has stated that *Dodecaceria concharum* and *D. pulchra* are parthenogens, as males are not found and 70% of the mature worms have oocytes. However, he has not provided any additional evidence to confirm that these polychaetes are parthenogens.

2.4.1 Parthenogenic Types

In annelids, the origin of parthenogenesis can be traced to two sources: (i) the repeated polyphyletic origin of new parthenogens from sexual ancestors (e.g. *Fridericia striata*, Christensen et al., 1989) and (ii) monophyletic origin through accumulation of new mutations within an existing lineage. Cytological studies have revealed the presence of three major parthenogenic types: 1. Ameiotic/apomictic parthenogens. In them, synapsis and bivalent formation do not occur (e.g. *Dendrobaena octaedra*, Suomalainen et al., 1987). 2. Automictic/gynogenic parthenogens retain features of normal meiosis including crossing over and so on, *albeit* no crossing over in polyploids. Whereas meiosis in diploids is chiasmatic, polyploids undergo synaptic meiotic division in the oocytes (Christensen, 1960, 1961, 1980). However, the haploid egg pronucleus restores diploidy by fusing with that of homologous second polar body (e.g. *Lumbricillus lineatus*). This is also true of the family Lumbricidae and in the tubificid genus *Limnodrilus* (Christensen, 1984). Interestingly, the eggs of triploid and other polyploid parthenogens do require activation by sperm from diploid sexual morph. Consequently, if sperm producing morphs are absent, there will be no triploids (Christensen et al., 1978). 3. In telytochorous parthenogens, in which pre-meiotic doubling of chromosomes at the last oogonial divisions result in endomitosis followed by the formation of chiasmatic bivalents and regular meiosis with extrusions of two polar bodies. The genetic consequences of this cytological mechanism are parallel to that of apomictics, as synapsis is limited between sister chromosomes that are exact molecular copies of one another (see Diaz-Cosin et al., 2011). As a result, diploid eggs are produced without recombination leading to production of genetically indistinguishable clones (e.g. *Tubifex tubifex*, Christensen, 1984). Nevertheless, electrophoretic analysis of 3n parthenogenic *Fridericia galba* has revealed the presence of 13 different clonal phenotypes (Christensen et al., 1992). Similarly, allozyme analysis can also discriminate selfers from parthenogens by comparing heterozygous loci of an individual and its related uniparental progeny (see Baldo and Ferraguti, 2005). Notably, of 33 lumbricid species found in North America, 17 of them reproduce primarily or exclusively by parthenogenesis, i.e. they are obligate parthenogens (Jaenike and Selander, 1979). There are also facultative (e.g. *Allolobophora muldali*) and amphimictic cum parthenogens (e.g. *Diplocardia singularis, Dendrobaena rubida, Eisenia foetida*) (Reynolds, 1974).

Incidentally, despite his elaborative cytological study on oogenesis in parthenogenic worms (e.g. Christensen, 1980), Christensen has not considered the centrosomes and their derivative centrioles. Their function is important for assembling the spindle apparatus that organizes separation of chromatids. In most animals, centrioles are degraded during oogenesis and are inherited paternally through the sperm (Engelstadter, 2008). It is not clear whether centrioles are either not degraded during oogenesis or synthesized *de novo* in the parthenogenic eggs of annelids.

2.4.2 Parthenogenic Levels

The existence of facultative parthenogens and amphimictic parthenogens clearly indicates that components of male reproductive system are not stably maintained. In his summarizing note, Gates (1971) has reported the gradual but progressive reductions in components of male reproductive system of parthenogenic oligochaetes. Accordingly, the reductions are from four to three pairs of testes, two to one pair of seminal vesicles, two to one pair of prostate, two to one pair of spermatheca and disappearance of oviducts and male pores. Reynolds (1974) has indicated that despite retention of one or more of these male components, sperms may be absent. For example, sperms/spermatophores are absent in some parthenogenic earthworms, despite the presence of testes and seminal vesicles. In *Tubifex tubifex*, the function of the testes diminishes and the spermatogonia are not replenished with the development of ovary. In fact, the process of spermatogenesis is not completed and ceases at the spermatid stage (Poddubnaya, 1984). In the parthenogenic earthworms too, the spermatogenesis may be arrested at spermatid stage, despite the presence of one or more components of

TABLE 2.6

Levels of reduction in components of male reproductive system in the Taiwanish parthenogenic earthworms (condensed and compiled from Tsai et al., 2007)

Spermatheca	Prostate Glands	Seminal Vesicles
Amynthas shinanmontis		
Absent 10/20	Absent 3/20	Vestigial 5
One pair 1/20	Vestigial 3/20	Medium 3
Two pairs 2/20	Nodular 13/20	Normal 12
Three pairs 7/20	Normal 1/20	
A. chilanensis		
Absent	Absent	Small
A. sheni		
Absent	Small	Small

the reproductive system. Interestingly, Tsai et al. (2007) have described a spectrum of reductions or loss of spermatheca, prostate glands and seminal vesicles in Taiwanese earthworms. Irrespective of the presence of seminal vesicle in *Amynthas shinanmontis*, the number of spermatheca is reduced from four to two pairs in 2 out of 20 specimens collected (Table 2.6). Clearly, the parthenogens may have originated from sexual oligochaetes by a stroke of one or more mutations arresting spermatogenesis in parthenogenic oligochaetes but the components of male organs are reduced gradually with more and more mutations.

2.4.3 Ploidy Levels

In oligochaete parthenogens, polyploidy is the most common phenomenon. An advantage of polyploidy in parthenogens is the increased scope for genetic diversity (Diaz-Cosin et al., 2011). Table 2.7 indicates that the number of polyploid parthenogenic species is more in lumbricids than enchytraeids. Diaz-Cosin et al. (2011) have listed the following advantages for polyploid oligochaetes, especially the earthworms: 1. High levels of heterozygosity and exceptionally fit genomes that are sustained and inherited by avoiding recombination and segregation. 2. Manifestation and maintenance of a single reproductive system facilitate relatively higher growth and reproductive rates. The questionable parthenogenic singletons of *Octolasion cyaneum* grow faster at the rate of 204 mg/g/d, in comparison to the twins generated by cross fertilizers (177 mg/g/d). With diploid and haploid eggs in them, cocoon production is 0.8 and 1.55 cocoon/w for the singletons and twins, respectively. The singletons generate also 32 mg weighing hatchling, in comparison to cross fertilizing twins producing 18 mg weighing hatchling (Lowe and Butt, 2008). In *Aporrectodea trapezoides*, parthenogenic singletons attain sexual maturity at the age of 153 days and 1 g size, and produce 2.03 cocoon per

w. 3. The no need to search for a mate and mating expedite colonizing ability and 4. Polyploid and polymorphic morphs are generated from selection at the level of genomes. The level of polyploids ranges from triploid, tetraploid, pentaploid, hexaploid and heptaploid in lumbricids and enchytraeids (Table 2.7). Of 319 parthenogens collected from Australia, Canada and Europe, 54.5, 6.0, 39.5 and 1.0% are diploids, triploids, tetraploids and pentaploids, respectively (Coates, 1995). Briefly, the incidence of polyploid parthenogens is more fecund in enchytraeids than in lumbricids (Table 2.8).

TABLE 2.7

Polyploids in parthenogenic oligochaetes (compiled from Jaenike and Selander, 1979, Coates, 1995)

Species	Ploidy	Chromosomes (no.)
Lumbricidae		
Aporrectodea trapezoides	2n	36
	3n	54
Octolasion tyrtaeum	2n	38
	3n	54
	4n	72
A. rosea	3n	54
	5n	90
	6n	108
Dendrodrilus rubidus	2n	34
	4n	68
	6n	102
D. octaedra	6n	108
	7n	124
Enchytraeidae		
Lumbricillus lineatus	2n	26
	3n	39
	4n	52
	5n	65

TABLE 2.8

Polyploid distribution in Lumbricidae and Enchytraeidea (modified from Coates, 1995)

	Lumbricidae	Enchytraeidea
Species (no.)	56	79
2n species	37	47
Parthenogens (no.)	0	1
In + Polyploidy species (no.)	9	9
Polyploidy species (no.)	10	23
Parthenogens (no.)	8–9	4

2.5 Fecundity

Fecundity is a decisively important factor in recruitment and sustaining a population. By providing space to accommodate the oocytes/eggs, body size becomes a critically important factor in determining fecundity. Sexuality and egg size are also important factors in deciding the level of fecundity. Beside these, food supply (e.g. *Dinophilus gyrociliatus*, diet quality, Prevedelli and Vandini, 1999), temperature (e.g. *D. gyrociliatus*, Akesson and Costlow, 1991), salinity (e.g. *D. gyrociliatus*, Akesson and Costlow, 1991, *Neanthes limnicola*, Fong and Pearse, 1992) and photoperiod (e.g. *Streblospio benedicti*, Chu and Levin, 1989) also play a role in deciding fecundity.

The total number of oocytes contributing to fecundity is assured by waves of oogonial proliferation and subsequent oocyte recruitment (see Pandian, 2013). With regard to fecundity of annelids, the terms used by fishery biologists have to be introduced. Firstly, Batch Fecundity (BF) or a clutch is the number of eggs produced per spawning or brood. It is a function of body size (Length, L or Weight, W). BF is related to the volume of space available in the (intra-ovarian) ovary or (extra-ovarian) coelomic cavity. To accommodate maturing oocytes, geometry ($F = aL/W^b$) suggests that the length/weight exponent would be 3.0. Unlike broadcasting oviparous fishes, ~ 50% of polychaetes brood their eggs and the clitellates encapsulate them in cocoons. Though these features complicate BF relation to body size, BF increases with increasing body size in many annelids (Figs. 1.10D, E, F, 1.11A, B). In annelids, growth is expressed mostly in length, girth in some (e.g. *Hediste japonica*) and breadth in others (e.g. *Hirudo medicinalis*). With these morphological differences, the fecundity-body size relation may tilt the expected exponent of 3.0. However, no information is yet available on this aspect.

Secondly, Potential Fecundity (PF) is the maximum number of oocytes commencing to differentiate and develop. However, due to one or other environmental factors like food supply, a fraction of these developing oocytes may not be developed or resorbed through atresia. For example, ~ 42% of the pre-vitellogenic oocytes released into the coelom in *Spirorbis spirorbis* are matured representing the Realized Fecundity (RF) (Fig. 2.4B). Life time Fecundity (LF) is a cumulative number of eggs produced in different batches/seasons. For example, *S. spirorbis* produces five batches of broods during its LF totaling to 633 eggs (Daly, 1978b). Relative Fecundity (RF) is the number of eggs/oocytes per unit weight of the body. For example, it is ~ 1.0 egg/μg dry body weight of *S. spirorbis* (Daly, 1978b). It provides a scope for analysis of the reproductive performance with reference to body size within a species, populations from different geographical areas and different years within a geographical area as well as between comparable species.

Thirdly, a distinction must also be made between species with determinate and indeterminate fecundity. In the former, the PF is fixed prior to the commencement of spawning or a spawning season. It depends on the stored energy reserves from the body and/or storage organs, as the chloragogue in oligochaetes and blood in leeches; hence the clitellates are all capital breeders (see Pandian, 2015). In them, development of oocytes is likely to be synchronized. In the indeterminates, the PF, however, is not fixed and the development of oocytes is asynchronized (e.g. *Perkinsiana* spp, Gambi et al., 2000). Their PF depends on incoming energy resource and are not capital income breeders (see Pandian, 2015). The determinate annelids adopt two different spawning strategies: (i) synchronous determinate total spawning (e.g. semelparous *Nereis virens*, Fig. 2.7A) and (ii) synchronous determinate iteroparous spawning (e.g. *Marphysa sanguinea*, Fig. 2.7C). On the other hand, consequent to low incoming energy, oocytes are asynchronously developed in those characterized by indeterminate fecundity, resulting in successive spawning during a spawning season (e.g. *Nephelopsis obscura*, Peterson, 1983). In them too, fecundity decreases with advancing time/age and increasing body weight. Though these fecundity traits are applicable to annelids, they remain to be tested in more number of species.

2.5.1 Sexuality

Within hermaphroditic oligochaetes, incidence of parthenogenics is not uncommon. A fact that is known but not adequately recognized is that the establishment and maintenance of dual sex within a single individual cost resources. In them, resource allocation for reproduction is limited with the consequence of reduced fecundity. But the reduced fecundity is compensated by (i) internal fertilization ensuring a higher level of fertilization success and (ii) embryonic development within a hard cocoon affording fairly high level of protection. However, these two clitellate features may further reduce fecundity. Within the genus *Ophryotrocha*, there are gonochoric and hermaphroditic species, providing a rare but excellent opportunity to study the effect of sexuality on fecundity. Fecundity values range from 80 eggs in gonochoric *O. macrovifera* to 200 eggs in *O. robusta* but from 11 eggs in hermaphroditic *O. gracilis* to 50 eggs in *O. maculata* (Table 2.9). Hence, the mean fecundity is 9.4 eggs/d and 4.5 eggs/d in gonochoric and hermaphroditic *Ophryotrocha* spp, respectively, despite the interspawning period averaging to 13.6 days and 4.8 days in gonochorics and hermaphrodites, respectively. Not surprisingly, maintaining the dual sex within an individual hermaphrodite reduces the batch fecundity to half of that gonochorics.

To reduce the cost of maintaining dual sex, many earthworms have 'sacrificed' one or more organs in the male reproductive system (Table 2.6). Consequently, some of these 'female hermaphrodites' switched to parthenogenesis (Diaz-Cosin et al., 2011). To enhance genetic diversity, ploidy level has been increased up to heptoploidy in some of these parthenogenic clones (Table 2.7). Hence, elevated ploidy is also expected to express in their eggs. For example, diploid and triploid eggs are produced by *Aporrectodea*

TABLE 2.9

Effect of sexuality on fecundity in gonochoric and hermaphroditic species in polychaete genus *Ophryotrocha* (from Premoli and Sella, 1995, modified) and hermaphroditic and parthenogenic earthworms (c = cocoon)

Species	Fecundity (egg no./spawn)	Inter-spawning Interval (d)
	Gonochoric polychaetes	
O. labronica	130	11.1
O. macrovifera	80	11.3
O. notoglandulata	115	14.0
O. robusta	200	18.1
Mean	**9.4 eggs/day**	
	Hermaphroditic polychaetes	
O. diadema	25	2.9
O. gracilis	11	6.2
O. hartmanni	30	5.3
O. maculata	50	–
O. socialis	25	–
Mean	**4.5 eggs/day**	
	Hermaphroditic earthworms	
Eudrilus eugeniae (Mba, 1988)		10.27 c/w
Eisenia foetida (Siddique et al., 2005)		9.0 c/w
Apporrectodea longa (Lowe and Butt, 2005)		0.33 c/w
Allolobophora chlorotica (Lowe and Butt, 2005)		0.20 c/w
Mean		**4.95 c/w**
	Parthenogenic earthworms	
Aporrectodea trapezoides (Fernandez et al., 2010)		2.03 c/w
Pontoscolex corethrurus (Bhattacharjee and Chaudhuri, 2002)		2.30 c/w
Drawida nepalensis (Bhattacharjee and Chaudhuri, 2002)		0.46 c/w
Aporrectodea caliginosa (Lowe and Butt, 2005)		0.61 c/w
Lumbricus terrestris (Lowe and Butt, 2005)		0.46 c/w
Mean		**1.17 c/w**

trapezoides (Table 2.7). Considering the relatively more frequent incidence of parthenogens in earthworms, available information on fecundity of hermaphrodites and parthenogens is summarized in Table 2.9. Expectedly, the mean cocoon production is 1.17/week (w) and 4.95/w in parthenogens and hermaphrodites, respectively. Hence, the switch to parthenogenesis and polyploidy may have facilitated the survival of these earthworms in 'patchy' habitats but certainly at the cost of fecundity. Briefly, both hermaphroditism and parthenogenesis have significantly reduced batch fecundity.

2.5.2 Oogenic Anlage

Being a key factor, food quality and availability may affect the age at sexual maturity. In the sediment feeding *Capitella* sp I, the age is significantly increased from 28 days in those feeding sediment containing 3.0% Total Organic Matter (TOM) to 56 days in those feeding sediment containing only 0.25% TOM (Ramskov and Forbes, 2008). Oogenesis can be intraovarian, as in *Streblospio* spp and *Branchipolynoe seepensis* with 10–300 brooded eggs involving indirect development in the former and direct one in the latter (Table 2.11) or extraovarian, as in most polychaetes. In the dorvilleid *Veneriserva pygoclava meridionalis*, vitellogenesis involves progressive reduction in the number from about 100 pre-vitellogenic oocytes of 50 μm to 25 eggs each measuring > 150 μm (Fig. 2.4A). Presumably, oogenesis is intra-ovarian in *V. pygoclava meridionalis*, in which only 25% oocytes are matured. In the extraovarian oogenesis, about 50% oocytes are developed into mature eggs, as in the serpulid *Spirorbis spirorbis* (Fig. 2.4B). With comparatively far less space available in the intraovary, less number of pre-vitellogenic oocytes is expected to mature; hence, the anlage, in which oocytes undergo vitellogenesis (see p 31) and maturation, imposes a profound effect on fecundity.

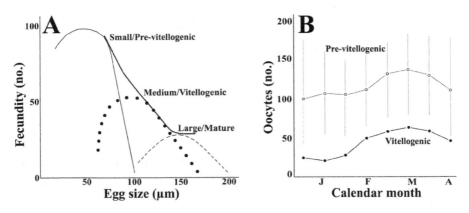

FIGURE 2.4

Progressive decrease in the number of oocytes from pre-vitellogenic to vitellogenic/mature egg stage in the presumably intraovarian A. *Veneriserva pygoclava meridionalis* with increasing egg size and B. extraovarian *Spirorbis spirorbis* during succesive calendar months. Note the levels of variance indicated by thin vertical lines in Fig. 2.4B (modified and redrawn from Micaletto et al., 2002, Daly, 1978b).

2.5.3 Egg Size

In polychaetes, egg size ranges from 30 μm (e.g. *Protodrilus albicans*, Westheide, 1984) to 600 μm (e.g. *Thelepus cincinnatus*, Garraffoni et al., 2014) and to

1,000 µm (e.g. *Myxicola* cf *sulcata*, Gambi et al., 2001). It has a profound effect on fecundity and recruitment. From his survey, Giangrande (1997) has grouped them into semelparous and iteroparous forms: the latter is divided into short- and long-living forms. The short living forms may produce two broods, the fecundity of the second brood being strongly influenced by resource availability (e.g. *Harmathoe imbricata*, see Cassai and Prevedelli, 1998a). But the long living forms generate about five broods (e.g. *Spirorbis spirorbis*, Daly, 1978b). Of 32 families (sabellids, spionids and syllids represented in two groups) considered by Giangrande, 7, 9 and 16 families comprise semelparous, short- and long-living iteroparous forms, respectively. Clearly, long living iteroparity is the most common reproductive strategy of polychaetes. On the basis of their reproductive mode, these polychaetes are considered in this analysis under 1. planktotrophic (PLK), 2. lecithotrophic (LEC) and 3. direct developers (DIR). The latter is considered in two subgroups, in which the egg size measures > 100 µm (DIR$_1$) and < 100 µm (DIR$_2$), which obviously draw nutrients from the mother. From Table 2.10, the following may be inferred: 1. expectedly, the mean egg size increases from 113 µm in PLK to 196 µm and 209 µm in LEC and DIR$_1$, respectively but decreases to 83 µm in DIR$_2$. The variations from the respective mean are very wide in the first three groups but the least in DIR$_2$ subgroup, 2. within the families considered, the choice for reproductive strategy decreases in the following order: long living iteroparity > short living iteroparity > semelparity; for the reproductive mode, it is PLK > LEC > DIR$_1$ > DIR$_2$. Notably, the long living iteroparous species do not invest on viviparity.

Within a species, egg size remains equal in worms of different body weights (e.g. *Galeolaria caespitosa*, Fig. 1.12A). However, contrasting trends become apparent, when fecundity-egg size relation is considered at interspecific level. Expectedly, fecundity is decreased with increasing egg size in semelparous and iteroparous polychaetes characterized by indirect life cycle with holoplanktonic or meroplanktonic (brooder) larva (Fig. 2.5A). With small oocyte size (112 µm), the semelparous broadcasters spawn

TABLE 2.10

Egg size (µm) in planktotrophic (PLK), lecithotrophic (LEC) and directly (DIR) developing polychaetes. Values in brackets represent the number of families and square brackets the range of egg size (compiled from data reported by Giangrande, 1997)

Developmental Mode	PLK	LEC	DIR$_1$, egg size > 100 µm	DIR$_2$, egg size < 100 µm
Semelparous	98 (2)	205 (2)	178 (2)	96 (1)
Iteroparous				
short living	113 (4)	175 (1)	142 (1)	178 (2)
long living	123 (8)	196 (5)	241 (4)	–
Total	112 (14)	196 (8)	209 (7)	83 (3)
Range	[64–155]	[122–250]	[142–290]	[72–96]

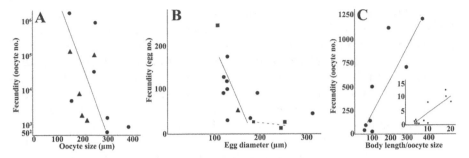

FIGURE 2.5

A. Effect of oocyte/egg size on fecundity of (●) semelparous and (▲) iteroparous polychaetes (drawn using data reported for *Nereis virens* (Creaser and Clifford, 1982), *Glycera dibranchiata* (Creaser, 1973), *Perinereis cultrifera* (Cassai and Prevedelli, 1988a), *Hediste japonica* (Sato, 1999), *Nicolea zostericola* (Eckelbarger, 1973), *Eupolymnia crescentis, Neoamphitrite robusta, Thelepus crispus* and *Lanice conchilega* (McHugh, 1993). B. Effect of egg size on fecundity in dorvilleid polychaete embryos with (●) maternal, (■) biparental and (▲) communal care (drawn from the data reported by Sella and Ramella, 1999), C. Effect of body length/oocyte size on fecundity of interstitial polychaetes. The relationship in smaller interstitial worms is shown in a window (drawn using data reported by Westheide, 1984).

millions and thousands of eggs, while the iteroparous brooders generate a few thousands or hundreds of larger (196, 209 µm) eggs. Interestingly, a slightly negative but linear relation is also apparent between egg size and fecundity in *Ophryotrocha* spp that exhibit biparental care (Fig. 2.5B). Hence, egg size significantly imposes reduction in fecundity of broadcasters and others displaying parental care. In some subgroups of sabellid brooders a positive linear relation between log body size and log fecundity is apparent (Fig. 1.10E). However, this positive relation may be altered, when normal values are considered (cf Fig. 2.5A, B, Fig. 2.7H, I).

On the other hand, increasing body size may also increase egg size, when the relation is considered at interspecies level. For example, the oocyte/egg size increases with increasing body length in broadcast spawning nereidids, dorvilleidid and dinophilid as well as extra- and intra-tubular brooding polychaetes (Fig. 1.10C, D, E, F), although the positive linear relation in Fig. 1.10E, F are based on log values. Unusually, planktotrophic duration increases with increasing egg size in spionids (Fig. 1.10A). In them, with increasing egg size, however, the chaetigers stage, at which they are released, also increases (Fig. 1.10B).

At this point, a brief account on the interstitial polychaetes becomes necessary. Represented by as many as 16 families (Westheide, 1971), these polychaetes constitute one of the most species-rich and number-rich communities of the marine fauna. In an impressive concept, Westheide (1984) has characterized their life history traits. Inhabiting within the interstitial space of ~ 125 µm, (i) they are small (1–2 mm body length, the smallest

Diurodrilus sp measuring 300 µm), (ii) mostly hermaphroditic (e.g. Nerillidae, *Microphthalmus*), (iii) fertilization by pseudocopulation (*Ophryotrocha gracilis*)/hypodermic injection (e.g. *Protodrilus albicans*) of limited sperms (with considerable morphological alterations including the lack of flagellum and loss of mid-piece, *Pisione remota*) and (iv) brood/gestate their eggs (*O. vivipara*). In the interstitial polychaetes, body size is measured in number of times of the egg diameter (Fig. 2.5C). Fecundity also increases with increasing body length/oocyte diameter from 45 times in *Hesionides arenaria* with 70 oocytes to 400 times in *P. albicans* with 1,200 oocytes in fairly large worms. In smaller polychaetes, it is, however, from four times in *Ikosipodus carolensis* with one–two oocytes to 20 times in *M. sczelkowii* with 11–15 oocytes.

2.5.4 Body Size

In a rare investigation, Daly (1978b) has made two important observations in the serpulid *Spirorbis spirorbis*. The extent of individual variations in the output of pre-vitellogenic oocytes by the 2.9 mm size class ranges between 25 and 35% (Fig. 2.4B). The oocytes mature into vitellogenic ones in the coelomic cavity. Of the pre-vitellogenic oocytes released from the 'ovary', ~ 50% of them mature into vitellogenic ones during the period from January to April. From his estimates on changes in the mean number of pre-vitellogenic and vitellogenic oocytes during successive broods, Daly has estimated that the gonadal release of pre-vitellogenic oocytes compensates the number of oocytes entering vitellogenesis. Secondly, the oocyte output in the first brood is positively correlated with body size and increases from 24 in 2.1 mm coil diameter body size (= 75 mg dry body weight) to 59 but from ~ 16 to ~ 28 in 3.1 mm body sized (= 225 mg body weight) (Fig. 2.6A). However, these values decrease to lower levels from 64–25, 54–24 and 24–20 in the first, second and fifth broods, respectively (Fig. 2.6B). Hence, fecundity decreases with successive broods in relation to increasing body size. In other words, it decreases from 1.2 eggs/µg dry body weight in the smallest worm to 0.97 egg/µg dry body weight in the largest worm.

In polychaetes, reproductive modes are greatly diverse. Not surprisingly, the patterns of spawning and fecundity are also equally diverse. Spawning is a critically important event to release the time-long gametic investment and polychaetes, as other animals, select a strategically important location and time for spawning to ensure higher fertilization success and potential offspring survival. Some of these reproductive modes are listed below: (i) Oogenesis can be synchronized in broadcasters involving epitoky and semelparity, as in *Glycera dibranchiata* with a few thousands/millions oocytes or (ii) iteroparous synchronized atokous spawner, as in *Marphysa sanguinea* with a few thousand oocytes or (iii) asynchronous broadcast spawning, as in *Eupolymnia crescentis* involving indirect development of 128,500 oocytes (Table 2.11). In iteroparous brooders, (iv) development can be indirect with discrete synchronized spawning, as in *Nicolea zostericola* with 665 eggs/y, (v) asynchronized spawning over a period of 6 months, as in *Thelepus crispus* with 515,555 eggs, (vi) asynchronous oocyte maturation with continuous spawning of 44 eggs/brood throughout the year and brooding 11 sequential broods at a time, as in *Ramex californiensis* and (vii) direct development of self-fertilized ~ 150 gestated embryos parturited once in 6 months, as in viviparous *Neanthes limnicola*. In others, (viii) spawning is continuous and lasts almost throughout the year with 11 eggs/w and 3.5 egg masses/d, as in *Dinophilus gyrociliatus* and *Ophryotrocha adherens*, respectively. (ix) In oligochaetes, cocoons are generated at the rate of 2.25–3.14/w in earthworms, 0.15/w in *Branchiura sowerbyi* but 10 eggs per w in *Tubifex tubifex*. (x) In hirudineans, the inverted 'U'-shaped spawning trends within a reproductive bout are presumably regulated by blood ingestion in sanguivorous *Hirudo medicinalis* and actually by temperature in carnivorous *Nephelopsis obscura*. The large *H. medicinalis* can gorge itself sucking ~ 10 g blood at a time (Davies and McLoughlin, 1996).

Clearly, the reported values for the observed fecundity are so diverse that it is difficult to consider them comparable for following reasons: 1. The number of gametogenic (ovarian) segments varies from two in *Spirorbis spirorbis* (Daly, 1978a) to a large number in nereidids. Hence, the fecundity–body size relationship becomes not comparable with such wide variations in the number of ovarian segments. 2. The few eggs developed and matured intraovarially cannot be compared with iteroparous brooders having thousands of eggs and semelparous broadcasters having millions of oocytes

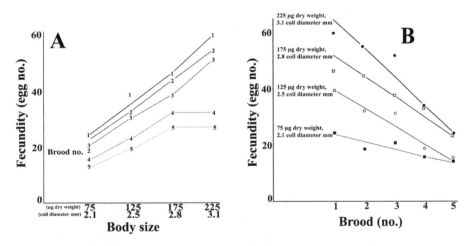

FIGURE 2.6

Fecundity in *Spirorbis spirorbis*: A. increasing fecundity with increasing body size, B. decreasing fecundity as function (of body size) in successive broods from 1 to 5 (drawn from data reported by Daly, 1978b).

TABLE 2.11

Fecundity (F) and modes of oogenesis, spawning and breeding by annelids, as reported by respective authors. coc = cocoon

Intraovarian oogenesis

Streblospio shrubsolii	*S. benedicti*	*S. benedicti*	*Branchiopolynoe seepensis*
Anterior gametic segments extend from 21 to 30 chaetigers. 10–34 eggs (Lardicci et al., 1997).	F = 11, 4 and 5.5 eggs per ovary in PLK, LEC and hybrid morphs, respectively (Levin and Bridges, 1994).	F decreases from ~ 60 eggs on 60th d to 10 eggs on 161st d and 500 d in PLK and LEC (Bridges and Heppell, 1996).	100–300 eggs/♀ brooded in I (125 µm) and II (300–350 µm) cohorts. Direct development (Jollivet et al., 2000).

Extraovarian oogenesis: Broadcast spawners

Semelparous, synchronous epitoky	Iteroparous, synchronous atoky	Iteroparous, Asynchronous
Glycera dibranchiata	*Marphysa sanguinea*	*Eupolymnia crescentis*
F increases from 0.7 to 9.5 million oocytes in 17 and 47 cm body size, respectively. Indirect development (Creaser, 1973).	F increases 8,500 to 24,300 oocytes in 0.7 and 3.0 g size, respectively. Indirect development (Cassai and Prevedelli, 1998b).	128,500 oocytes spawning lasts for 3 months. Indirect development (McHugh, 1993).

Iteroparous brooders

Asynchronous

Direct	Viviparous
Ramex californiensis	*Neanthes limnicola*
44 eggs/coc. Asynchronized oocyte maturation and sequentially laid 11 coc. Breeding throughout y (McHugh, 1993).	~ 150 embryos/brood. Breeding twice/y (Fong and Pearse, 1992).

Semelparous, synchronous

Indirect	Direct
Thelepus crispus	*Nicolea zostericola*
515,555 eggs/brood spawned during 6 months (McHugh, 1993).	Male and female pair prior to spawning. Single discrete spawning of 665 eggs/y (Eckelbarger, 1973).

Continuous spawners

Dinophilus gyrociliatus ~ 11 eggs/w (30‰ S, tetramin) F, as coc no. decreases with increasing age (Prevedelli and Simonini, 2000).

Protandric *Ophryotrocha adherens* ~ 3.5 egg mass/d. F, as coc no. decreases with increasing age (Paavo et al., 2000).

Clitellates

Earthworms: *Lumbricus terrestris* 3.14 coc/w. F, as coc no. decreases with increasing age (Lowe and Butt, 2005) (Fig. 2.8A).

Hyperiodrilus africanus 2.25 coc/w. in paired worms; 1.12 coc/w. in singletons (Tondoh and Lavelle, 1997).

Eisenia foetida 2.6 coc/w. (Siddique et al., 2005).

Naidid: *Branchiura sowerbyi* 0.15 coc/w. F, as coc no. decreases with increasing age (Lobo and Alves, 2011) (Fig. 2.8B).

Tubificid: *Tubifex tubifex* ~ 10 eggs/w. F, as coc no. decreases with increasing age (Pasteris et al., 1996) (Fig. 2.8D, E).

Hirudinea: *Hirudo medicinalis*. F increases with increasing age, body weight and blood ingestion (Davies and McLoughlin et al., 1996) (Fig. 2.7F, G, H).

Nephelopsis obscura. An inverted U trend for F within a reproductive bout (Holmstrand and Collins, 1985).

developed and matured in the coelom. In terms of space to accommodate eggs/oocytes, there can be a vast difference between the ovary and coelom. 3. The thousands-millions of oocytes spawned by semelparous epitokous broadcasters are considered as batch (BF) as well as lifetime fecundity, in contrast to the PF consisting of few thousand oocytes of iteroparous, atokous broadcasters. 4. Within iteroparous brooders, the counts for fecundity in synchronous and asynchronous spawners are not comparable. In iteroparous asynchronous spawners characterized by indirect development like *T. crispus* produce > 500,000 eggs, which may not be comparable to the 150 embryos gestated by self-fertilizing viviparous *N. limnicola*, a biannual breeder. 5. *R. californiensis* is not comparable with any of the above, as it simultaneously broods 11 sequentials broods, each consisting of 44 eggs.

A second complication is that different morphological features have been considered to describe fecundity-body size relationship: (i) body length (e.g. *Nereis virens*, Fig. 2.7A, body length/oocyte length, Fig. 2.5C), (ii) body width (e.g. *Hediste japonica*, Fig. 2.7B), (iii) number of chaetigers (e.g. *Ophryotrocha puerilis puerilis*, Fig. 2.7D), (iv) number of segments (e.g. *Enchytraeus variatus*, Fig. 2.8C), (v) body weight (e.g. live body weight: *Tubifex tubifex*, Fig. 2.8E; dry body weight *Spirorbis spirorbis*, Fig. 2.6A, B), and (vi) age (e.g. *T. tubifex*, Fig. 2.8D). Notably, similar trends for the fecundity-body size relationship have been reported, when body size is considered in units of body weight

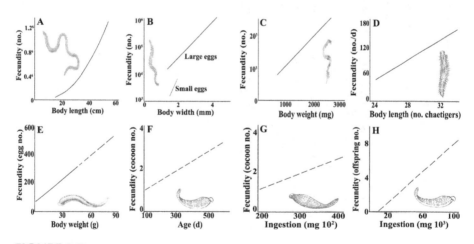

FIGURE 2.7

Fecundity in broadcast spawning polychaetes as function of A. body length in semelparous epitokous *Nereis virens* (Creaser and Clifford, 1982), B. body width in estuarine (large eggs) and marine (small eggs) forms of semelparous epitokous *Hediste japonica* (Sato, 1999), C. body weight in iteroparous *Marphysa sanguinea* (Cassai and Prevedelli, 1998b) and D. chaetiger number in protandrus continuous spawner *Ophryotrocha puerilis puerilis* (Berglund, 1991). Fecundity in sanguivorous hirudineans as function of body weight in E. the giant leech *Haementeria ghilianii* (Sawyer et al., 1981) and in *Hirudo medicinalis* as functions of F. age and G, H. ingestion of blood (Davies and McLoughlin, 1996).

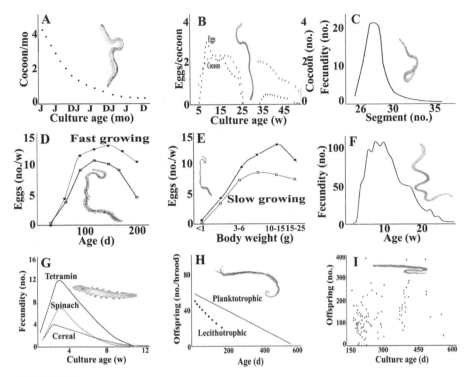

FIGURE 2.8

Hermaphrodites: Fecundity as function of age in A. earthworm *Lumbricus terrestris* (for a period of 3 y, Lowe and Butt, 2005), B. in naidid *Branchiura sowerbyi* (Lobo and Alves, 2011), C. segment number in *Enchytraeus variatus* (Bouguenec and Giani, 1989), D. age and E. body weight in fast and slow growing morphs of *Tubifex tubifex* (Pasteris et al., 1996). Polychaetes: Fecundity as function of age in F. *Ophryotrocha adherens* (Paavo et al., 2002), G. digamic *Dinophilus gyrociliatus* fed on different diets (Prevedelli and Vandini, 1999), H. tube-dwelling, egg brooding planktotrophic and lecithotrophic morphs of *Streblospio benedicti* and I. self-fertilizing viviparous *Neanthes limnicola* (Fong and Pearse, 1992). All figures are modified and redrawn.

and age. Obviously, growth in body weight proceeds along with age in *T. tubifex*. But in others like *S. spirorbis*, age may significantly vary for a given body size. Some scientists (Britayev and Belov, 1994, Plyuscheva et al., 2004) have unsuccessfully attempted to correlate jaw length with age in polychaetes.

In most hirudineans, a third complication is the dorso-ventrally flattened body to an almost rectangular shaped mid-body, in which a pair of ovary is accommodated. Contrastingly, the cylindrical body shape in all other annelids differs distinctly from that of the hirudineans. The flattened rectangular body provides 1.5–2.0 times larger surface area to nourish more number of eggs to undergo vitellogenesis than that of cylindrical body in typical annelids limiting the number of oocytes entering vitellogenesis. Of course, there are exceptions to this generalization. For example, the body shape of continuously breeding polychaetes like *Dinophilus gyrociliatus* (Fig. 2.8G) and *Ophryotrocha puerilis puerilis* (Fig. 2.7D) tends to be rectangular and that of annually breeding hirudinean *Nephelopsis obscura* (Fig. 2.7E) is more of cylindrical than rectangular.

Despite the diversity and complications, two major trends emerge for the fecundity body size relation. In the first one, the relation is positive and linear, and is observed in broadcasters and sanguivorous leeches. Irrespective of the morphological features (taken to represent body size) like body length in semelparous epitokous *Nereis virens* (Fig. 2.7A), body width in *Hediste japonica* (Fig. 2.7B), body weight in iteroparous *Marphysa sanguinea* (Fig. 2.7C) and number of chaetigers in the protandrous continuous breeders *Ophryotrocha puerilis puerilis* (Fig. 2.7D), the linear relationship remains, *albeit* at different levels. In the semelparous *Perinereis cultrifera*, the reported correlation *per se* is weak. About 26% of the variances of fecundity are influenced by body size and the remaining 74% by environmental factors like food supply (Rettob, 2012). However, BF decreases in the following order: semelparous epitokous spawners (e.g. *Nereis virens*, Fig. 2.7A) < iteroparous spawners (e.g. *Marphysa sanguinea*, Fig. 2.7C) < iteroparous brooders (e.g. *Nicolea zostericola*, Table 2.11). In them, the number of gametogenic segments also progressively decreases from almost all the abdominal segments in the epitokous spawners to a few in the iteroparous brooders. Apparently, semelparous species allocate majority of the available resources to gamete production, whereas iteroparous species invest the bulk of available resources on somatic growth (see Cassai and Prevedelli, 1998a). In broadcasters, the durations required to complete oogenesis are 10, 12 and 12 or 19 months in *Glycera dibranchiata* (Creaser, 1973), *Neoamphitrite robusta* (McHugh, 1993) and *N. virens* (Creaser and Clifford, 1982), respectively. With dorso-ventrally flattened body and ovaries in the sanguivorous leeches *Hirudo medicinalis* and *Haementeria ghilianii*, fecundity also increases linearly and positively with advancing age (Fig. 2.7F), increasing body weight (Fig. 2.7E) as well as the quantum of ingested blood (Fig. 2.7F, G). Apparently, more than age and body weight, the assured supply of nutritional source ensures increase in fecundity with increasing quantum of ingested blood. However, within a single reproductive season/bout, an inverted 'U'-shaped trend for the fecundity-time relations has been reported for the glossiphoniid *Helobdella californica* (Kutschera, 1989) and *Nephelopsis obscura*, in which temperature regulates the spawning pulses (Peterson, 1983, Holmstrand and Collins, 1985). Notably, the trend is linear for the continuous spawner *Ophryotrocha puerilis puerilis* (Fig. 2.7D), in which eggs are spawned once a day. But it is an inverted 'V' trend in *Dinophilus gyrociliatus* (Fig. 2.8G), which spawns egg masses once a week.

About 50% of polychaetes brood their eggs/egg sacs attaching them on suitable substratum (e.g. mucous nesting site in some sabellids, Gambi et al., 2001), on dorsal elytra (e.g. *Harmothoe imbricata*, Daly, 1972) or within the branchial crown (e.g. *Potamilla antarctica*, Knight-Jones and Bowden, 1984). Encountering problems of safety and space, extra- or intra-tubular structures have been developed. Being cylindrical, these tubes also limit space with increasing body size. It must also be noted that only the number of brooded eggs are counted; however, none has reported the number of fertilized eggs that have been accommodated within the tube. In all the brooding polychaetes, an inverted 'U' or 'V' shaped trends is noted for the fecundity-body size relation. The trends are, however, linear and negative but are parallel with different levels in PLK and LEC morphs of *Streblospio benedicti* (Fig. 2.8H). Notable is the shorter lifespan of LEC morph. In viviparous *Neanthes limnicola*, no trend is apparent (Fig. 2.8I). Almost all the oligochaetes with cylindrical body also exhibit an inverted 'V' shaped trend, as in *Enchytraeus variatus* (Fig. 2.8C) or an inverted 'U' shape, as in *Branchiura sowerbyi* (Fig. 2.8B) and *Tubifex tubifex* (Fig. 2.8D, E). In the earthworm *Lumbricus terrestris*, fecundity also decreases linearly with advancing age (Fig. 2.8A).

Arguably, almost all the oligochaetes and brooding polychaetes exhibit declining trends for fecundity with advancing age or increasing body size. In contrast, broadcast spawning polychaetes with more but variable number of gametogenic segments provide cumulatively a larger coelomic surface area to nourish more number of eggs to undergo vitellogenesis. Similarly, the flattened ovaries of the sanguivorous leeches provide relatively more surface area than that in the saccular ovaries of oligochaetes. Consequently, the broadcast spawning polychaetes and sanguivorous leeches do not limit the number of oocytes entering vitellogenesis. In them, fecundity increases with advancing age and/or increasing body size. However, it decreases *per se* in oligochaetes and brooding polychaetes, as their relatively smaller ovarian/coelomic surface area in old worms limit the number of pre-vitellogenic oocytes entering vitellogenesis.

In this context, the maintenance of Oogonial Stem Cells (OSCs) is relevant. The OSCs are responsible for the sustained production of oogonia (see Pandian, 2012). In an important publication, Daly (1978b) has shown that the number of pre-vitellogenic oocytes released from the 'ovary' for the ensuing brood is compensated by that entering vitellogenesis in the preceding brood of the serpulid *Spirorbis spirorbis*. This rare observation is likely to be confirmed in other annelids. Hence, it is likely that: 1. The compensatory release of oogonia indicates that the OSCs in annelids do not undergo ageing and senescence (however see Martinez and Levinton, 1992). 2. The senescence is rather imposed in oligochaetes and brooding polychaetes by the progressively decreasing ovarian/coelomic surface area in the old/large worms by limiting the number of pre-vitellogenic oocytes entering vitellogenesis. As in annelids, broadcast spawning crustaceans also do not suffer reproductive senescence but the brooders suffer from it (Pandian, 2016).

2.6 Poecilogony and Dispersal

Poecilogonics exhibit multiple-patterns of dichotomy in egg size/egg type and larval development. They alter the existence and efficiency of selection force for alternative and coexisting reproductive and developmental strategies (Fischer, 1999). It is a rare but interesting reproductive mode that occurs in half a dozen opsithobranch molluscs (Pandian, 2017). In polychaetes, it occurs in *Hediste japonica* (Sato, 1999) and in six spionid species: *Streblospio benedicti* (Levin and Huggertt, 1990), *Boccardia acus*, *B. andrologyna*, *B. chilensis* (Read, 1975), *B. proboscidae* (Gibson, 1997), *B. semibranchiata* (Guerin, 1991) and *Pygospio elegans* (Jenni, 2012). In *H. japonica* and *S. benedicti*, it includes planktotrophic (PLK) and lecithotrophic (LEC) morphs. The PLKs produce many small eggs that develop into feeding larvae with 2–3 weeks stay as plankton. But the LECs generate a few but larger eggs that develop into non-feeding larvae with or without planktonic phase. An advantage of the PLK larval development includes (i) greater fecundity, (ii) reduced parental investment, (iii) enhanced dispersal and gene flow, (iv) ability to colonize new habitats and (v) greater potential to delay metamorphosis to select a better habitat. On the other hand, the potential advantages of LEC larval development includes (i) higher larval survivorship, (ii) independence from external food supply, (iii) utilization of potential habitat and resources resulting in denser density and greater production (Levin and Huggett, 1990 see also Table 2.12).

Available information on the cluster of life history traits of the poecilogonic polychaetes reveals dichotomy in *S. benedicti* and *H. japonica* but trichotomy in *B. proboscidae* (Table 2.13). Production of nurse eggs and adelpophagy occurs in *B. proboscidae* but not in *S. benedicti* and *H. japonica*. In the rivers of Aomori and Kagoshima (Japan), *H. japonica* PLK and LEC morphs co-occur but may not interbreed. The free spawning small sized (65 mm) PLK adult produces ten thousands to one million small (150 µm) eggs (Fig. 2.7B) and its neochaete larva with long chaete migrates to the sea and returns as juvenile. But the larger (73 mm) LEC adult spawns a few thousand eggs (230 µm) and its direct development is completed within the estuary (Sato, 1999). Electrophoretic studies have revealed the homogenous genetic structure in PLK morph but genetic differentiation in LEC morph.

In the coast of North Carolina (USA), the brooding *S. benedicti* PLK and LEC morphs within a deme co-occur, interbreed and generate viable hybrids (Table 2.12). With available hybrids, the composition of the PLK within the deme consists of mostly PLK (64%) and LEC (10%) and hybrids (20%). Within the same deme also, it consists of predominantly 74% LEC but only 10% PLK and 16% hybrids. The PLK embryos seem to be more dependent

TABLE 2.12

Dichotomic poecilogonic developmental traits of *Streblospio benedicti* (compiled from
Levin and Huggett, 1990, Levin and Bridges, 1994)

Trait	PLK	LEC	Hybrid
Ova			
Size (µm)	65	152	74
Volume (µl × 10^{-3})	0.5	3.1	–
Carbon/embryo (µg)	0.11	0.85	–
Nitrogen/embryo (µg)	0.023	0.017	–
No./ovary	11	4	5.5
Larvae			
No./brood	200	40	80
No./pouch	10	3	5
Released chaetiger stage (no.)	3–5	9–12	–
Released larval length (µm)	250	600	–
Swimming setae	Present	Absent	Variable
Larval nutrition	Feeding	Non-feeding	Facultative
Larval duration (d)	12–20	0–9	7–9
Adults			
Life span (w)	38	30–75	–
Age at Ist spawning (w)	9–10	13–14	–
Brood (no./life time)	6	6	–
Composition (%)			
PLK deme	64	10	20
LEC deme	10	74	16
Colonizing PLK (%)	82	20	–
Mean density (no./m^2)	5,030	12,935	–
Production (g/m^2)	2.57	3.68	–

on protein (0.023 µg N/embryo), whereas the LEC on carbohydrate
(0.85 µg C/embryo). With relatively more number of broods and brood
pouches, the PLKs release smaller (250 µm) feeding larvae at ~ 4 chaetiger
stage lasting for 12–20 days, in comparison to the large (600 µm) non-feeding
larvae released at 9–12th chaetiger stage with planktonic duration of 0–9
days. Though the colonizing ability of PLK is higher (82%), their density
(5,030 no./m^2) and production (2.6 g/m^2) are lower than (12,935 no./m^2,
3.7 g/m^2) those of LEC. Briefly, the PLK is more an explorative morph, while
LEC is more a productive morph in the favorable parental habitat. Reciprocal
mating between PLK and LEC reveals a tendency for perpetuation of the
dichotomy (Levin et al., 1991).

TABLE 2.13

Dichotomic poecilogonic developmental traits of *Boccardia proboscidae* (condensed from Gibson, 1997)

Trait	Type 1	Type 2	Type 3	
Egg size (μm)	94.5	92.7	108.8	
Capsules/brood	41.7	46.8	42.9	
Eggs/brood	2180	1637	2309	
Larvae/capsule	52.4	35.0	53.8	
Nurse eggs/capsule	0.6	6.7	34.4	
Nurse egg/larva	0.01	0.014	8.7	
Encapsulated period (d)	6	6	11	
			Type 3A	Type 3B
Post-hatchling	PLK	PLK	PLK	Benthic
Larval size (μm)	206	251	209	484
Adelpophagic	No	Some	Some	Yes
Planktonic duration (d)	30	19	15	0
Age at sexual maturity (mo)	4	4	4	3

In contrast, the morphs of *B. proboscidae* are not recognizable as PLKs and LECs but are typed as Types 1, 2 and 3; Type 1 is characterized by the absence of or presence of a very few nurse eggs and is developed into planktotrophic dispersive larvae but Types 2 and 3 by the presence of abundant nurse eggs consumed by adelpophagous viable number of embryos (Table 2.13). In Types 2 and 3, the nurse eggs appear similar to the viable oocytes prior to cleavage but fail to cleave. Females (78%) dominate the population of Type 1, but they are much less abundant in Type 2 (6%) and Type 3 (16%). Type 1 females brood eggs in the absence or presence of 0.01 nurse egg/viable egg; the hatched offspring (200 μm) is planktotrophic for 30 days. Type 2 females are similar to that of Type 1 but for every viable larva, there is 0.14 nurse egg and the hatched (251 μm) larva is planktotrophic for 19 days only. Type 3 females are further subdivided into A and B groups. In Type 3A, adelpophagy may occur but its larva is planktonic for 15 days. In Type 3B, adelpophagy is most common and direct development results in the release of benthic juvenile. Remarkably, the poecilogonic patterns vary among broods produced by different females as well as within a single brood and even in an egg capsule. Type 1 and Type 3 adults can successfully be mated. All the offspring produce broods characterized by maternal breeding type, regardless of paternal origin. Apparently, its sex is determined by ZZ-ZW female heterogametic system. In Type 1, female produce no or very few nurse eggs and release 3-chaetiger larva on 6 days after spawning. Type 3 females spawn both viable and a large number of nurse eggs. In

B. proboscidae, poecilogony represents a gradual shift from planktotrophy to adelpophagy.

2.7 Mating Systems

Egg production is costlier than that of sperm (Charnov, 1982). Consequently, reproductive success of the female is limited by access to resources, whereas that of a male is limited by availability of females (Bateman, 1948). The sex allocation theory (Charnov, 1987) explains the differences in resource allocation to male versus female reproduction. Fishes have proved as an excellent animal system to test a number of hypotheses on sex allocation theory (Fischer and Petersen, 1987). For the following reasons, annelids, especially polychaetes as an ancient group can also serve as an excellent invertebrate model: 1. The presence of different forms of sexuality ranging from gonochorism to sequential, serial and simultaneous hermaphroditism (SH). 2. About 50% of polychaetes have free swimming larval stages(s), while their adults are sessile. Hence, they allow a study on the effects of a range of population structure and selection pressure on mating system. 3. The ease with which they can be reared in the laboratory as well as culture system (see Chapter 8) allows precise estimates on individual reproductive success (Premoli and Sella, 1995). *Ophryotrocha* sp collected from 1,500 m depth is readily amenable for culture in laboratory, where it has completed three successive generations within 7 years of the experimental study (Mercier et al., 2014).

Annelids display the three recognized mating systems. However, monogamy is limited to very few (e.g. *Nereis acuminata*, Weinberg et al., 1990; enforced monogamy, *Ophryotrocha puerilis puerilis*, Berglund, 1986, *O. diadema*, Cannarsa et al., 2015). A microsatellite marker study has revealed the presence of a near monogamy in *Homogaster elisae*, in which all the four spermatheca stores sperm from the same male (see Diaz-Cosin et al., 2011). However, *H. elisae* individuals select partners with similar body size (Novo et al., 2010). In-breeding and out-breeding pairs of *Eisenia andrei* produce 30 and 19%, respectively fewer cocoons than intrapopulation mating pairs (Velando et al., 2006). From their observations, Dominguez et al. (2003) have reported that 88 and 10% of matings in *E. foetida* are reciprocal and unilateral, respectively. Body shape and locations of male and female pores minimize the scope for self-fertilization in earthworms, though rare such events are recorded in some species.

For experimental study on mating systems, a few species belonging to *Ophryotrocha* have served as an ideal model. They are small marine polychaetes and amenable to rearing. In them, some are SH (e.g. *O. diadema*), others are sequentials (e.g. *O. puerilis puerilis*) and yet others are gonochoric (*O. labronica*). *O. diadema* commences its sexual life very early with a progametic male phase lasting for 21 days and sperms are produced from the 3rd and 4th segments in a worm with six segmented body. When body size reaches a length 15 chaetigers, it commences to generate oocytes from the last five segments onward. On becoming SH, the mating pairs regularly take turns to assume either the male or female role. Assuming the female role, it allows only a smaller number of oocytes within a parcel to mature at a time and limits the parcel to hold 29 eggs/cocoon and lay parcels more frequently during the 30–40 d-SH phase (Sella, 1988), in comparison to gonochores and sequential hermaphroditic *Ophryotrocha* species that lay hundred eggs/cocoon.

2.7.1 Simultaneous Hermaphrodites

As it is cheaper to produce sperms, male assuming SH tends to cheat the other partner. To guard against the non-reciprocating mating partners, three mechanisms have been developed: 1. The female acting worms can distinguish whether 'her' partner is an adolescent male on the verge of sex change or a hermaphrodite or an young/small male. The female-acting SH can regulate the clutch size by releasing more or less number of oocytes to mature and subsequently spawn. Accordingly, successive spawnings are regulated at intervals of 3.0, 5.2 and 5.4 days, with the partner being a young/small male, adolescent male and hermaphrodite, respectively (Sella, 1988). 2. The female acting SH parcels 29 eggs per cocoon (Akesson, 1973). This parceling strategy enables her to regulate the number of parcels, according to her mate being a young male, adolescent male or SH. 3. A mating partner from a bigger group invests more on sperm production to escape from male-male competition than a male acting partner from a small group. Hence, the chances for the male acting SH from a larger group to escape from reciprocation are less, in comparison to the male acting SH from a smaller group (Premoli and Sella, 1995). In another SH *O. gracilis*, mating occurs in pairs of ovigerous hermaphrodites, which sequentially alternate sex roles more than once. No male competition occurs. Protandrous males are not involved in pair formation. Hence reciprocal insemination is safeguarded (Sella et al., 1997).

Female acting SH can assess the group size through a water-borne lipidic pheromone and adjust the reproductive output appropriately (Schleicherova et al., 2010). On the other hand, males can also regulate the quantum of sperm donation. For example, *Eisenia andrei*, an epigeic earthworm, detects the virgin status of the mating partner and appropriately adjusts the quantum of donated sperms (Velando et al., 2008). Not only sex allocation is adjusted by SH according to social conditions, but can also adjust the body size of sex change. In *O. diadema*, protandrous males from isolated and intermediate groups attain sexual maturity with significantly larger number of segments than those from a larger group. In *O. diadema*, the female acting monogamaous SH spawns more number of eggs than the SH exposed to promiscuity (Cannarsa et al., 2015). Characterized by strongly female-based allocation, the ovarian biomass of *O. diadema*, *O. gracilis* and *O. hartmanni* is ~ 80% of the gonadal biomass. Hence, the reciprocating hermaphrodites of *Ophryotrocha* produce eggs equivalent to 0.8 times of the eggs produced by a hypothetical mutant pure female. With reciprocation of equal amount by the partner, the reproductive success of SH shall be $0.8 \times 2 = 1.6$ times more than that of a pure female (Sella and Ramella, 1999).

2.7.2 Sequential Hermaphrodites

The majority of sequential hermaphrodites are protandrics. Schroeder and Hermans (1975) recorded 32 out of 67 hermaphroditic species as protandrics. In them, male ratio decreases from 0.80 in young worm to 0.25 in older ones (e.g. *Hydroides elegans*, Qiu and Qian, 1997). The size advantage hypothesis of Ghiselin (1969) and others predicts that reproductive success will increase less with increasing body size for males than for females. In *Ophryotrocha puerilis puerilis*, an individual commences sperm production at the nineth segment size and changes sex, when it attains a body length with 19 chaetigers. From a series of experiments, in which the potential mating pairs consisted of all possible combinations of different body sizes of males and females of *O. puerilis puerilis* in mate choice experiment, Berglund (1990) has brought experimental proof for the size advantage hypothesis. He has observed that the pair consisting of a small male and a large female produces the highest daily egg output, a measure of reproductive success. In an another experiment, Berglund (1991) has assessed the fecundity in isolated male-female pairs of protandric *O. puerilis puerilis* and gonochoric *O. labronica* and concluded that the former are not benefited from large body size. Hence, they will be better off by changing sex, as females prefer to mate with small males. However, the larger *O. labronica* males have greater access to more females. Hence, they gain nothing by changing sex.

In *O. puerilis puerilis*, the time cost for sex change is only ~ 6 days. Interestingly, Monahan (1988) has shown that within 48 hours, individuals in male phase can commence oocyte production. Amazingly, the mating partners in a pair are able to simultaneously change sex several times in their life time (Premoli and Sella, 1995). *Trypanosyllis zebra* and *Syllis amica* are also reported to undergo this type of serial sex change (Policansky, 1982). Berglund (1986) has recorded that 4 out of 14 pairs in *O. puerilis puerilis* have become simultaneous hermaphrodites, after they have reproduced for a month by performing simultaneous sex change. Subsequently, these pairs require assuming an alternate male and female role alone. Indeed, that has saved much of the resource and time required to change sex from male to female or female to male. Berglund explained this phenomenon in the light of sex allocation theory. As eggs are costlier to produce than sperm, a female is quickly exhausted all of its resources. The sex changed to male phase provides adequate time to recover quickly and store adequate resource to act as female.

2.7.3 Labile Gonochorics

In a few gonochorics, either the sex determination or differentiation process is labile. For example, the obligate need for the presence of male pheromone for the mature female to spawn is demonstrated in *Brania clavata*. In the absence of the male pheromones, the female resorbs the oocyte and changes sex to male (Hauenschild, 1953). The protandric *Typosyllis prolifera* is a stolonizing broadcast spawner with a planktonic larval phase and a sedentary adult phase. A small or larger portion of females in a population undergo irreversible change to male sex at one of its subsequent reproductive cycles, indicating the labile sex differentiation (Franke, 1986b). Consequently, resource allocation for reproduction is altered in these sex changing males. In *Capitella* sp I, the processes of sex determination and differentiation remain stable in heterogametic (ZW) female but labile in homogametic (ZZ) male. When the worm is reared in isolation or in a small group, some males become ZZ hermaphrodites and function as either sex. The development of hermaphroditism is regarded as an 'emergency adaptation' to low density. Not only in the absence of females but also excess food is required to induce hermaphroditism. Isolated males require only 22 days to develop ova. Interestingly, these hermaphrodites do not self-fertilize. But their ability to mate as a male or female depends on density and frequency of hermaphrodites. However, the hermaphrodite is unsuccessful as a male. On crossing with a normal ZZ male, the female-acting ZZ hermaphrodite produces only males (Petraitis, 1985a, b). Estimates on the fertility of these female-acting hermaphrodites range from 0.1 to 0.3, assuming that of a true female as one (Petraitis, 1988).

On the other hand, the male heterogametic *Dinophilus gyrociliatus* is gonochoric, dimorphic and reproduces iteroparously. Females lay transparent egg capsules and in its life time; she may spawn a minimum of 40–50 eggs and a maximum of 120–130 eggs. In it, a single ovary simultaneously generates one small (XO) male egg of 40 μm size and many larger female eggs each measuring about 80 μm in a single batch. Consequently, a female egg (by its larger volume) receives ~ 8-times more resources than the male egg with its smaller volume. With sex ratio of 3 ♀ : 1 ♂, the resource allocation is ~ 24-times more to daughters, in comparison to the sons (Minetti et al., 2013). In *D. gyrociliatus*, the unique mechanism, that allows the mothers to overcome the chromosomal mechanism of sex determination, is the selective fertilization of larger eggs by X barring sperms and of small egg by sperms without sex chromosome. The dwarf male arising from a small egg possesses no digestive system, inoculates sperms into his many immature sisters and dies prior to hatching (Traut, 1969a, b, 1970). When reared under stressful temperature (Simonini and Prevedelli, 2003, Akesson and Costlow, 1991) and/or salinity (Akesson and Costlow, 1991) or diet (Prevedelli and Vandini, 1999), more number of small eggs are produced resulting in more number of dwarf males.

3

Regeneration

Introduction

Regeneration is the potency to repair and replace the voluntary loss of cells, tissues and organs of an animal to escape from (i) sub-lethal predation, (ii) self-inflicted spontaneous autotomy and (iii) intolerable shock (e.g. encounter frequency in *Diopatra aciculata*, Safarik et al., 2006, hypo-osmotic stress in *Marenzellaria viridis*, David and Williams, 2016). In 1898, Morgan has classified regeneration into two types: 1. Morphallaxis involving the remodeling of existing tissues into missing ones without extensive cell proliferation (e.g. regeneration of mid-body segments in the oligochaete *Lumbriculus variegatus*, Martinez-Acosta and Zoran, 2015). 2. Epimorphosis involving massive proliferation of undifferentiated cells/stem cells and formation blastema (e.g. head and tail regeneration in *L. variegatus*). Morphallaxis may be less energy-intensive than epimorphosis, as it involves remodeling some pre-existing structure, whereas epimorphosis requires *de novo* formation of missing body parts from blastema. Like echinoderms (Pandian, 2018), annelids represent another phylum with exceptional prodigius potency for regeneration. The head of *L. variegatus* can be regenerated as many as 21 times and the tail 42 times and both together 20 times (Muller, 1908). However, unlike the radially symmetrical echinoderms, the bilaterally symmetrical body of annelid is composed of metameric segments, each consisting of the same organs and so on. Hence, amputation at any axial position along the body of annelids results primarily in the removal of different quantum of the same organ system rather than the removal of a different organ system with a unique structure of echinoderms. Once a small or larger body part is lost, the individual suffers from locomotor function and feeding activity (e.g. *Dipolydora quadrilobata*, Lindsay et al., 2007), growth (e.g. *Eisenia foetida*, Xiao et al., 2011) and longevity (e.g. *D. commensalis*, Dualan and Williams, 2011), reproduction (e.g. *Polydora ligni*, *P. cornuta*, Zajac, 1985, 1995) and behavior (e.g. *Pseudopolydora kempi japonica*, Lindsay and Woodin, 1995). Not surprisingly, only a few polychaete and oligochaete species (~ 250 species, Zattara and Bely, 2016) can regenerate the lost body parts. Further, the regenerative potency varies greatly from species that can regenerate every part of the body from even a single isolated segment (e.g. polychaetes: *Chaetopterus variopedatus*, Berrill, 1928; oligochaetes: *L. variegatus*, Morgulis, 1907) to those that cannot regenerate even a single lost segment (e.g. all hirudineans).

3.1 Regeneration and Reproduction

Embryogenesis, regeneration and clonal reproduction are similar but not identical developmental processes. For example, wound healing and blastema formation are not part of embryogenesis (Myohara, 2004). No blastema is formed during clonal reproduction (see Martinez-Acosta and Zoran, 2015). In *Amphipolydora vestalis*, for example, morphogenetic regeneration requires only 8 days but that of embryogenesis as long as 22 days (Gibson and Harvey, 2000). Table 3.1 represents a comparative summary of major events in regeneration, and architomic and paratomic reproduction. The differences in many of these events are explained in the ensuing sections and Chapter 4, as well. On the other hand, regeneration and clonal reproduction share extensive similarities, especially in development of peripheral nervous system and thereby provides a strong support for the hypothesis that clonal reproduction is derived from regeneration (Zattara and Bely, 2011). From a more detailed analysis, Zattara and Bely (2016) have noted that of 87 species with potency for anterior regeneration, 32% of species alone are capable of clonal reproduction (however, see Table 3.8). But only 7% of (67) species with potency for posterior regeneration can reproduce clonally. Hence, clonal reproduction may have evolved from those with potency for anterior regeneration. Still, there are differences between these two processes. For example, four anterior segments of the naidid *Paranais litoralis* are generated in each clonal reproductive cycle but regeneration fails to produce these four anterior segments (Bely, 1999). In *Lumbriculus variegatus*, these two processes share common cellular and molecular (Lan 3-2 epitopic expression) mechanisms during temporal and spatial developmental events. However, the Lan 3-2 epitopic upregulation is confined to regenerative segments and occur prior to clonal reproduction (Martinez et al., 2005). Differences between the trajectory of regeneration and clonal reproduction occur throughout these developmental processes suggesting that the divergence has occurred all along the developmental course of these trajectories (Zattara and Bely, 2016).

TABLE 3.1

Comparison of major events during regeneration and two types of clonal reproduction: architomy and paratomy (compiled from Kostyuchenko et al., 2016)

Event	Regeneration	Architomy	Paratomy
Wound/Fission	Occurs before	Occurs after	Terminates the process
Wound healing	Commences after	Poorly apparent	Not apparent till end
Cell migration	Commences after	Commences after	Precedes at fission zone
Blastema formation	Occurs after cell proliferation	Precedes fission	Precedes but continues until segmentation
Segmentation	Hypomorphic. Accompanied by morphallaxis	Hypomorphic, morphallactic, old segments reorganized in new antero-posterior (A–P) axis in polychaetes (Boilly et al., 2017) and oligochaetes (Myohara, 2004)	Limited morphallactic old segments precede and persist through the process
Muscular system	Circular muscle fibers formed *de nova*	Changes in muscles facilitate fission	Changes in muscles facilitate fission of new fragments
Digestive system	Transient reorganization. Stomach formed by morphallaxis	Morphallactic reorganization of digestive system along A–P axis	Emergence of morphallactic digestive system at late stage of fission
Nervous system	Following rupture, new fibers grow out from old ventral nerve cord (VNC) to form cereberal commissures and VNC ganglia	New nerves, cerebral commissures and VNC ganglia developed after fission	New nerves, cerebral commissures and VNC ganglia are formed prior to fission. VNC is ruptured at fission (see also Muller, 2004)
Molecular changes	Commences immediately after an injury/amputation	Precedes fission from 1–7 day	Precedes since formation of fission zone and persists through growth and differentiation of new zooids

In the light of the definition for clonal reproduction, the term of regeneration may have to be extended. Clonal reproduction is bi- (or multi-)directional, when an animal (genet) divides to produce two or more fully functional progenies (ramets). But it is unidirectional, when only a single (half of the ramets) progeny is developed (Rychel and Swalla, 2009, see also Bely, 1999), as in enteropneusts and solitary ascidians (Pandian, 2018). In principle, reproduction, whether sexual (including self-fertilizing hermaphrodite involving a single parent) or asexual (clonal), is expected to generate more than one offspring. Hence, a fission or amputation that generates a single ramet, as a product of unidirectional cloning, can be considered more as regeneration than reproduction. For some reason, many authors/reviewers (e.g. Berrill, 1952, Zattara and Bely, 2011, 2016) have not taken this into consideration. This chapter includes annelid species that are characterized by unidirectional cloning as regeneration. The following examples may explain it.

The siboglinid vestimentiferan polychaetes live in chitinous tubes and inhabit primarily in hydrothermal vents. They lack a mouth, an anus and a digestive tract but harbor chaemoautotrophic symbiotic bacteria in the trophosome and derive metabolic needs from the bacteria (see also Chapter 1.5). The body of *Lamellibrachia satsuma* is divisible into (i) tentacular, (ii) vestimental, (iii) trunk and (iv) ophisthosoma regions (Table 3.2). The vestimental region contains the brain, heart, excretory organs and reproductive systems. The longest trunk region harbors the trophosome with symbiotic bacteria and reproductive organs. The ophisthosoma is the only segmented region of the body. From the description and Fig. 1 of Miyamoto et al. (2014), the following inferences can be made: 1. Amputated into two,

TABLE 3.2

Stages of regeneration in two series of experiments in *Lamellibrachia satsuma* (compiled from information reported by Miyamoto et al., 2014; freehand drawing of *L. satsuma* is included, ten = tentacular, ves = vestimental, tr = trunk and op = ophisthosoma regions)

Stage	Post-amputation (d)	External Features	Internal Features
Amputation between anterior and posterior trunk			
1	0–10	Wound healing	Trophosome evaginated
2	14	Wound sealed	Bacteriocytes beneath the healed epidermis
3	20	White/pink blastema formed	Mesodermal cells aggregated
4	30	No sign of segmentation	Septa formed
5	40	Blastema projection Continuous nerve cord and blood vessels formed	Ophisthosoma segmented

Amputation between anterior and posterior ophisthosoma

Anterior	Posterior
80% of ophisthosoma Single row of setae	20% of ophisthosoma No setae

21 days after amputation, trophosomal structures appeared
40 days after amputation, regeneration completed

the anterior fragment containing the tentacular, vestimental and an anterior part of the trunk regenerates the entire posterior region. 2. However, neither an anterior fragment containing only the tentacular and vestimental regions nor the posterior fragment containing the posterior fraction of the trunk and ophisthosoma is able to regenerate the missing body regions. The stages of regeneration in two series of the experiment are summarized in Table 3.2. Clearly, (i) neither the so called 'anterior' nor 'posterior' fragment of *L. satsuma* is able to regenerate, (ii) the stem cells responsible for regeneration remain in the mid-body, namely the anterior trunk and (iii) potency for regeneration limited to the mid-body and represents an example for regeneration of unidirectional cloning polychaetes.

In oligochaetes, earthworms are a good model for the study of regeneration because of their availability and economic importance. However, their regeneration is complicated by the chloragogue, position and size of clitellum (Fig. 3.1) as well as aestivation/hibernation (e.g. aestivation in the tubificid *Rhyacodrilus hiemalis*, Narita, 2006, hibernation in *Lumbricus terrestris*, Liebmann, 1942). In earthworms, the amputed head and tail are regenerated by the mid-body; however, the head and tail do not have the potency to regenerate the mid-body. Amputation at different positions along the mid-body length of *Eisenia foetida* is followed by regeneration toward anterior or posterior direction in immature worms. In fact, the middle amputee containing 30–66 segments in immature *E. foetida* and mid-body segments of *Perionyx excavatus* (Fig. 3.1) regenerate bidirectionally. But regeneration is limited to the head (six segments) alone in sexually mature *E. foetida* (Fig. 3.1). In sexually mature *E. foetida* and *L. terrestris*, regeneration completely ceases (Liebmann, 1942). As the amputated fragment dies but the amputee regenerates in many earthworms, it is arguably clear that amputation at any level results in regeneration and not in clonal reproduction. Understandably, despite low survival (see Xiao et al., 2011), regenerative potency for anterior regeneration is common among these investigated earthworms, as they emerge with the head first from the burrows. Notably, the potency for the head is retained in both immature and mature worms. Recently, S. Sudhakar and his team (pers. comm.) have found that any fragment consisting of six segments in the post-clitellar region of *P. excavatus* has the potency to clonally reproduce, a feature experimentally shown for the first time (Fig. 3.1). While neoblasts may have been retained in almost all the post-clitellar segments, the chloragogue to provide adequate nutrients for clonal reproduction may have to originate from the minimum of six segments. It is not clear whether neoblasts are also present in other earthworms and if so, the need for research is obvious to know the overriding role played by chloragogue on regeneration and clonal reproduction.

Lumbricus terrestris (Fielde, 1885)

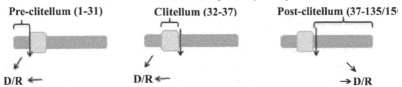

Pre-clitellum (1-31) Clitellum (32-37) Post-clitellum (37-135/15

D/R ← D/R ← →D/R

Eudrilus eugeniae (Sudhakar et al., unpublished)

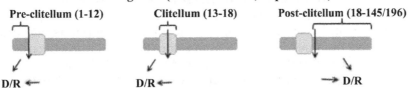

Pre-clitellum (1-12) Clitellum (13-18) Post-clitellum (18-145/196)

D/R ← D/R ← → D/R

Eisenia foetida (Xiao et al., 2011)

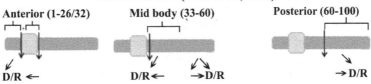

Anterior (1-26/32) Mid body (33-60) Posterior (60-100)

D/R ← D/R← →D/R →D/R

Perionyx excavatus (Gates, 1941)

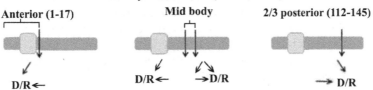

Anterior (1-17) Mid body 2/3 posterior (112-145)

D/R← D/R← →D/R → D/R

Perionyx excavatus (Sudhakar et al., unpublished)

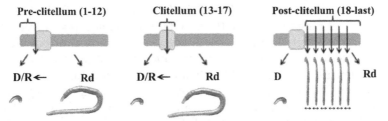

Pre-clitellum (1-12) Clitellum (13-17) Post-clitellum (18-last)

D/R← Rd D/R← Rd D Rd

FIGURE 3.1

Death (D) and regeneration (R) of different fragments following amputation at pre-clitellum, clitellum and post-clitellum positions in selected earthworms. Long vertical arrows indicate the position of amputation. Short horizontal arrows indicate the direction of regeneration. Rd = regenerated worm. Note the experimental clonal reproduction in *Perionyx excavatus* (compiled from different sources including Sudhakar et al., unpublished).

3.2 Incidence and Prevalence

Incidence: Among oligochaetes, terrestrial species emerge the head first from their burrow. In contrast, most aquatic oligochaetes live with their heads burrowed into substrate but their tail is extended up into the water (Martinez-Acosta and Zoran, 2015). Hence, it is likely that the former suffer more frequently from the loss of anterior head (e.g. earthworms) but the latter from the posterior tail tip. Due to the loss of brain ganglia and reproductive structures in the anterior region, regeneration of anterior region can be costlier than the posterior one (Zattara and Bely, 2013). Understandably, polychaete species belonging to Arenicolidae and Opheliidae are incapable of anterior regeneration (see Weidhase et al., 2014). Sub-lethal predation has been a selective force responsible for the sustenance of regeneration of missing body parts (Bely, 2010). Not surprisingly, many authors believe that "Almost all polychaetes can regenerate appendages such as palps, tentacles, cirri and parapodia, and most are capable of regenerating the posterior end of the body" (Pires et al., 2012). However, an intense search has revealed that there are reports for ~ 40 species only (Table 3.3). Description in these reports is also limited to palps, head, a few anterior and/or posterior segments including the tail. This is also true for the > 2,000 speciose aquatic oligochaetes (Table 1.3), as reports available on structural regeneration are limited to ~ 10 species (Table 3.3). On this aspect, more information is provided later. The need is obvious for reports in species with feeding tentacles. Incidentally, a few polychaetes, which are unable to regenerate one or another body part, undertake clonal reproduction, whose products namely the ramets regain them (e.g. *Paranais litoralis*, Bely, 1999).

Prevalence: On prevalence of regenerated structures, limited information is available from field collected species. Prevalence for the palp(s) ranges from 7–8% in *Rhynchospio glutaeus* to 20–50% in *Pygospio elegans* (Table 3.4). The values for the tail range from 29% (head + tail) in *Lumbriculus variegatus* to as high as 90% in *Arenicola marina*. Notably, a larger fraction of population undergoes posterior regeneration than anterior regeneration (e.g. *Euclymene oerstedi*, Clavier, 1984). Clavier (1984) has also estimated that the biomass production from regeneration of *E. oerstedi* is in the range of 2 $g/m^2/y$. Notably, there are more losses of the tail (45%) than the head (28%) in earthworm *Perionyx excavatus*. Yet, that the field collected 2.6 and 1.1% of the worms can regenerate the tail and head, respectively indicate that the worm, *albeit* in very smaller proportion, has retained the bidirectional potential to regenerate the tail and head (see also Fig. 3.1).

Regeneration: Our understanding of regeneration of body parts is based mostly from experimental studies. In spionids, the deciduous palps serve as food gathering organs (Lindsay and Woodin, 1992). But they (each measuring

TABLE 3.3

List of annelid species reported to undergo regeneration (compiled from many sources)

Organ and segmental regeneration

Nereididae: *Nereis virens, Perinereis nuntia, Platynereis dumerilii*

Phyllodocidae: *Eulalia* spp

Amphinomidae: *Eurythoe complanata*

Nephtyidae: *Nephtys caeca*

Dorvilleidae: *Dorvillea bermudensis*

Chaetopteridae: *Chaetopterus variopedatus, Spiochaetopterus oculatus*

Cirratulidae: *Cirratulus cirratus*

Capitellidae: *Capitella* sp I

Arenicolidae: *Abarenicola, Arenicola marina*

Maldanidae: *Clymenella torquata, Euclymene oerstedi*

Oweniidae: *Owenia fusiformis*

Spionidae: *Boccardia syria, Dipolydora commensalis, D. quadrilobata, Marenzellaria viridis, Pseudopolydora kempi japonica, Polydora caulleryi, P. ciliata, P. cornuta, P. flava, P. ligni, Pygospio elegans, Rhynchospio glutaeus, Scolelepis squamata, Spio benedicti, S. setosa*

Sabellidae: *Branchiomma luctuosum, B. nigromaculata, Myxicola aesthetica, Sabella pavonina, S. melanostigma, S. spallazanii*

Serpulidae: *Hydroides dianthus, Kirkegaardia*

Onuphidae: *Diopatra amboinensis, D. dexiognatha, D. neapolitana*

Lumbriculidae: *Lumbriculus variegatus*

Lumbricidae: *Allolobophora molleri, Eisenia foetida, Eophila*

Enchytraeidae: *Dero, Limnodrilus hoffmeisteri, Pristina leidyi, Paranais litoralis, Tubifex tubifex*

Hirudinidae: *Hirudo medicinalis*

0.9 mm in length) are difficult to be removed in *Dipolydora commensalis*, which receives food particles from its host hermit crab, on which it lives within the shell. On being reared in isolation from its host, the palp grows to a length 1.4 mm to enlarge the food gathering structure with cirri (Dualan and Williams, 2011). Ablation at different positions of the gill-bearing anterior part of the tubiculous onuphid *Diopatra neapolitana* indicates that the ablated first few (15) segments die and are not regenerated from the anterior. However, the ablation of anterior fragments from the 10th and those from the 25th to 45/55th chaetigers (except that of 20 chaetigers) are all regenerated from the cut ends of the remaining anterior zone (Fig. 3.2A). Notably, the ablated/lost anterior segments are regenerated by the remaining anterior segments. However, the posterior is not able to regenerate the anterior (Pires et al., 2012). The regenerative growth of the errant *Eurythoe complanata* is estimated as 0.14 segments per day. And, it is completed within 105 and

270 days for the 16-segmented anterior and 37-segmented posterior, respectively (Kudenov, 1974).

In polychaetes, segmental regeneration ranges from simple extension of the last abdominal segment by *Arenicola marina* (De Vlas, 1979) to complete formation of segments, as in many species. The regenerative process in *Pygospio elegans*, for example, proceeds as follows: on the 3rd day after ablation, the wound is sealed and blastema is formed, on the 6th day segments without setae are formed, on the 9th d prostomium, nuchal organs, mouth, short palps and setae are formed and on the 16th day the head is formed, chaetiger number is increased and palp length is increased (Lindsay et al., 2007). However, survival and normal growth of polychaetes decrease with increasing number of chaetigers ablated. For example, ablation of > 30% chaetigers is followed by 25% mortality and 51% abnormal regeneration in *Marenzellaria viridis*. But that of < 20% chaetigers results in 13% mortality and 9% abnormal regeneration (Whitford and Williams, 2016). In *Polydora caulleryi* too, only 10 chaetigers are regenerated, when ablation exceeds 14 chaetigers. A climax seems to be *Spio setosa* (83–126 chaetigers), in which regeneration ceases, when > 32 chaetigers are ablated. The response of *Autolytus pictus* is a shade different. In response to ablation of 1 to 5 and 5 to 13 segments, a head + 3 or 4 chaetigerous segments are regenerated but for that of 13 to 42 segments, a head alone is regenerated and for that involving > 42 segments, only a stump is developed (Okada, 1929). However, amputation, a half of the frontal segment (a half of prostomium and another half of prostomium in *Ophryotrocha notoglandulata* is followed by an increase of 0.7 (female) and 1.0 (male) segment (Pfannenstiel, 1974). Whereas only 14 anterior segments are regenerated in the field, the number of regenerated segments in *Eurythoe complanata* progressively decreases from 40–24 for the loss of ~ 80 segments. Notably, the anterior fragment regenerates a constant number of 14 segments in this worm with 130 segments (Kudenov, 1974). The oligochaetes also display similar reductions with increasing ablated number of segments. In *Lumbriculus variegatus*, ablation upto eight segments virtually from any position of the body are all replaced. However, the maximum of only eight segments are replaced, when ablation exceeds eight segments (Von Haffner, 1928). In *Eisenia foetida* too, regeneration ceases with ablation of 20 segments, although that for four to eight segments, three or four segments are regenerated (Moment, 1950). Hence, the degree of anterior regeneration has an upper limit, as in *M. viridis* and is reduced with increasing number of segments lost, as in *Eisenia foetida* (Fig. 3.2B). Conversely, the speed of regeneration is accelerated with increasing number of segments lost, especially in gill-bearing chaetigers, as in *Pygospio elegans* (Fig. 3.2C).

An important factor that controls regenerative potency is the cerebral hormone in errant polychaetes. In *Nereis diversicolor*, sexual maturation stages are assessed by the oocyte size and their synchronous development.

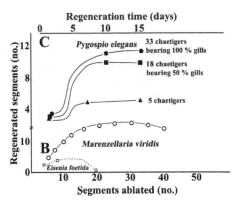

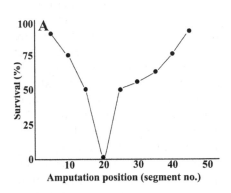

FIGURE 3.2

A. Survival of *Diopatra neapolitana* following amputation at different positions in the anterior region. The anterior amputee dies following ablation. The remaining anterior fragment survives and regenerates the missing parts, but the posterior is unable to regenerate the anterior (modified and redrawn from Pires et al., 2012). B. The number of segments regenerated as a function of the number of ablated segments in *Marenzellaria viridis* (drawn using data reported by Whitford and Williams, 2016) and *Eisenia foetida* (drawn using data reported by Moment, 1950). C. Number of regenerated segments as a function of time in *Pygospio elegans*, in which 5, 18 (50% gill bearing-) and 33 (100% gill bearing-) segments have been ablated (simplified and redrawn from Lindsay et al., 2007).

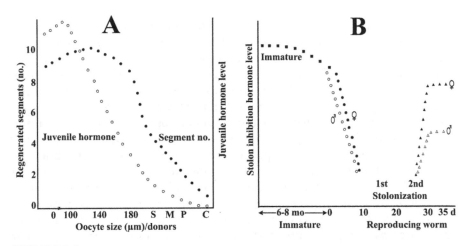

FIGURE 3.3

A. Effect of declining juvenile hormone level on oocyte size and number of regenerated segments in *Nereis diversicolor* S = submature, M = mature, P = postspawing, C = control (compiled, modified and redrawn from Golding, 1983). B. Schematic representation of declining level of ootrophic hormone in females and males and its effect on first and second stolonization in mature *Typosyllis pulchra* (modified schematic representation from Heacox and Schroeder, 1982).

Females with occyte measuring < 100 µm, those with asynchronous population of oocytes measuring ~ 180 µm and those with synchronously developing oocytes of > 200 µm are considered as immature, submature and mature, respectively. Golding (1983) has estimated the number of regenerated segments in a 3 weeks period after amputation of 30 segments in decerebrated immature worms implanted with the 'brain' from a matured one. Figure 3.3A shows progressively decreasing number of regenerated segments with increasing oocyte size, a marker of sexual maturity. The decerebrated recipient control without the 'brain' implantation has almost lost the regenerative potency. The figure also indicates the reported decreasing trend for the cerebral Juvenile Hormone (JH) as a function of oocyte size (see Chapter 7). In the errant syllids, which reproduce through stolon, regeneration is also eliminated with the commencement of stolonization (e.g. *Typosyllis pulchra*, Heacox and Schroeder, 1982). Figure 3.3B also shows the reported declining trend for Ootrophic Hormone (OH) (see Chapter 7) with the commencement of stolonization. With the progressive decrease in JH/OH, reproduction is commenced but regeneration is ceased. Understandably, these two processes competing for resource allocation mutually eliminate each other. However, sedentary polychaetes, which are mostly suspension feeder with continuous food supply, may undertake regeneration even after sexual maturity. Incidentally, regeneration in errant polychaetes depends on cerebral neurosecretion, whereas the sedentary polychaetes on posterior segmental ganglia (see p 111).

Polychaete vs oligochaete reproduction: Limited information is available for the loss and regeneration of body parts on growth and/or reproduction. Nevertheless, the available information clearly indicates the adoption of contrasting strategies by oligochaetes with the chloragogue and sedentary polychaetes without it to meet the cost of regeneration. Whereas the former mutually eliminates regenerative and reproductive processes, the latter simultaneously undertake both of them but at a cost of reproductive output. In oligochaetes, the chloragogue functions as an organ equivalent to the liver in vertebrates. Derived from the peritoneum, it is located around the intestine and serves as the major center for synthesis and storage of glycogen and fats. It is colored due to the inclusion of green-yellow colored lipids. On release into the coelom, the chloragocytes become eleocytes (Barnes, 1974). Abundance of the chloragocytes and eleocytes depends on the amount of food consumed. The processes of regeneration and reproduction require eleocytes for their functioning. In *E. foetida*, the occurrence of one of these processes retards or arrests the other (Table 3.5), as the quantum of eleocytes is evidently not adequate to meet the simultaneous costs of these two resource-demanding processes (Liebmann, 1942). This is also true of *Lumbricus terrestris* and *Pheretima* (*indica* ?) and *Tubifex tubifex* (Liebmann, 1942). In the naidid *Stylaria lacustris*, Harper (1904) reports the mutual elimination between regeneration on one hand, and sexual and asexual reproduction on the other.

TABLE 3.4

Prevalence of the loss of body parts in some annelids

Species, Reference	Reported Observations
Sedentary polychaetes	
Arenicola marina (de Vlas, 1979)	90% worms have lost the tail; regeneration is limited to extension of the last segment but no new segment is added
Rhynchospio glutaeus, Pseudopolydora kempi japonica (Lindsay and Woodin, 1992)	7–8 and 10% of worms have lost I and II palps, respectively
Pygospio elegans, Polydora cornuta (Woodin, 1982, Miller and Jumars, 1986)	50–20% worms have lost palps
Marenzellaria viridis (Whitford and Williams, 2016)	7% worms have lost and are regenerating anterior segments
Euclymene oerstedi (Clavier, 1984)	8 and 44% worms are regenerating anterior and posterior segments, respectively
Clymenella torquata (Sayles, 1936)	0.1 and 3.9% of worms are regenerating anterior and posterior segments, respectively
Errant polychaetes	
Eurythoe complanata (Kudenov, 1974)	More headless fragments than tailless ones. 35–80% (mean ~ 50%) regenerating fragments have been recorded
Oligochaetes	
Lumbriculus variegatus (Morgulis, 1907)	Loss of palps + head, head + tail and posterior segments is 19, 29 and 41%, respectively
Perionyx excavatus (Gates, 1927)	Of 533 worms in 4 collections, 77% have lost tail or head and regenerating it. Of them, 45 and 28% of the worms have lost tail and head, respectively. About 2.6 and 1.1% of the worms are regenerating tail and head, respectively

Clonal reproduction and regeneration intensely compete with sexual reproduction for resource allocation. Not surprisingly, both these reproductive modes are temporally separated. In most asexually reproducing annelids, clonal reproduction occurs prior to sexual maturity but sexual reproduction after sexual maturity. However, these two modes may also alternate, as in the sedentary *Polydorella* spp (Radashevsky, 1996). In other sedentary polychaetes like the spionid *Amphipolydora vestalis*, architomic fragmentation is the dominant form of reproduction but occurs concurrently with sexual reproduction. About 46% of asexual propagules or ramets contain gametes in the parental abdominal chaetigers (Gibson and Paterson, 2003). The other ramets gain the germline by the migrating *piwi*-positive ventral cells through the ventral nerve cord (Ozpolat and Bely, 2015).

TABLE 3.5

Effect of regeneration of body parts on reproduction in some annelids

Species, Reference	Reported Observations
Regeneration and reproduction: Sedentary polychaetes	
Polydora ligni (Zajac, 1985)	Palp regeneration in 8% worms reduces fecundity by 10–29%. Regeneration of 71% posterior segment reduces 49–80% fecundity
Polydora cornuta (Zajac, 1995)	In 30-, 60- and 70-segmented worm, regeneration of gametic segments is 13.8, 23.4 and 17.2% of the worms. Consequent reduction in reproductive performance is 27, 32 and 28% for gametogenic segments, egg capsules and cumulative fecundity, respectively
Regeneration and reproduction: Oligochaetes	
Eisenia foetida, Lumbricus terrestris, Pheretima (indica?), Tubifex tubifex (Liebmann, 1942)	Regeneration, asexual and sexual reproduction processes are mutually excluded, as they depend on reserve tissue, chloragogue. Regeneration of head is not affected by reproduction. During hibernation, not even posterior regeneration occurs, as eleocytes/chloragogue are exhausted
Stylaria lacustris (Harper, 1904)	Regeneration, asexual and sexual reproduction processes are mutually excluded
Enchytraeus variatus (Bouguenec and Giani, 1989)	Simultaneously occur for a brief period at the cost of cocoon production

In general, oral structures like palps are more vulnerable in epifaunal polychaetes than the infaunal ones. Fecundity may be decreased with a loss of one or two oral palps, which are important oral structures for food collection and thereby decrease fecundity. But the loss of gametogenic or posterior segments profoundly reduces fecundity. Zajac (1985, 1995) has provided valuable information to show that regeneration of palps and segments occurs at the cost of reproduction. From Fig. 3.4, the following may be inferred: 1. The loss of one or two palps, followed their regeneration within 15 days, does not affect the number of gametogenic segments. 2. However, fecundity is significantly reduced, when gametogenic segments or posterior segments are lost. The loss of fecundity in the sedentary *Polydora ligni* ranges between 10 and 29% for palp(s) regeneration as well as 49 and 80% for regeneration of posterior segments (Table 3.5). In *P. cornuta*, the reduction in reproductive performance is 27, 32 and 28% for the loss of gametogenic segments, egg capsule and cumulative fecundity, respectively. Incidentally, from both field and experimental observations, Zajac (1986) has also reported that not only regeneration but also limited food availability and increased intra-specific density also slows growth, postpones sexual maturity, prolongs inter-brood and reduces fecundity.

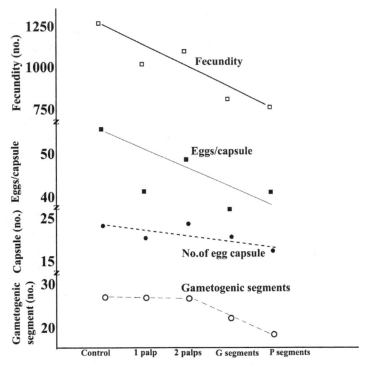

FIGURE 3.4

Effect of loss of one and two palps, as well as gametogenic segments (G segments) or posterior segments (P segments) on fecundity of the spionid polychaete *Polydora ligni* (compiled and simplified from Zajac, 1985).

3.3 Regenerative Process

In annelids, regeneration has been studied for well over a century and in an array of representative species across the phylum (e.g. Randolph, 1892, Hyman, 1940, Hill, 1970, Bely, 2006). Earlier studies have focused on cellular and tissue-level dynamics of regeneration. They have laid the foundation for the present molecular level studies. This account has also consulted Bely's (2014) review, which has considered a range of publications on histological and descriptive studies, experimental surgeries and graftings, and proliferation assays. The collated information is sequenced under the following headings: 1. wound-healing, 2. blastema formation and differentiation, 3. segmental reorganization, 4. growth and elongation (Weidhase et al., 2014). Briefly, regeneration is a fairly complex but an orderly process. In annelids, it ranges from species that are completely incapable of any regeneration, as in leeches and species that has no ability to regenerate even a single anterior segment, as in *Streblospio benedicti* (Lindsay et al., 2007) to species that are capable of regenerating an entire animal from a single mid-body (14th) segment in *Chaetopterus variopedatus* (Berrill, 1928). Not surprisingly, it is a highly variable trait.

3.3.1 Wound-healing

In annelids, wound-healing includes (i) muscle contraction, (ii) autolysis, (iii) cell migration and (iv) re-epithelialization. Immediately following an amputation, most annelids seal the wound effectively by rapid longitudinal muscular contraction (e.g. Hyman, 1940, see also Samuel et al., 2012), stemming the loss of body fluids and possibly preventing infection. The acanthobdellid leeches are exception to it. With the presence of an outer stiff cuticle, they fail to seal the wound (Bely, 2006). Other leeches, however, respond to surgical lesion with a same sequence of events, as in vertebrates. In *Hirudo medicinalis*, for example, the newly synthesized collagen granulates the tissue and remodels the scar tissue (Tettamandi et al., 2005). In other annelids, wound-healing triggers early regenerative signaling events (King and Newmark, 2012). In *Paranais frici, P. litoralis, Chaetogaster diaphanus, C. diastrophus* as well as *Amphichaeta raptisae* B and C, the wound healing presumably fails to trigger the signaling. Hence, the sealed wound neither forms a blastema nor replaces any missing structures (Bely and Sikes, 2010). In *Enchytraeus japonensis* (Kawamoto et al., 2005) and *Lumbriculus variegatus* (Lesuik and Drewes, 1999), muscular contraction precedes autotomy. Within hours of an inflicted injury, autolysis is evident both in ectoderm and mesoderm (Cornec et al., 1987). Migration of an array of cells to the wound site is a general feature of regenerating annelids. The migrant cells phagocytize the damaged cells and tissue debris and form a tissue plug, while others serve as a source for the formation of new cells (Bely, 2014). The phagocytes are of different types and are called coelomocytes and splanchnopleural cells in *Limnodrilus hoffmeisteri*, chloragocytes/eleocytes and amoebocytes type I in *L. variegatus* as well as hyalocytes and macrophages in *Eisenia foetida*. Usually, the severed epidermal edges fuse with each other and the same occurs between the gut epithelia, as well (e.g. Hill, 1970). Hence, re-epithelialization is completed prior to cell proliferation at the wound.

3.3.2 Cell Migration

An injury on any part of the body elicits an extensive migration of different cell types toward the wound. The origin of new cells and tissues is one of the fundamental problems in regeneration of annelids. Three hypotheses have been postulated concerning the origin of new tissue (Paulus and Muller, 2006). They are: 1. the epidermis proliferate new tissues including new epidermis or mesoderm (e.g. Stone, 1933), 2. the new tissue is generated by specific pluripotent mesodermal cells (e.g. Randolph, 1892) and 3. each germ layer retains its identity and proliferates material for the respective tissues (Hill, 1970). In a shade different from these hypotheses, Goss (1974) has postulated that the blastema is a mixture of differentiated cells and stem cells that migrate to the wound site.

In a seminal publication, Randolph (1891) has discovered the 'quiescent specialized embryonic cells called "neoblasts" from the posterior surface of the septa adjacent to the ventral nerve cord of *Lumbriculus variegatus*. Subsequently, her discovery is extended not only to some oligochaetes like *Tubifex tubifex* (Stone, 1932), *Nais* (Herlant-Meewis, 1947) and *Limnodrilus hoffmeisteri* (Cornec et al., 1987) but also to polychaetes like *Aricia*, *Chaetopterus* and *Diopatra* (see Berrill, 1952). Counting the number of neoblasts following posterior transection in the naidid *Ophidonais serpentina*, Bilello and Postwald (1974) have brought evidence for migration of neoblasts toward the wound site. Activated by an injury, these cells do migrate to the wound along the ventral nerve cord in oligochaetes (e.g. *Enchytraeus japonensis*, Tadokoro et al., 2006). Incidentally, the neoblasts of annelids are multipotent stem cells. Because of their pluripotency, the term neoblasts are more appropriate to planarians than to annelids (Randolph, 1892, Paulus and Muller, 2006). However, the term neoblasts (in multipotent *sensu*) here is continued to be used.

Besides the mesodermal neoblasts, the injury-activated dedifferentiated stem cells are considered to be present in all the three germ layers (Jamieson, 1981). From their BrdU pulse-chase experiment in the amputated (at the 7th and others at every five segment up to the pygidium) *Parougia bermudensis*, Paulus and Muller (2006) have inferred the following: 1. Sealing of the wound is a pre-requisite for regeneration (see also Kato et al., 1999). 2. Following amputation, the amputees lose the gut contents (cf Subramanian et al., 2017). The everted gut epithelium forms an outer layer and closes the wound at both ends. The muscle fibers adjacent to the wound contract and constrict the tissue around the wound to ~ 20% of its size within 2 hours post-operation. Within a day, the entire wound is covered by the epithelium. 3. The BrdU-labeled cells appear 16 hours post-operation. Incidentally, this observation also confirms that of Hill (1970), who has reported that tritiated thymidine [³H] T-labeled cells of *Nereis* sp, *Nephtys* sp and *Sabella melanostigma* appear at the wound site 24 hours post-operation. 4. The delayed appearance of the epidermal stem cells indicates the time required for dedifferentiation of already existing epidermal cells. 5. The descendants of epidermal cells proliferate centrally but those of the gut peripherally and 6. The anterior region of the blastema transforms into the head, while the posterior forms the pygidium and persists as proliferation zones. 7a. Proliferation of the BrdU-labeled epidermal cells fills ~ 67% of the blastema within 250 hours of regenerative period. 7b. These BrdU-labeled epidermal cells appear in nearly almost all the areas of the body. 7c. But most of them are detectable at the posterior end and their number gradually decreases anteriorly. 7d. More importantly, the proliferation ceases earlier in the 'older' anterior segments than in the 'young' posterior segments. Incidentally, this may explain why less number of species regenerates anterior. Briefly, the interesting investigation by Paulus and Muller suggests that activated by the injury, the epidermal cells are dedifferentiated to become ectodermal stem cells. Incidentally, employing histochemical techniques, dedifferentiated mesodermal cells are shown to arise from the muscle cell tissues in the syllid *Autolytus pictus* (Okada, 1929) and *Owenia* (Probst, 1931).

In this context, Prof. S. Sudhakar and his team at Manonmaniam Sundaranar University, Thirunelveli, India has brought a series of publications in recent years. Samuel et al. (2012) have traced the origin of mesodermal stem cells and blastema formation in *Eudrilus eugeniae*. The earthworm was amputated between the 11th and 12th segments passing through the clitellum. The wound was sealed 24–48 hours after amputation and soft transparent blastema was formed between 48 and 72 hours following amputation. A section through the worm 48 hours after amputation indicated reduction in the thickness of circular muscle layer from 170 to 98 μm, and an increase in the longitudinal layer from 100 to 188 μm, when the blastema was being formed. BrdU-labeling retention assay in the blastema revealed the presence of the BrdU-positive longitudinal cells in the blastema, presumably arising from the muscles. The immunochemical-BrdU antibody staining of the 2-day old blastema also confirmed the presence of BrdU-positive cells in the blastema. In another interesting study, Kalidas et al. (2015) demonstrated that Lamin A, an intermediate filament protein, is absent in the BrdU-positive stem cells of *E. eugeniae*.

In the blastema of *E. eugeniae*, the cytoplasm but not the nuclei of the BrdU-positive cells is autofluorescent (Fig. 3.5A) due to the accumulation of riboflavin. In the amputated worm, functional mouth and anus are formed 6 days after amputation. The adaptive starvation during the first 5 days

TABLE 3.6

Riboflavin content (μg/g) in different organs in intact and regenerating *Eudrilus eugeniae*. Right panel A and B shows the milky white and yellow riboflavin synthesizing *Bacillus endophyticus*, respectively (modified from Subramanian et al., 2017)

Organ	Intact Worm	Regeneration on	
		3rd day	6th day
Intestine	305	966	321
CML + LCL	203	640	277
Coelomic fluid	368	386	415
Prostate gland	395	279	216
Setae	284	112	103

CML = Circular muscle layer, LCL = Longitudinal cell layer

Upper panel showing autofluorescence

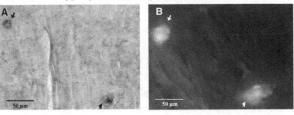

Lower panel (I) showing regeneration in intact worm on

3 d   5 d   7 d

Lower panel (II) showing regeneration in riboflavin supplemented worm on

3 d   5 d   7 d

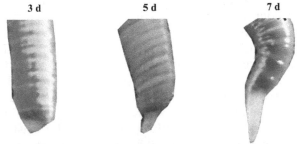

Lower panel(III)showing regeneration in worm inhibited by antibiotic

3 d   5 d   7 d

FIGURE 3.5

Eudrilus eugeniae: Upper panel shows A. BrdU-positive cells marked by arrows and B. shows autofluorescence of BrdU-positive cells under red filter. Lower panel showing regeneration on 3rd, 5th, and 7th d in intact worm (lower panel I), riboflavin supplemented worm (lower panel II) and antibiotic injected worm (lower panel III). Note the transparent regenerative blastema in these worms (modified from protocol record of Dr. S.C.J.R. Samuel).

triggers the gut microflora to rapidly multiply and synthesize riboflavin. The presence of an efficient riboflavin synthesizing *Bacillus endophyticus* has been confirmed by 16Sr RNA sequencing. The microbial synthesis of riboflavin has also been confirmed as the colony of *B. endophyticus* changes its color from milky white to yellow on the second day. Table 3.6 shows that effectively employing the symbiotic microflora, *E. eugeniae* draws riboflavin for its regeneration between the 3rd and 5th day following the amputation. The intestine is the major location, in which much of riboflavin is synthesized by the riboflavin synthesizing symbiotic microbes. Other organs like prostate gland and setae do neither harbor the symbiotic microbes nor synthesize riboflavin. The presence of riboflavin in the circular muscle layer and longitudinal cell layer clearly indicates the stem cell rich tissues have abundant riboflavin in these layers to augment regeneration (Subramanian et al., 2017). The riboflavin supplementation accelerates blastemal length (Fig. 3.5C), a marker for the blastemal growth and possibly differentiation. But its suppression by injection of antibiotic reduces the blastemal growth (Samuel et al., 2012). Yet it remains to be known whether the mesodermal stem cells differentiate into mesoderm and its derivatives alone or they can also differentiate into epidermal and gut cells.

Thanks are due to the Japanese scientists for establishing the source and sequence of mesodermal regeneration in *Enchytraeus japonensis*, a small (10–15 mm), almost transparent worm with the life span of < 2 weeks (Yoshida-Noro and Tochinai, 2010). Analyzing ~ 400 clones, Takeo et al. (2010) have isolated five genes, whose expression level is altered during the regenerative process. RT-PCR studies have shown that three of them *Ejpsmd*, *EjTuba* and *grimp* are upregulated, while the two others *horu* and *minor* are downregulated. Of the three, *grimp* is transiently but specifically expressed only in mesodermal cells just beneath epidermis at the blastema tip and entire amputees from 3 to 12 hours. But it is never expressed in the epidermis and digestive tract. The suppression of its expression inhibits mesodermal proliferation and anterior differentiation. Cross section analysis has shown that *grimp* is expressed in the mesodermal cells and neoblasts, characterized by large nuclei and nucleolei, and a large nucleo-cytoplasmic ratio. It must, however, be noted that Myohara (2012) has shown that the neoblasts are dispensable for regeneration in annelids. The annelid 'neoblasts' are peritoneal stem cells and may not be equal to those of planarians (Bilello and Postwald, 1974). Hence, activated by the injury, the dedifferentiated mesodermal stem cells are responsible for regeneration of mesoderm and its derivatives.

Regarding regeneration of endoderm and its derivatives, there are only passing remarks (e.g. Paulus and Muller, 2006, Takeo et al., 2008). Briefly, available information lends support to Hill's (1970) hypothesis that each germ layer has retained its identity. Activated by an injury, the differentiated stem cells from the respective germ layers regenerate their respective organs and systems. This view is further supported by the fact that annelids undergo mosaic development. In *Tubifex rivulorum*, blastomere deletion at four cell stage has revealed the loss of totipotency by all the three blastomeres (A, B, C) except the D blastomere (Penners, 1922). Subsequent to the formation of germ layers in spiralian embryos of annelids, each of the differentiated germ layer retains its identity but has already lost the potency to revert back (Needham, 1990).

3.3.3 Blastema and Differentiation

Shortly after wound healing, a blastema is developed with the contribution from all three germ layers (e.g. ectoderm: Paulus and Muller, 2006, mesoderm: Randolph, 1892, endoderm: Takeo et al., 2008). Cellular contribution to the blastemal tissue arises from dedifferentiated cells beneath the wound site (Balavione, 2014).

Differentiation: A large number of publications (Clark and Bonney, 1960, Clark et al., 1962, Hauenschild and Fischer, 1962, Nayar, 1966) have experimentally shown that posterior regeneration is induced by neurosecretion arising by the suboesophageal ganglia, the brain in many errant polychaetes like *Nereis diversicolor*, *Platynereis dumerilii* and *Nephtys*. Recently, Muller et al. (2003) have also found that the cephalic ganglia are essential for regeneration in the errant amphinomid *Eurythoe complanata*. However, the ventral ganglia in each posterior segment of the sedentary polychaetes are capable of posterior/caudal regeneration in the complete absence of a brain (e.g. *Branchiomma nigromaculata*, *Chaetopterus variopedatus*, Hill, 1972). Correlated with the reduction in sense organs, the brain of sedentary tube-dwelling polychaetes is quite simple, in comparison to that of the errant polychaetes (Bullock and Horridge, 1965). As the projected head of the sedentary worms is often subjected to sublethal predation, it may be advantageous for them to have the neurosecretory ability distributed along the nerve cord rather than localized in the cephalic ganglia. Not surprisingly, barring *Arenicola marina*, all the investigated sedentary tubiculous worms are capable of regeneration.

Investigations on the Antero-Posterior Axis (APA) of the errant polychaete *Nereis latescens* have found that the Ventral Nerve Cord (VNC) assigns the APA. In the absence of VNC, the regenerate fails to develop APA especially the parapodia. But in the sedentary sabellids *Dasychone infrata* and *Branchiomma nigromaculata*, parapodial inversion occurs along the APA in the absence of VNC and it is correlated with the position of nerve cord (Boilly et al., 2017, see also Table 3.1).

In the blastema, the system induces self-proliferation initially and regeneration subsequently. In *Eurythoe complanata*, extirpation of a single ganglion retards regeneration and that from more ganglia inhibits regeneration completely, unless the affected regenerate is autotomized (Muller et al., 2003). Understandably, the nervous system develops remarkably earlier than

A 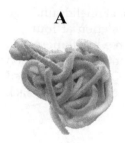   B 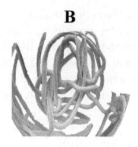   C

FIGURE 3.6

Self assemblage of *Eudrilus eugeniae*, when kept in water. A. Intact worm, B. Worm in which the brain harboring cephalic region up to 4th segment has been amputated, C. Worm, in which regions beyond 70th segment has been amputated (modified from protocol record of Dr. G. Daisy).

musculature in *Cirratulus* cf *cirrus* (Weidhase et al., 2014). Using phalloidin-labeling, anti-body staining and confocal laser scanning microscopy, Weidhase et al. (2014) have described the differentiation processes of nerves and musculature in the blastema of *C. cirrus*. In blastema, the nervous system is a typically a rope-ladder like arrangement and circumpharyngeal connectives exhibit two separate roots leading to the brain. As a general pattern of annelids, the regeneration of circular and longitudinal musculature originates from different group of cells. During regeneration, the longitudinal musculature commences with diffuse growth and subsequent structuring into the blastema. In contrast, circular musculature develops independently inside the blastema.

In many oligochaetes, self assemblage is a behavioral response to stress conditions. For example, an exposure to hypoxic water (1.5 mg O_2/l), the self assemblage of *Tubifex tubifex* forms a ball enabling the harvest of the worm in culture system (Marian and Pandian, 1984). In earthworm *Eudrilus eugeniae*, stressed by immersion into water, Daisy et al. (2016) have observed that the self assemblage behavior remarkably changes in the presence and absence of the brain. Even in the absence of the brain, the worm, in which the first four segments with brain are amputed can survive but is unable to self assemblage (Fig. 3.6B) and requires the intact brain for the self-assemblage within an area of 2.3 cm^2 per worm. However, in the presence of the brain, the worm amputated at the 70th segment is able to self assemblage (Fig. 3.6C) and requires an area of just 0.9 cm^2 per worm.

3.3.4 Segmentation and Reorganization

In annelids, the number of segment ranges from a few (e.g. ~ eight segments, 400 μm long *Apodotrocha progenerans*, see Westheide, 1984) to as many as 300–700 in *Eunice siciliensis* (Hofmann, 1974). The segments are almost uniform throughout the body. Yet, based on the parapodial structure and function, they can be grouped into cephalic, trunk and caudal regions. In annelids, segments are added successively forward from the growth zone of the structurally well developed pygidium, although the number is limited in a few exceptional polychaetes (e.g. *Clymenella torquata*, Sayles, 1932) and oligochaetes (e.g. *Eisenia foetida*, Moment, 1946, 1950). Following amputation at the position of 50th or 80th segment in *E. foetida*, new segments are added posteriorly by regeneration until the original total numbers of 100 segments are restored. Considering only the publications by Moment, Berrill (1952) has wrongly concluded that in all oligochaetes, the addition of segments is ceased early and is precociously attained. In oligochaetes, the segment number varies widely from 80 to 100 in *Eudrilus eugeniae* (*shodhganga.inflibnet. ac.in*), 135 to 150 in *Lumbricus terrestris* (The Robinson Library) and 145 to 195 in *Perionyx excavatus* (*shodhganga.inflibnet.ac.in*) as well as 34 to 120 in *Tubifex tubifex* (*http://eol.org/pages/620440/overview*). Barring the exceptional polychaetes and oligochaetes, the segment number of annelids is a labile trait and is regulated mostly by food availability and temperature (Van Cleave, 1937) as well as photoperiod (Kharin et al., 2006).

In the segmentation process, polychaetes and oligochaetes display following contrasting differences: 1. In oligochaetes, the processes of segmental regeneration and reproduction, irrespective of sexual or clonal, are competitive to draw resources from the chloragogue. Consequently, they mutually eliminate each other (Liebmann, 1942, Okrzesik et al., 2013). 2. In majority of oligochaetes (e.g. enchytraeids), the neoblasts are obligately required for asexual reproduction but not for segmental regeneration. In polychaetes, the presence of neoblasts is reported only from *Chaetopterus variopedatus*, in which the posterior (without the 14th segment) is not capable of anterior regeneration (Berrill, 1928) and *Diopatra amboinensis* (see Berrill, 1952), which is not known to regenerate. These neoblasts are not known to obligately require for asexual reproduction. 3. In polychaetes, expression of the genes *hedgehog* (*hh*) in the segment-addition zone (SAZ) of *Perinereis nuntia* (Niwa et al., 2013) and *Hox* genes in the central nervous system in regenerating *Nereis virens* (Novikova et al., 2013) is reported to manifest segmentation. But in oligochaetes, *grimp* expression in the neoblasts and mesodermal cells is reported to manifest segmentation. 4. More than 50% polychaetes are characterized by indirect life cycle involving external fertilization (see p 44) and one or more larval stages lasting for some hours to several months (Giangrande, 1997). But the oligochaetes display direct life cycle involving internal fertilization and completion of entire embryogenesis within the relatively safer cocoon. Consequent to these contrasting features, regeneration in polychaetes and oligochaetes markedly differ from each other. Though this has been overlooked by many reviewers (e.g. Berrill, 1952, Zattara and Bely, 2016), this account considers them separately.

Polychaetes: Experimental amputation studies have indicated that the minimum number of segment(s) required to regenerate segments both anteriorly and posteriorly is only the 10th in the maldanid *Clymenella torquata* and 14th in *Chaetopetrus variopedatus* (Fig. 3.7). In the syllid *Procerastea halleziana* too, it is limited to the 17th and 18th segments, i.e. the seminal segments, harboring stem cells (a newly coined term) responsible for regeneration, are positioned between the cephalic and anterior region of the trunk (cf *Lamellibrachia satsuma*, p 96). In these worms, anterior regeneration commences with formation of the head, from which new segments are added posteriorly. The posterior regeneration begins with differentiation of the pygidium, from which new segments are added anteriorly (see Licciano et al., 2012). Remarkably, these seminal segments retain the same respective position in the newly regenerated worms indicating that they direct the reorganization of newly added segments. Whereas the seminal 10th and 14th segments can regenerate anteriorly and posteriorly, their posterior (in the absence of the seminal segment) is unable to regenerate anterior in *C. torquata* (Moment, 1951) and *C. variopedatus* (Berrill, 1928) and are not capable of

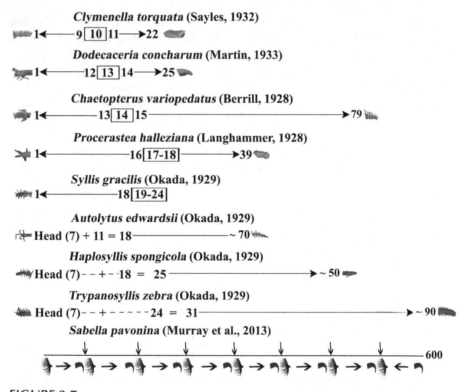

FIGURE 3.7

Regenerative potency of representative polychaetes. Seminal segments are boxed. Arabic numerals indicate the segment number. Note the heads are decorated with food acquiring structures in polychaetes. Arrows indicate the directions of regeneration. Continuous lines indicate regenerative segmental regions.

clonal reproduction. However, *P. halleziana*, in which the posterior containing 17th and 18th seminal segments regenerate the anterior, is capable of clonal reproduction by architomy (see Table 4.2). The intact cirratulid *Dodecaceria concharum* is also able to naturally regenerate both anterior and posterior from the 13th seminal segment and is also capable of clonal reproduction.

But it is difficult to comprehend why the syllids like *Autolytus edwardsii*, *Haplosyllis spongicola* and *Trypanosyllis zebra* do require (7 + 11) 18 to (7 + 18) = 25, anterior segments to regenerate the posterior, as the head consists of only seven anterior segments in syllids (Fig. 3.8). In *S. gracilis*, a fragment consisting of 6, i.e. the 19th to 24th segments regenerate all the 18 anterior segments.

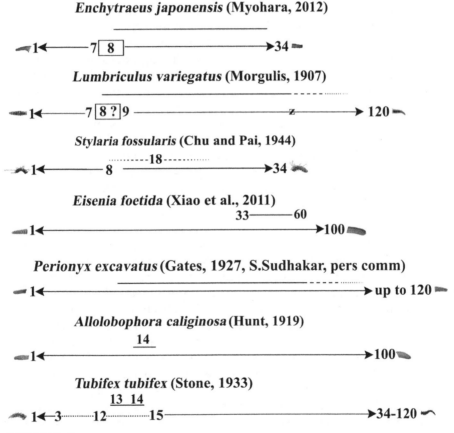

FIGURE 3.8

Regenerative potency of representative oligochaetes. Straight lines with arrow heads indicate the number of body segments and directions of regeneration. The straight lines above the arrow headed lines indicate the number of segments harboring neoblasts. The neoblasts are present in all the segments beyond the boxed number. The dotted lines on either side of the Arabic number indicates the seminal segment, on either side of which the neoblasts harboring segments progressively decrease. Note the narrower head in the earthworms.

These observations on syllids indicate that the positioned seminal segments between 18th and 34th harbors stem cells. The presence of 17 seminal segments in the mid-body markedly differs from the presence of one or two seminal segments alone in the identified species listed above. Further, any fragment containing ~ 75 segments from any part of the body in the 600-segmented *Sabella pavonina* is reported to add regenerative segments both anteriorly and posteriorly (Murray et al., 2013). In the spionids too, the fragment consisting of a few segments in *Pygospio elegans* (Gibson and Harvey, 2000) and 2–3 chaetigers in *Amphipolydora vestalis* (Gibson and Paterson, 2003) drawn from any part of the body is able to clonally reproduce. This is also true of *Potamila torelli* (Watson, 1906). These observations imply the presence of regenerative stem cells almost throughout the body. Hence, the polychaetes seem to fall in one group including maldanid, chaetopterid and cirratulid possessing one or two seminal segments and the others including syllids, spionids and sabellids harboring many regenerative seminal segments positioned in almost all the body segments. Although the identified seminal segment(s) in the first group of polychaetes are capable of adding segments both anteriorly and posteriorly, their posteriors (without the seminal segments) are unable to regenerate and add segments anteriorly (e.g. *C. variopedatus*). Hence, in the first group the position of seminal segment(s) harboring stem cells is positioned in the anterior part of the trunk region immediately behind the cephalic region and the second group by almost all the segments are seminal. This may imply that the regenerative stem cells harbored in all the seminal segments are lost from the second group and are progressively limited to one or a few segments in the first group and thereby confirm the postulation by Bely (2006, 2010).

Oligochaetes: In general, the 'head' in an annelid is recognizable from the rest of the trunk. Using Methyl Green-Pyronin (MGP) staining technique, Myohara (2012) recognized the presence of a pair of neoblasts located in all the trunk segments but not in the first seven 'head' segments and the 8th, i.e. the 1st trunk segment in a 50 to 70 segmented oligochaete *Enchytraeus japonensis*. Thereby the presence of a head and trunk in this worm can be distinguished. This finding seems to hold true for a range of oligochaetes from the small *E. japonensis* to a large 120 segmented *Lumbriculus variegatus*. With reference to segmentation, Myohara (2004) has reported an important observation that in the 7th segmented head of oligochaetes, segmentation occurs in sequence from anterior to posterior in embryogenesis but simultaneously in regeneration. Hence, segmentation and its reorganization during regeneration in annelids assume a great importance.

A prelude on the segmentation process in *E. japonensis* is required for a better understanding of segmentation in different taxonomic groups. Molecular level studies have shown that regeneration occurs through the epimorphic regeneration of the head and tail, and morphallactic transformation of old segments into the appropriate middle segments.

In these studies, different combinations of techniques are employed; for example, MGP staining and BrdU-labeling and detection (Myohara, 2004) or simple hematoxylin and eosin staining, sequence analysis of cDNA clones and expression of marker genes (Takeo et al., 2008). Confirming the finding of Randolph (1892), Myohara (2004) has shown that each segment from the 9th to the last 34th contains neoblasts in asexually reproducing *E. japonensis* and is capable of regenerating a complete head and a complete tail, irrespective of its position in the body, from which it has been derived. Therefore, no antero-posterior gradient of regeneration potential exists in the trunk. Yet, the number of regenerated segments is dependent on the position of amputation along the body axis, when amputation is made within the head segments of 1 to 7, where no neoblast is present. Conversely, *E. buchholzi*, which reproduces sexually alone and possesses no neoblasts in any of its segment, does not regenerate a complete head but displays an antero-posterior gradient, when subjected to amputation. Hence, the presence of neoblasts correlates with an absence of antero-posterior gradient of regeneration ability along the body axis. The presence of neoblasts is obligatory for asexual reproduction but not for regeneration of missing body parts. This observation also distinctly delineates regeneration from asexual reproduction. Incidentally, of 27 aeolosomatid clonal species, five alone alternate clonal with sexual reproduction (Falconi et al., 2015). A few lumbriculid species (e.g. *Lumbriculus variegatus*, Morgulis, 1907) are capable of regeneration, besides clonal reproduction. In enchytraeids, eight species (Christensen, 1959, Collado et al., 2011) out of 670 (Schmelz and Collado, 2012) are capable of either regeneration or clonal reproduction at a time. In naidids, the majority of them are capable of clonal reproduction. However, five species (e.g. *Chaetogaster*) can clonally reproduce but are not capable of regeneration (Bely and Sikes, 2010). On the other hand, all the investigated lumbricid and megascolecid earthworms (except *Perionyx excavatus*) do not clonally reproduce but are capable of regeneration. Hence, regeneration and clonal reproduction are quite independent processes in oligochaetes.

In *E. japonensis*, Takeo et al. (2008) have isolated three region-specific genes namely *EjTuba*, *mino* and *horu*. In the normally growing worm, *EjTuba* and *mino* are expressed in the head and trunk, respectively. The expression areas for *EjTuba* and *horu* in the trunk are proportionate to the total number of segments. In a regenerating worm, the expression of these genes is suppressed but subsequently restored on the 7th day, when regeneration is completed. Using these molecular markers, Takeo et al. have found that the epimorphic regeneration is completed on day 4. Subsequently, morphallactic regeneration occurs between the 4th and 7th day. As a result, the epimorphically regenerated head reorganizes the subsequent morphallactic segmental organization. The morphallactic regulation arises from the balance of molecular gradient derived from the head and tail. Briefly, the morphallactic regeneration regulates the segmentation process but the epimorphic regeneration of head and tail regulates segmental reorganization through the newly formed antero-posterior axis. Using immunohistochemistry and immunoblotting with antibodies against these proteins, Martinez et al. (2005) have shown the existence of a gradient distribution of mannose-rich neural glycoproteins with varying molecular masses along the antero-posterior body axis and overlapping in the architomic fission zone of *Lumbriculus variegatus*.

It is known that the neoblasts are firmly adhered to the posterior surface of septa of *E. japonensis*. Sugio et al. (2012) have discovered a second type of smaller neoblast-like cells designated as N-cells. The N-cells are located dorsal to the neoblasts on the septa. A BrdU labeled pulse chase study has shown that neoblasts are slow cycling in the growing intact worm and possess stem cells characteristics, as evidenced by the expression of *Ej-vlg2* and by the telomerase activity during regeneration. Both neoblasts and N-cells of are mesodermal origin, and actively proliferate and migrate toward the autotomized site to form the mesodermal region of blastema.

In *Lumbriculus variegatus* too, any segment beyond the 8th is capable of regenerating a complete head and a complete tail. However, the regenerative potential progressively decreases posteriorly, indicating the increasing loss of neoblasts from the posterior end (Morgulis, 1907). In the naidid *Stylaria fossularis*, the distribution is limited between the 8th and 22nd segments with the maximum on the 18th segment and progressive decreases on either direction (Fig. 3.8). In *Nais commensalis* too, the maximum is on the 17th segment with progressive decreases from 16th to 12th segment anteriorly and 18th to 23rd posteriorly (Kharin et al., 2006). Investigations on the naidids *Dero limosa* (Hyman, 1916), *Pristina longiseta* (Van Cleave, 1937) and *Nais paraguayensis* (O'Brien, 1946) suggest that the distribution of neoblasts is a labile feature. It shifts toward posterior, with favorable food availability and temperature but anterior during unfavorable winter temperature conditions. In *Tubifex tubifex*, the amputations upto the 12th anterior segment results in regeneration of the hypomorphic 3-segmented head but the posterior beyond the 15th segment is unable to regenerate the anterior. Evidently, the number of segments harboring neoblasts is limited between the 13th and 14th segments alone. However, the tubificids, which are unable to clonally reproduce, have opted to parthenogeneic mode of reproduction (Table 2.3).

The trend for the loss of neoblasts from posterior is almost total in the megascolecid (except in *P. excavatus*) and lumbricid earthworms, as they are unable to reproduce clonally. The following descriptive surgical studies on the easily available earthworms have formed the basis for the segmentation process. In *Eisenia foetida*, Gates (1950) has noted that regeneration potency decreases with increasing number of amputated segments. In 1943, Moment also recorded that the rate of regenerative growth is faster, when the position of amputation is closer to anterior segment. In a key publication, Liebmann (1946) reported that the rate of head regeneration in *E. foetida* is progressively decreased with amputation increasing from the 9th segment toward posterior segment. Incidentally, his observation confirms that the first 8th anterior segments do not possess neoblasts (see Randolph, 1892, Myohara, 2004). Xiao et al. (2011) have made amputation at selected positions (head, up to 8th segments, pre-clitellar 9th to 25th segment, clitellar from 26th to 32nd segments and post-clitellar segments from the 33rd segment onwards and have observed survival and regeneration of *E. foetida*. Their observations may briefly be summarized: 1. Regeneration commences with blastema formation at the amputated end, followed by epimorphic cellular proliferation and differentiation of the head and tail bud. At 4 hours following amputation, the wound is sealed, 5–7 days after, a conical, unpigmented tail bud is developed, 7–11 days after, segmental delineation begins and 12–16 days after, segments are formed followed by increasing segmental dimension but not segmental number. 2. Survival of amputated worms is linearly and significantly increased with the number of remaining segments but not with position of amputation. 3. However, the anteriorly regenerated body length is correlated with the position of amputation but not with the remaining segments. 4. The posteriorly regenerated body length is correlated neither with a number of remaining segments nor with a position of amputation. Clearly, there are remarkable differences between the formation of anterior head and posterior tail. Mostly, anterior regeneration is achieved, *albeit* at low survival. In another earthworm *Allolobophora caliginosa*, the anterior is generated by a fragment containing the 16th segment, indicating the presence of regenerative stem cells in the 16th segment. Briefly, the presence and distribution of the stem cells in oligochaetes and seminal segments or its equivalent stem cells in polychaetes vary from family to family and in families like the syllids from species to species. Clearly, regeneration in annelids is a labile process and has independently originated and/or lost multiple numbers of times (see Bely, 2006, 2010).

3.4 Anterior vs Posterior

A large volume of literature concerning anterior and/or posterior regeneration in polychaetes and oligochaetes has been accumulated. Surprisingly, no author has defined the anterior and posterior fragments. A reason for it seems to be the fact that despite the uniformity of segments throughout the body, different functional regions with varying number of segments have been recognized. For example, the polychaete body is considered by different authors as divisible into (i) two regions, the anterior with 1–8 segments and posterior with 8–12 segments in the maldanid *Euclymene oerstedi* (Clavier, 1984), (ii) three regions, the head, thorax (12 segments) and abdomen (21) in the sabellid *Bispira brunnea* (Davila-Jimenez et al., 2017) and (iii) into four regions, head, thorax (10–12 segments), abdomen (20–35) and tail (6–12) in

TABLE 3.7

Ability of annelids to regenerate anterior or posterior segments (compiled from Bely, 2006, also added from other sources)

Anterior regeneration occurs

Maldanidae: *Clymenella torquata, Euclymene oerstedi, Petaloproctus socialis*

Phyllodocidae: *Eulalia viridis*

Syllidae: *Autolytus pictus, Procerastea halleziana, Syllis* spp, *Streptosyllis websteri, Typosyllis prolifera, T. pulchra, Trypanosyllis asterobia*

Amphinomidae: *Eurythoe complanata*

Dorvilleidae: *Dorvillea bermudensis*

Eunicidae: *Lysidice* sp, *Nematonereis unicornis*

Onuphidae: *Diopatra* spp

Oweniidae: *Owenia fusiformis*

Sabellidae: *Branchiomma nigromaculata, Myxicola aesthetica, Sabella* spp

Serpulidae: *Hydroides dianthus*

Spionidae: *Dipolydora quadrilobata, Pygospio elegans*

Criodrillidae: *Criodrilus lacuum*

Enchytraeidae: *Enchytraeus dudichi, E. fragmentosus, E. japonensis*

Lumbricidae: *Eisenia foetida, Lumbricus terrestris*

Lumbriculidae: *Lumbriculus lineatus, Rhynchelmis vagensis*

Megascolecidae: *Metaphire peguana, Perionyx excavatus, Pheretima* sp

Naididae: *Allonais paraguayensis, Dero digitata, Limnodrilus claparedianus, Nais elinguis, Stylaria* spp

Tubificidae: *Tubifex tubifex*

Posterior regeneration occurs

Eunicidae: *Eunice fucata, E. siciliensis, E. viridis* (Hofmann, 1974)

Syllidae: *Autolytus cornuta, Brania pusilla, Eusyllis blomstrandi, Exogone naidina, Grubeosyllis clavata, Haplosyllis spongicola, Haplosyllides floridana, Myrianida pachycera* (Franke, 1999), *Odontosyllis enopla, Pionosyllis lamelligera, P. procera, P. pulligera, Odontosyllis prolifera, O. polycera* (Fischer and Fischer, 1995), *O. phosphorea* (Tsuji and Hill, 1983)

Anterior regeneration does not occur

Arenicolidae: *Arenicola marina*

Capitellidae: *Capitella* sp I, *Capitella* sp II, *Mediomastus* sp

Opheliidae: *Polyphthalmus pictus*

Nereididae: *Platynereis dumerilii*

Polynoidae: *Harmothoe imbricata*

Dorvilleididae: *Ophryotrocha notoglandulata, O. puerilis puerilis*

Eunicidae: *Eunice afra, E. schizobranchia, E. siciliensis, E. torquata, E. viridis*

Spionidae: *Streblospio benedicti*

Species/taxa, in which posterior regeneration does not occur

Dinophillidae: *Dinophilus gardnieri*

Arenicolidae: *Arenicola marina*

Opheliidae: *Polyophthalmus pictus*

Hirudinea: Almost all species

the spionid *Pygospio elegans* (Gibson and Harvey, 2000). In the siboglinid *Lamellibrachia satsuma*, it is divisible into the three non-segmented tentacular, vestimental and trunk regions and the fourth segmented opisthosoma region (see p 95). Considering the head alone, this consists of seven segments in oligochaetes and can be recognized by the absence of chloragogue, metanephridia (e.g. *Nais communis*, Kharin et al., 2006) and neoblasts (e.g. *Enchytraeus japonensis*, Myohara, 2012). Regeneration potency of the head ranges widely from *Paranais litoralis*, which is unable to regenerate the head to *Lumbriculus variegatus*, which stereotypically regenerate a restricted but precise number of 7–8 head segments, regardless of the position of amputation (see Martinez-Acosta and Zoran, 2015).

Nevertheless, it is possible to distinguish the anterior from posterior in natural fission from field and experimental observations. In the cirratulid *Dodecaceria concharum*, natural fission occurs between the 1–13 anterior segments and 13–34 posterior segments (Martin, 1933). In the amphinomid *Eurythoe complanata*, a field study has shown that the anterior consists of 16 segments and posterior 40 segments (Kudenov, 1974). Experimental studies have identified the 10th, 14th and 17–18th seminal segments in *Clymenella torquata*, *Dodecaceria concharum*, *Chaetopterus variopedatus*, *Procerastea halleziana* in polychaetes (Fig. 3.7) and *Stylaria fossularis* in oligochaetes (Fig. 3.8). All the segments in front of the respective seminal segments constitute the anterior and those behind them the posterior. Similarly, all the segments ahead of the 20-segmented mid-body constitute the anterior and those behind it posterior in *Perionyx excavatus* (Fig. 3.1). Pigmentation has been used as a marker of pre-fragmentation stage in *Dodecaceria pulchra* (Gibson, 1977) and post-pigmentation stage in *Bispira brunnea* (Davila-Jimenez et al., 2017). In *Typosyllis antoni*, anterior can be identified by its distinct red lines on the dorsal and posterior by the presence of bendantate chaetae and a tiny thin acicula (Aguado et al., 2011). Besides them, the identified seminal segments or the mid-body enable the identification of anterior and posterior fragments. However, the fact that any segment beyond the 8th can regenerate the entire body in 34-segmented *E. japonensis* and 120-segmented *Lumbriculus variegatus* in oligochaetes and a few (2–3 chaetigers) segments in the spionids *Amphipolydora vestalis* (Gibson and Paterson, 2003), *P. elegans* (Gibson and Harvey, 2000) and in the 600-segmented sabellids *Sabella pavonina* (Murray et al., 2013) in polychaetes complicate the identification of anterior and posterior. Despite these constraints, it is still possible to understand the differences between anterior and posterior fragments and their distribution across the phylogenetic groups of polychaetes and oligochaetes.

Thankfully, Zattara and Bely (2016) have accomplished an onerous task of assembling the relevant information on regeneration in 247 species belonging to 28 families of polychaetes and 129 species belonging to 7 families of oligochaetes. Zattara (2012) must be complimented for accomplishing this job earlier and promptly sending the appendix of Zattara and Bely (2016) on email request (*ezattara@gmail.com*). For immediate reference, Table 3.7 lists some of these annelids with potency for anterior, posterior and/or anterior cum posterior regeneration. The said appendix provides an opportunity for further analysis. In view of the contrasting features, regeneration and clonal reproduction in polychaetes and oligochaetes are separately considered. From Table 3.7, the following inferences can be made: 1. The number of species capable of anterior, posterior and anterior cum posterior regeneration is 149, 206 and 143, respectively (Table 3.8). A calculation of the said numbers as fractions of 16,931 annelid species (Table 1.2) indicates that only 0.88, 1.22 and 0.85% of annelids are capable of anterior, posterior and anterior cum posterior regeneration, respectively. Comparing even the 1.22% incidence for the posterior regenerative potency of the bilaterally symmetrical annelids with 2.95% of 7,000 speciose radially symmetrical echinoderms (Pandian, 2018), the potency of annelids is far less than that of echinoderms. It is not clear, whether radial symmetry facilitates a greater regenerative potency. Comparative studies on the incidence frequency of regeneration between the bilaterally symmetrical non-segmented, acoelomate Turbellaria and the segmented coelomate annelids as well as bilaterally symmetrical hydrozoan cnidarians and radially symmetrical schiphozoan cnidarians may clarify it. 2. In annelids, posterior regeneration occurs (incidence frequency 1.2%) more frequently than the anterior (0.85%) indicating that the former costs less than that of the latter. The reasons for it are listed: (i) in most polychaetes and oligochaetes, the anterior consists of the brain, heart and metanephridia as well as reproductive organs in oligochaetes. (ii) regenerative potency diminishes faster in the 'old' anterior segments than in the 'young' posterior ones (see Paulus and Muller, 2006), (iii) in oligochaetes, the loss of neoblasts progresses from the posterior end (e.g. *Lumbriculus variegatus*) toward the mid-body (e.g. *Stylaria fossularis*) and (iv) amputation causes a strong shift

TABLE 3.8

Regeneration in annelids. Percentage values in brackets. For details see text (estimated from Appendix of Zattara and Bely, 2016)

Regenerating Fragment(s)	Total (no.)	Polychaeta (no.)	Oligochaeta (no.)
Anterior only	149 (0.88)	104 (0.80)	45 (1.57)
Posterior only	206 (1.22)	146 (1.12)	60 (1.89)
Anterior + posterior	143 (0.85)	98 (0.75)	45 (1.42)
No anterior + posterior	3	2	1
Anterior + posterior and cloning	61 (43)	32 (33)	29 (64)
Anterior + posterior but no cloning	34 (23.8)	31 (31.6)	3 (6.7)
Anterior without cloning	19 (13.2)	18 (17.3)	1 (2.2)
Posterior without cloning	15 (10.5)	10 (6.9)	5 (11.2)
No anterior + posterior but cloning	6	0	6

in resource allocation from growth to regeneration and the shift is strong and persists longer during anterior regeneration than the posterior (Zattara and Bely, 2013). The values of 149 for anterior and 143 for anterior cum posterior indicate that the former is usually accompanied by the posterior (Table 3.8). In the absence of anterior, the posterior regeneration occurs only in the naidids *Amphichaeta raptisae, Chaetogaster diaphanus, C. diastrophus, Paranais frici* and *P. litoralis* (see Bely and Sikes, 2010). 3. When the number of anterior, posterior and anterior cum posterior regeneration is considered as a fraction of 13,012 polychaete species and 3,175 oligochaete species (Table 1.2), the percentage values (1.57, 1.80, 1.42) obtained indicates that the prevalence of regenerative potency is 1.5 times greater in oligochaetes than the respective ones (0.76, 1.15, 0.76) of polychaetes. 4. Only 61 species (out of 143, i.e. 42.6%) possess clonal potency; of them, 33% (i.e. 32 species out of 98) polychaetes and 64% (i.e. 29 species out of 45) oligochaetes are characterized by anterior cum posterior regeneration along with clonal reproduction. Surprisingly, all the 32 polychaete species *per se* undergo clonal reproduction by architomy. Of 29 oligochaete species *per se*, 19 and 21 species undergo clonal reproduction by paratomy and architomy, respectively. When these 32 and 29 species numbers are related to 98 polychaete species and 45 oligochaete species characterized by *per se*, it is only 32% of the polychaetes undergo architomic clonal reproduction (with a couple of exceptions), in comparison to 64% of oligochaetes. 5. As indicated earlier, regeneration and clonal reproduction are quite independent processes. Hence, it is not correct to consider that the latter is derived from the former (Zattara and Bely, 2016). 5a. In clonal oligochaetes, the presence of neoblasts is obligatorily required to manifest the clonal potency. In the absence or loss of the neoblasts, almost all the lumbricid and megascolecid earthworms (except *Perionyx excavatus,* see p 96, Fig. 3.1), some enchytraeids (e.g. *Lumbricillus lineatus, Enchytraeus buchholzi* [Myohara, 2012]) and naidids (e.g. *Limnodrilus claparedianus, Tubifex rivulorum, T. tubifex*) are unable to clonally reproduce. Even in the absence of anterior cum posterior regenerative ability clonal reproduction does occur in *Chaetogaster diastrophus*. Clearly, the clonal potency of oligochaetes arises from the neoblasts and not from anterior cum posterior regenerative potency. 5b. In polychaetes, however, 60% species characterized by anterior (18.3%), posterior (10.2%) and anterior cum posterior (31.6%) regeneration are unable to clonally reproduce. Only 33% of them characterized by anterior cum posterior regenerative ability are able to clonally reproduce. Researches in polychaetes to identify the factor responsible for clonal reproduction are urgently required. 6. Despite the absence of anterior cum posterior regeneration, *C. diaphanus* and *C. diastrophus* are unique being capable of clonal reproduction by paratomy. 7. Both anterior and posterior regeneration is absent *Dinophilus gardnieri, Polyophthalmus pictus* (Opheliidae) and *Arenicola marina* in polychaetes as well as *C. diaphanus* and *C. diastrophus* in oligochaetes.

For immediate reference Table 3.9 lists the families, in which anterior, posterior and anterior cum posterior regeneration occurs. Whereas the

TABLE 3.9

Families in which anterior, posterior or anterior cum posterior regeneration occurs. Continuous underline indicates the incidence of regeneration in almost all the investigated species of the family. Dotted underline indicates the incidence in a few of investigated species of the family (compiled from Appendix of Zattara and Bely, 2016)

Anterior

Polychaetes: Oweniidae, Chaetopteridae, Amphinomidae, Saccocirridae, Dorvilleidae, Eunicidae, Onuphidae, Syllidae (exception: *Brania pusilla*), Polygordiidae, Nephtyidae, Hesionidae, Phyllodocidae, Orbiniidae, Siboglinidae, Cirratulidae, Ctenodrilidae, Sabellidae, Serpulidae, Spionidae (exception: *Streblospio benedicti*), Terebellidae, Maldanidae

Oligochaetes: Aeolosomatidae, Naididae, Enchytraeidae, Glossoscolecidae, Lumbricidae, Megascolecidae, Lumbriculidae

Posterior

Polychaetes: Oweniidae, Chaetopteridae, Amphinomidae, Saccocirridae, Dorvilleidae, Eunicidae, Onuphidae, Syllidae, Polynoidae, Polygordiidae, Nephtyidae, Hesionidae, Nereidae, Phyllodocidae, Tomopteridae, Orbiniidae, Siboglinidae, Cirratulidae, Ctenodrilidae, Sabellidae, Sabellariidae, Serpulidae, Spionidae, Captellidae, Terebellidae, Maldanidae

Oligochaetes: Aeolosomatidae, Naididae, Enchytraeidae, Glossoscolecidae, Lumbricidae, Megascolecidae, Lumbriculidae

Anterior cum posterior

Polychaetes: Oweniidae, Chaetopteridae, Amphinomidae, Saccocirridae, Dorvilleidae, Eunicidae (only in *Lysidice collaris, L. ninetta, Nematonereis unicornis*), Onuphidae, Syllidae, Polygordiidae, Orbiniidae (only in *Proscoloplos cygnochaetus, Scoloplos armiger*), Cirratulidae, Sabellidae, Serpulidae, Spionidae (exception: *Myxicola aesthetica*), Maldanidae

Oligochaetes: Aeolosomatidae, Naididae, Enchytraeidae, Lumbricidae, Megascolecidae, Lumbriculidae

incidence of posterior regeneration is spread over as many as 33 families, that of anterior and anterior cum posterior is limited to 28 and 15 families, respectively.

4

Asexual Reproduction

Introduction

Besides the amazing potency of regenerating the missing body parts from a small feeding structure to larger fragment of the entire anterior and/or posterior body, some annelids are capable of agametically clone and reproduce asexually. As indicated earlier, regeneration and reproduction are quite independent processes. In oligochaetes, the presence of 'multipotent' neoblasts is obligately required to manifest clonal reproduction. Sex is costly, and demands time and resource but clonal reproduction saves them and avoids the risks involved in sexual reproduction. It also provides a mechanism for potential rapid amplification of a genotype known for its fitness. However, it involves no gametogenesis and recombination. In the absence of recombination and fusion of gametes in clonally reproducing polychaetes and oligochaetes, adaptation can be impeded and deleterious mutations may be accumulated, due to Muller's ratchet (Engelstadter, 2008). Understandably, the incidence of clonal reproduction is limited to ~ 190+ species, i.e. 1.1% of annelids; it is also limited to a few polychaete and oligochaete species alone. In them, the clonal reproduction can broadly be grouped into architomy and paratomy. In architomy, fission is followed after completion of regeneration and formation of progenies (ramets). But, it occurs even before the ramets are fully formed in paratomy. Further, a fission zone(s) is formed, prior to fragmentation in paratomics but not in architomics.

4.1 Obligate Cloners?

About a dozen polychaete and oligochaete species are reported to survive and flourish by clonal reproduction alone for periods from 3 years (*Aeolosoma hemprichii*) to 60 years in *Zeppelina monostyla* and *Pristina leidyi* (Table 4.1). In *Polydorella kamakamai*, sexual reproduction is rare or transient in the field and covers only 0.3% of the population (Williams, 2004). These observations may arguably question the general conclusion that no animal species exclusively reproduce asexually (see Pandian, 2016). While the institutions at Bologna,

TABLE 4.1

Obligately asexually reproducing polychaetes and oligochaetes

Species/Reference	Reported Observations
Polychaetes	
Cirratulidae: *Zeppelina monostyla* (see Akesson and Rice, 1992)	No sexual reproduction observed for 60 years in an Aquarium Freiburg, Germany
Dorvilleidae: *Parougia albomaculatus* *P. bermudensis* (Akesson and Rice, 1992)	On breeding in the laboratory (University of Goldberg, Sweden), no indication of sexual reproduction observed over 15–17 years
Spionidae: *Pygospio elegans* (Anger, 1984)	A single population entirely relies on asexual reproduction
Polydorella kamakamai (Williams, 2004)	In the field, sexual reproduction is rare or transient and covers only 0.3% population
Sabellidae: *Perkinsiana milae* (Gambi et al., 2000)	In the field collected specimens, no gametes in the coelom. Indications for fission within a single tube
Oligochaetes	
Enchytraeidae: *Cognettia sphagnetorum* (Christensen, 1959)	Sexually mature worm is very rare at any season in Denmark. Sexual eggs never hatch
Enchytraeus higentius (Christensen, 1984)	Asexual reproduction totally suppressed only at high density
Aeolosomatidae: *Aeolosoma hemprichi* (Stolc, 1903)	Survive in laboratory for 3 years by only paratomic fission
A. viride (Falconi et al., 2015)	Reproduce only by paratomic fission for > 20 years in the laboratory of Bologna University, Italy
Naididae: *Nais communis* *Potamothrix bedoti* *P. vejdoskyyaneum* (Timm, 1984)	Survive in laboratory (Estonia) for 6 years only by paratomic fission
Pristina leidyi (Ozpolat and Bely, 2015)	A clone flourishes for 8 years without any sign of senescence (cf Martinez and Levinton, 1992)
	Clonal reproduction by only paratomic fission for 20 years (University of College Park, USA) and earlier for 40 years (Carolina Biological Supply Co, USA)
Stylaria lacustris (see Schierwater and Hauenschild, 1990)	At LD = 16 hours and T = 20°C, the naidid reproduced asexually alone for > 6 years

Freiburg, Goteburg and College Park deserve admiration for accomplishing the arduous task of maintaining clonally reproducing annelids over long years, it must be noted that these clonally reproducing worms have been maintained at optimal rearing conditions enabling them to reproduce clonally alone. The absence of gonads and gametes (see Table 4.8) and failure of spawning in some seasons/years (e.g. *Streblospio benedicti, Nephtys hombergii*) may not indicate that the species concerned is asexual. For, the failure of spawning may be a consequence of poor nutritional conditions (see Kolbasova et al., 2013). When stressed/induced, these worms can switch over to sexual reproduction. Indeed, *P. leidyi* is capable of becoming sexual after ~ 1,000–3,000 agametic rounds of cloning over a prolonged period of 60 y (Ozpolat and Bely, 2015).

4.2 Incidence and Prevalence

Table 4.2 summarizes the incidence of clonal reproduction in polychaetes and oligochaetes. Some species like the oweniid *Myriochele heeri* is not included by Zattara and Bely (2016). However, Oliver (1984) reported 30% prevalence of clonal reproduction in it. Publications by Lohlein (1999) and Naidu (2005) have shown the need for inclusion of another 35 species. Hence, this list includes all of them. It has formed the base for further analysis and lists many new findings. 1. Of 100 and odd annelid families, clonal reproduction is limited to 12 polychaete families and five oligochaete families (Table 4.3). This estimate on clonal incidence reveals that the incidence is limited to 79 polychaete and 111 oligochaete species. Together, they make 190 species, i.e. only 1.1% of annelids are capable of clonal reproduction, which may be compared with 1.90% for echinoderms (Pandian, 2018). 2. When the values are related to the respective number of polychaetes (13,012) and oligochaete (3,175) species, the incidence of clonal reproduction in oligochaetes is ~ 4.6-times greater (3.18%) than that (0.68%) in polychaetes. 3. Further analysis of the incidences of architomy and paratomy has revealed the following: Among polychaetes, the incidence of architomy is widespread over 12 families, but paratomy is limited to four families alone (Table 4.3). Architomy occurs exclusively in eight polychaete families but paratomy exclusively in Aeolosomatidae in oligochaetes and Dinophilidae in polychaetes. The incidence within a family ranges from < 2% (13 out of 700 species, for species number see Franke, 1999) in Spionidae to 54% (69 out of 175 species, for species number see Ferraguti et al., 1999) in Naididae. Secondly, in terms of species number, the incidence is higher (75.6%, 60 species out of 79) for architomy in polychaetes but 77% (81 species out of 100) for paratomy in

TABLE 4.2

Annelid species reported to asexually reproduce; species in bold letters were not included in Zattara and Bely, 2016

1. Polychaetes
Architomy

(i) Oweniidae: ***Myriochele heeri*** (prevalence: 30%, Oliver, 1984)

(ii) Chaetopteridae: *Phyllochaetopterus prolifica* (see Purscke, 2006), *P. socialis* (see Kudenov, 1974), *Spiochaetopterus costarum costarum, Spiochaetopterus solitaries* (Zattara and Bely, 2016)

(iii) Amphinomidae: *Eurythoe complanata* (see Kudenov, 1974), *Linopherus canariensis* (Zattara and Bely, 2016)

(iv) Dorvilleidae: *Parougia albomaculatus, P. bermudensis* (see Akesson and Rice, 1992)

(v) Syllidae: *Odontosyllis gibba* (see Kudenov, 1974), *O. ctenostoma* (Zattara and Bely, 2016), *Procerastea halleziana* (Allen, 1923), *Syllis gracilis* (see Franke, 1999)

(vi) Orbiniidae: *Proscoloplos cygnochaetus* (Zattara and Bely, 2016)

(vii) Cirratulidae: *Caulleriella viridis, Cirratulus cirratus* (Zattara and Bely, 2016), *Dodecaceria berkeleyi* (see Kudenov, 1974), *D. concharum* (Gibson, 1977), ***D. coralii*** (Gibson, 1978), *D. fistulicola,* (see Kudenov, 1974), *D. fewkesi, D. fimbriata* (Berkeley and Berkeley, 1954), *D. pulchra, Timarete filigera, T. punctata* (Zattara and Bely, 2016), *Protocirrineris chrysoderma* (Purscke, 2006), *P. antarctica* (Zattara and Bely, 2016)

(xiva) Ctenodrillidae: *Raphidrilus nemasoma, Raricirrus beryli, Zeppelina monostyla* (see Akesson and Rice, 1992), ***Raricirrus maculatus, R. arcticus*** (A?) (Buzhinskaja and Smirov, 2017)

(viii) Sabellidae: *Bispira brunnea* (Davila-Jimenez et al., 2017), *Branchiomma bairdi* (Arias et al., 2013), ***B. curtum*** (see Tovar-Hernandez and Knight-Jones, 2006), *Megalomma cinctum* (Yuan, 1992), *Myxicola aesthetica* (Knight-Jones and Bowden, 1984), *Perkinsiana milae* (Gambi et al., 2000), *P. rubra, Potamilla torelli, Pseudobranchiomma emersoni, P. perkinsi* (Knight-Jones and Giangrande, 2003), *P. punctata, P. minima* (Nogueira and Kinght-Jones, 2002), ***P. schizogenica*** (see Davila-Jimenez et al., 2017), *Pseudopotamilla reniformis* (Kolbasova et al., 2013), *Sabella discifera* (Rioja, 1929), ***S. pavonina, Sabellastarte*** sp (Murray et al., 2013)

(xva) Serpulidae: *Josephella* sp, *Rhodopsis simplex* (Zattara and Bely, 2016)

(xvia) Spionidae: *Amphipolydora abranchiata* (Blake, 1983), *A. vestalis* (Gibson and Paterson, 2003), ***Dipolydora armata*** (Radashevsky and Nogueira, 2003), *Dipolydora caulleryi, D. socialis* (Allen, 1921), *Polydora colonia, P. elegantissima* (Allen, 1921), *Pygospio californica, P. elegans* (Gibson and Harvey, 2000)

Paratomy

(xii) Dinophilidae: *Dinophilus rostratus* (Zattara and Bely, 2016)

(xivb) Ctenodrilidae: *Ctenodrilus serratus* (Gibson, 1977), *Kirkegaardia* (Purscke, 2006)

(xvb) Serpulidae: *Filogranella elatensis, F. gracilis, Josephella marenzelleri, Rhodopsis pusilla, Spiraserpula snelli* (see Halt et al., 2006), *F. implexa, Salmacina amphidentata, S. australis* (Zattara and Bely, 2016), *S. incrustans* (Schroeder and Hermans, 1975), *S. dysteri* (Nishi and Nishihara, 1994)

(xvib) Spionidae: *Polydorella dawydoffi, P. kamakamai, P. tetrabranchia* (Gibson and Harvey, 2000), *P. smurovi, Polydorella prolifera* (Tzetlin and Britayev, 1985), *P. stolonifera* (Zattara and Bely, 2016)

Table 4.2 contd. ...

...Table 4.2 contd.

---

2. Oligochaetes
Architomy

(ix) Enchytareidae: *Buchholzia appendiculata, Cognettia sphagnetorum* (Christensen, 1959), *Enchytraeus variatus* (Bouguenec and Giani, 1989), *E. dudichi, E. fragmentosus, E. japonensis* (Collado et al., 2011), *Cognettia glandulosa, Enchytraeus bigeminus, Marionina* sp (Zattara and Bely, 2016)

(x) Lumbriculidae: *Lumbriculus variegatus* (Martinez et al., 2006)

(xi) Megascolecidae: *Perionyx excavatus* (S. Sudhakar, pers. comm.)

(xviia) Naididae: *Allonais paraguayensis* (Bely and Sikes, 2010), *A. inaequalis, A. lairdi, A. pectinata, Autodrilus japonicus, A. pluriseta, A.* sp, *Bothrioneurum righii, B. vedjioskyanum, Branchiodrilus menoni, B. semperi, Bratislavia unidentata, Dero bauchiensis, D. borelii, D. lutzi, D. malayana, Pedonais crassifaucis, Slavina evelinae, S. sawayai* (Zattara and Bely, 2016)

Paratomy

(xiii) Aeolosomatidae: *Aeolosoma travancorense* (Aiyer, 1926), ***A. hemprichi, A. quaternarum, A. singutare, A. titorale*** (see Falconi et al., 2015), *A. kashyapi, A. niveum, A.* spp, *A. viride* (Zattara and Bely, 2016), ***A. beddardi, A. headleyi, A. ternarium*** (Naidu, 2005)

(xviib) Naididae: *Arcteonais lomondi, Dero furcata, Dero* sp I, *Nais communis, N. elinguis, Piguetiella michiganensis, Pristina aequiseta, P. leidyi, Ripistes parasita, Slavina appendiculata, Specaria josinae, Stylaria lacustris* (Bely and Sikes, 2010), *Dero digitata, Pristina longiseta, Stylaria fossularis* (Bely, 1999), *Paranais litoralis* (Nilsson et al., 1997), *Amphichaeta raptisae, A. sannio, Branchiodrilus hortensis, Chaetogaster diaphanus, C. diastrophus, C. limnaei, Cruistipellis tribranchiata, Dero carteri, D. flabelliger, D. gravelyi, D. huaronensis, D. superterrenus, D. tonkinensis, D. vaga, Nais bretscheri, N. stolci, Ophidonais serpentina, Paranais frici, Stephenosoniana* sp, *Uncinais uncinata, Vejdovskyella* sp (Zattara and Bely, 2016), ***Chaetogaster langi, Nais* sp, *N. barbata, N. pseudobtusa*** (Lohlein, 1999), ***Allonais rayalaseemensis, Aulophorus carteri, A. flagellum, A. furcatus, A. gravelyi, A. hymanae, A. indicus, A. michaelseni, A. moghei, A. tonkinensis, Chaetogaster cristallinus, C. limnae bengalensis, Nais andina, N. andhrensis, N. pardalis, N. simplex, N. variabilis, Pristina breviseta, P. evelinae, P. macrochaeta, P. proboscidae, P. sperberae, P. synchites, Pristinella acuminata, P. jenkinae, P. menoni, P. minuta, Stephensoniana trivandrana*** (Naidu, 2005)

---

oligochaetes. Thirdly, 63 out of 79 clonal polychaete species are sedentary/ tubiculous, indicating that clonal reproduction is more frequent in sedentary than errant polychaetes. Fourthly, all the members of 10 families (i) Oweniidae, (ii) Chaetopteridae, (iii) Amphinomidae, (iv) Dorvilleidae, (v) Syllidae, (vi) Orbiniidae, (vii) Cirratulidae, (viii) Sabellidae, (ix) Enchytraeidae and (x) Lumbriculidae are architomic, but (xi) Dinophilidae and (xii) Aeolosomatidae are paratomic. Only, (xiii) Ctenodrillidae, (xiv) Serpulidae, (xv) Spionidae and (xvi) Naididae have representations for architomy and paratomy (Table 4.3). It is likely that each of the first 10 architomic family members underwent mutation for the clonal reproduction perhaps simultaneously at a time. It may also be true for the two paratomic families. However, it is not clear why different members of the last four families undergo architomy or paratomy.

TABLE 4.3

Incidence of clonal reproduction in polychaetes and oligochaetes. Estimations are based on Table 4.2

Family	Architomy Species (no.)	Paratomy Species (no.)
Polychaetes		
Sedentary polychaetes		
(i) Oweniidae	1	0
(ii) Chaetopteridae	4	0
(vi) Orbinidae	1	0
(vii) Cirratulidae	13	0
(viii) Sabellidae	17	0
(xv) Serpulidae	2	10
(xvi) Spionidae	9	6
Errant polychaetes		
(iii) Amphinomidae	2	0
(iv) Dorvilleidae	2	0
(v) Syllidae	4	0
(xii) Dinophilidae	0	1
(xiv) Ctenodrillidae	5	2
Subtotal	**60**	**19**
Oligochaetes		
(ix) Enchytraeidae	9	0
(x) Lumbriculidae	1	0
(xi) Megascolecidae	1	0
(xiii) Aeolosomatidae	0	12
(xvi) Naididae	19	69
Subtotal	**30**	**81**
Total	**90**	**100**

4.3 Observations and Characteristics

(i) *Publications*: (a) There are more publications on clonal reproduction in polychaetes than in oligochaetes. (b) On these annelids, there are more publications on experimental observations than field ones. Not surprisingly, there are limited publications on prevalence of clonal reproduction. (c) Only a few authors have reported observations from both field and experimental study (e.g. *Pseudopotamilla reniformis*, Kolbasova et al., 2013). Others have reported the findings from either experimental (e.g. *Parougia bermudensis*, Akesson and Rice, 1992) or field (e.g. *Pseudobranchiomma schizogenica*, Tovar-Hernandez and Dean, 2014) observation.

(ii) *Phylogeny*: (a) The oligochaetes, which have recolonized aquatic habitats, have retained a direct life cycle, although aquatic habitats can sustain halo- or mero-planktonic larvae, as in polychaetes. (b) For reasons not yet known, they are unable to manifest a sedentary/tubiculous mode of life, *albeit* a few inhabit within mucous or gelatinous tube (see Table 4.7). (c) Of 79 clonal polychaetes, 63 (architomic 47 species + paratomic 16 species), i.e. ~ 80% of them are sedentary and/or tubiculous. Of 290 tubiculous sabellids (McEuen et al., 1983), only 17 species, i.e. 5.9% of them are architomic cloners. Hence, it is difficult to consider that an intense sub-lethal predation alone have enforced clonal reproduction in the sedentary/tubiculous polychaetes (cf Oliver, 1984).

(iii) *Sexuality*: (a) In echinoderms, clonal reproduction occurs only in gonochoric oviparous species (Pandian, 2018). But the clonal polychaetes include protandric (e.g. *Salmacina australis*) and self-fertilizing simultaneous hermaphrodites (e.g. *Bispira brunnea*, Davila-Jimenez et al., 2017) as well as brooders. For example, the architomic cloning spionids *Dipolydora caulleryi*, *D. socialis*, *Pygospio californica* and *P. elegans* are all brooders (see Blake and Arnofsky, 1999). Similarly, the paratomic clonal serpulids *Salmacina amphidentata* and *S. dysteri* are brooders (Kupriyanova et al., 2001).

(iv) *Life stages*: (a) Barring crinoids, all other classes of echinoderms have representative species, in which one or other larva reproduces clonally (Pandian, 2018). In annelids, clonal reproduction is limited to immature and mature stages alone. Larval cloning is not so far reported for any annelid.

(v) *Reproduction*: Clonal species alter sex ratio and/or eliminate or reduce the number of gametes. For example, males are unknown in *Dodecaceria pulchra* (Gibson, 1977) and females in 12 clonal naidids (Table 4.8). Female ratio is also reduced to 0.03 in *Polydorella kamakamai* (Williams, 2004) and 0.09 in the oweniid *Myriochele heeri* (Oliver, 1984). Sperms are not observed in *Pseudobranchiomma schizogenica* (Tovar-Hernandez and Dean, 2014). Egg strings of *Amphipolydora vestalis* hold 69 eggs + 329 nurse eggs, i.e. 4.7 nurse eggs/egg (Gibson and Paterson, 2003). In *P. kamakamai*, the 13th, 14th and 15th gametogenic segments generate 39, 30 and 18 eggs, respectively (Williams, 2004).

(vi) *The trigger*: In polychaetes, food availability (e.g. *Parougia bermudensis*, Akesson and Rice, 1992) and density (e.g. *Pygospio elegans*, Wilson, 1985) as well as temperature (e.g. *P. elegans*, Rasmussen, 1953) trigger the initiation and proportion of architomic clonal reproduction. For example, increasing temperature during spring and consequent food availability initiates clonal reproduction in *P. elegans*. Increasing density may limit food availability and thereby suppress clonal reproduction. Contrastingly, none of these factors control clonal reproduction in naidids. It is photoperiod of < 12 hL that switches on sexual reproduction in *Stylaria lacustris* (Schierwater and Hauenschild, 1990). This (< 16 hL) is also true of other naidids *Nais communis* and *Pristina longiseta* (Kharin et al., 2006). However, clonal reproduction, at high densities above 300 and 400 worms, is suppressed in the architomic *Enchytraeus japonensis* (Myohara et al., 1999) and *E. bigeminus* (Christensen, 1973), respectively. Dependance on phytoplankton for food has limited the vertical distribution of naidids to euphotic zone (e.g. Martin et al., 1999, Hirabayshi et al., 2014). Understandably, decreasing photoperiod and the consequent diminishing phytoplankton in temperate and sub-arctic freshwater habitats initiate sexual reproduction and production of diapausing cocoons to overwinter. This may be an adaptive strategy to engage photoperiod as a trigger to sexual reproduction. Researches are required to know whether the photoperiod also serve to trigger sexual reproduction in the paratomic polychaetes and tropical naidids, to whom phytoplankton and/or detritus/sediment is available around the year.

(vii) *Clonal vs Sexual Reproduction*: These two reproductive processes intensely compete for resources. Understandably, they may temporally be separated in clonal polychaetes and oligochaetes. In the exceptional sabellid *Pseudopotamilla reniformis* from the sub-arctic White Sea, clonal reproduction dominates but lasts for (6 months) during autumn and winter. In it, sexual reproduction may rarely occur during spring and summer. Conversely, clonal reproduction dominantly occurs during the favorable spring and summer in the subtropical *Pygospio elegans*, over periods of 10 months in a year (Fig. 4.1). In them, sexual reproduction occurs only during the winter months. Sexual reproduction is limited to austral summer in the antarctic oweniid *Myriochele heeri*; but clonal reproduction occurs from austral autumn to spring. In the temperate naidid *Stylaria lacustris* too, clonal reproduction dominates over a period of 7 months during spring and summer. With the irreversible switch, sexual reproduction is limited to September only. In the enchytraeids, in which clonal reproduction dominates and sexual reproduction is suppressed until the critical densities are attained in the culture system.

Nevertheless, both clonal and sexual reproduction concurrently occurs in a few polychaetes and in enchytraeids. In the cirratulid *Dodecaceria pulchra* (South Africa), clonal reproduction dominates and occurs round the year with parthenogenic eggs being spawned only during late autumn. In it, sexual reproduction is virtually suppressed. Conversely, sexual reproduction dominates and occurs round the year in the sabellid *Bispira brunnea* (Mexico) with clonal reproduction concurrently occurring almost throughout the year but with a single ramet/fragmentation. In it, clonal reproduction does not also occur during spawning. Experimental study on the spionid *Amphipolydora vestalis* (New Zealand) indicates the concurrent occurrence of clonal and sexual reproduction. In fact, 46% of the ramets carry gametes and a fragment holding three chaetigers is adequate to accomplish bidirectional clonal reproduction. This is also true of *Lumbriculus variegatus* (Morgulis, 1907). It is not clear whether sexual and clonal reproduction in *A. vestalis* can

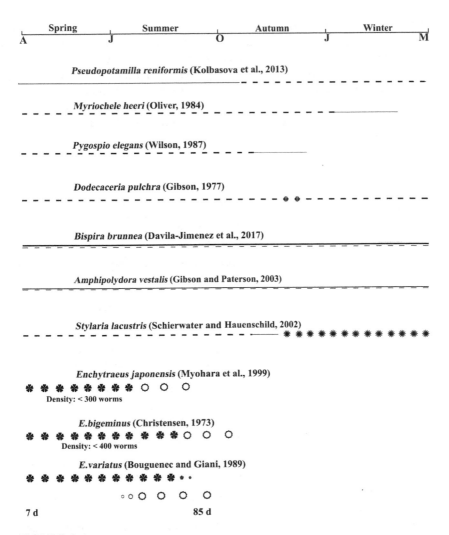

FIGURE 4.1

Temporally separated or concurrently synchronized clonal (– – –) and sexual (——) reproduction in architomic polychaetes and oligochaetes. Symbols marked by ❹ ❹ indicate parthenogenic reproduction. Thick lines indicate dominant frequency and thin line poor frequency. Lines marked by the symbols ✿ ✿ ✿ indicate clonal and ○ ○ ○ sexual reproduction at different densities in culture system. Lines marked by the symbols ✳ ✳ ✳ represent the period, during which cocoons diapause.

concurrently occur in the fields also. The concurrent occurrence of clonal and sexual reproduction is hinted in *Enchytraeus bigeminus* (Christensen, 1973) and *E. variatus* (Bouguenec and Giani, 1989). With a life span of ~ 85 days, clonal reproduction dominates initially in the 'young' worms but is replaced subsequently by sexual reproduction in the 'old' worms (Fig. 4.1).

Prevalence: Available information is limited to 11 architomic and one paratomic species. Among architomics, the values range from 4 to 13% for sedentary sabellid *Megalomma cinctum* to 90% in errant *Lumbriculus variegatus* and up to 100% in paratomic *Polydorella kamakamai* (Table 4.4). Of 297 tubiculous sabellid species, only 17 are cloners. Hence, the sublethal predation and the consequent prevalence of regenerates may be more related to the predatory pressure rather than errant/sedentary habit or architomic/paratomic fragmentation.

TABLE 4.4

Prevalence of clonal reproduction in field populations of some polychaetes and oligochaetes

Family/Species/Reference	Reported Observations
Errant polychaetes	
Amphinomidae *Eurythoe complanata* (Kudenov, 1974)	Architomy prevalent in 30% of population during spring from April to July
E. oerstedi (Clavier, 1984)	Architomy prevalent in 40 and 32% of anterior and posterior fragments, respectively
Oweniidae *Myriochele heeri* (Oliver, 1984)	Architomy prevalent in 30% of population
Cirratulidae *Dodecaceria pulchra* (Gibson, 1977)	Architomy prevalent in 100% of population round the year. Thirteen percent are at pre-fragmentary stage. 5, 21 and 54% are clonally regenerating at anterior, mid-body and posterior zones, respectively
Sedentary/Tubiculous polychaetes	
Sabellidae *Megalomma cinctum* (Yuan, 1992)	Architomy prevalent in 4–13% of population
Bispira brunnea (Davila-Jimenez et al., 2017)	Architomy prevalent in 23 and 13% of population during autumn and winter, respectively
Pseudobranchiomma schizogenica (Tovar-Hernandez and Dean, 2014)	Architomy prevalent in 82% of population
Pseudopotamilla reniformis (Kolbasova et al., 2013)	Architomy prevalent in 95% of population from October to March
Spionidae *Pygospio elegans* (Wilson, 1985)	Architomy prevalent in 80% of population
Polydora colonia (David and Williams, 2011)	Architomy prevalent in 16–33% of population from September to December
Polydorella kamakamai *P. stolinifera* (Williams, 2004)	Paratomy prevalent in 99.7% of population Paratomy prevalent in 83% of population
Errant oligochaete	
Lumbriculidae *Lumbriculus variegatus* (Morgulis, 1907)	Architomy prevalent in > 90% (19% anterior, 42% posterior and 29% anterior + posterior of population) during observation period of July–September

4.4 Architomy

In general, clonal reproduction is faster than that of sexual reproduction. For example, the former involves a fortnight but the latter a month in *Enchytraeus japonensis* (Yoshida-Noro and Tochinai, 2010). From field and experimental observations, duration from the commencement to the completion of clonal reproduction is reported for a few species. Irrespective of errant and sedentary habit, it requires a short duration of 3–8 days to avoid predation during this sensitive period (Table 4.5). A couple of natural cloners are indicated to require long duration of > 180 days. Understandably, inhabiting a highly unstable habitat, the White Sea with a low average yearly temperature and photoperiod, and consequent low phytoplankton productivity, *Pseudopotamilla reniformis* accumulates adequate resources throughout spring-summer and clonally reproduces 2–4 progenies during autumn-winter period of 180 days. However, the worm ensures 100% survival of all the two–four progenies. Besides, 70% lengths of its elastic tube strengthened by encrusting fine sand are spread horizontally over the substratum. Within this posterior end of the horizontally spreaded tube, the newly arising clones are maintained (Fig. 4.9). The newly emerged young worm, surrounded by ascidians and others, stands upto a height, at which its visibility is minimal (Kolbasova et al., 2013). The so called schizoparity or schizometamery occurs in a number of polychaetes like *Dodecaceria pulchra*, *Dipolydora armata*. In it, the fission involves three fragments including mid-body breaking off singly and cloning a complete progeny (Petersen, 1999). Following schizoparitic fission in *Pygospio elegans*, 9, 64 and 24% of progenies arising from anterior, mid-body and posterior fragments survive, respectively (Gibson and Harvey, 2000). In *Sabellastarte* sp, survival of eight fragments is 10, 20, 50 and 75% for the cephalic, mid 5th, mid 6th and posterior (8th) fragments, respectively. All the remaining 2nd to mid 4th fragments fail to survive (Murray et al., 2013).

The sequence of natural (e.g. *P. reniformis*) and artificial (e.g. *Pygospio elegans*) clonal reproduction is briefly summarized in Table 4.6. Notably, clonal reproduction involves blastema formation in naturally (e.g. *Polydora colonia*, David and Williams, 2011) and experimentally (e.g. *Pygospio elegans*, Gibson and Harvey, 2000, *Amphipolydora vestalis*, Gibson and Paterson, 2003) cloning spionids. Wound healing is noted in many architomics but blastema formation is not reported in any non-spionid species (cf Martinez-Acosta and Zoran, 2015). Interestingly, the minimum number of segments required for successful architomic cloning ranges from one seminal segment in *Dodecaceria concharum* (Martin, 1933) to two–three chaetigers in *A. vestalis* (Gibson and Paterson, 2003) and to 3.5 segments in *Lumbriculus variegatus* (Morgulis, 1907).

Short Life Span (LS) and high clonal frequency are typical features of some enchytraeids and dorvilleids. Elucidation of their clonal features becomes

TABLE 4.5

Progeny production and regeneration duration in clonally reproducing architomic polychaetes and oligochaetes

Family	Progeny (no.)	Regeneration (d)	Species, Reference
		Errant polychaetes	
Oweniidae	2	–	*Myriochele heeri* (Antarctica) (Oliver, 1984) (F)
Chaetopteridae	6	–	*Pseudochaetopterus prolifica* (see Purscke, 2006)
Amphinomidae	2	240†	*Eurythoe complanata* (USA) (Kudenov, 1974) (F)
Dorvilleidae	6–8	3	*Parougia bermudensis* (Akesson and Rice, 1992) (E)
Cirratulidae	2	–	*Dodecaceria concharum* (USA) (Martin, 1933) (F)
	–		*D. pulchra* (S. Africa) (Gibson, 1977) (F)
		Sedentary/tubiculous polychaetes	
Sabellidae	1	8	*Bispira brunnea* (Mexico) (Davila-Jimenez et al., 2017) (F)
	2–4	180	*Pseudopotamilla reniformis* (White Sea) (Kolbasova et al., 2013) (F)
	3	–	*Potamilla torelli* (UK) (Watson, 1906) (E)
	6–8	–	*Pseudobranchiomma schizogenica* (Gulf of California) (David and Williams, 2011)
Spionidae	8	> 28	*Sabella pavonina, Sabellastarte* sp (Murray et al., 2013) (E)
	2–3	–	*Dipolydora armata* (Brazil) (Radashevsky and Nogueria, 2003) (F)
	–	8	*Polydora colonia* (David and Williams, 2011)
	2–8	8	*Pygospio elegans* (Florida/Canada) (Watson, 1985/Gibson and Harvey, 2000) (F/E)
	3–6	8	*Amphipolydora vestalis* (New Zealand) (Gibson and Paterson, 2003) (E)
		Errant oligochaetes	
Enchytraeidae	3–6	6	*Enchytraeus variatus* (Bouguenec and Giani, 1989) (E)
	6–13	5	*E. japonensis* (Myohara et al., 1999) (E)
	7	–	*E. bigeminus* (Christensen, 1973) (E)
Lumbriculidae	51	–	*Lumbriculus variegatus* (Morgulis, 1907) (F)
Naididae	6		*Allonais paraguayensis* (Hyman, 1916) (E)

† for posterior, 90 days for anterior. F and E indicate field and experimental observations, respectively

important from culture point of view (cf Chapter 10). *Parougia bermudensis* grows to a length of 9.5 mm and consists of ~ 40 segments. A 20-day feeding and accumulation of reserves in coelomocytes is an obligate requirement to initiate schizoparity, yielding three fragments once every third day. Clonal regeneration involves anterior head with < eight segments, mid-body

TABLE 4.6

Morphogenesis during clonal reproduction in *Amphipolydora vestalis* (modified from Gibson and Paterson, 2003) and *Pseudopotamilla reniformis* (modified from Kolbasova et al., 2013)

Structure	Morphogenesis on								
	1st day	2nd day	3rd day	4th day	5th day	6th day	7th day	8th day	9th day

A. vestalis: Anterior blastema

Head
  Palps   (4th day ──────────────────→ 9th day)
  Mouth   (5th day ──────────────→ 9th day)
  Nuchal organ   (5th day ──────────────→ 9th day)
Thorax
  Gut   (4th day ──────────────────→ 9th day)
  Chaetigers   (4th day ──────────────────→ 9th day)
  Neuro chaetigers   (6th day ──────────→ 9th day)

A. vestalis: Posterior blastema

Tail
  Pygidium   (4th day ──────────────────→ 9th day)
  New chaetigers   (7th day ──────→ 9th day)
  Gut   (5th day ──────────────→ 9th day)

Duration (d)	Structure	Figure
	Pseudopotamilla reniformis	
1–4	Wound healing	
6–8	Formation of 2 anterior segments and pygidium	
11–30	Development of crown pinnules on radioles and fecal grove	
40–60	Further development of crown pinnules on radioles and fecal grove as well as reorganization of parapodia	
100	Regeneration completed	

fragment with ~ five or ~ nine segments and posterior with a few segments. Those with ~ nine mid-body segments regenerate faster than those with ~ five segments. Further, fragmentation in the mid-body and posterior zone can occur at any segment. However, based on the position and number of segments, the fragments are grouped into five classes. Due to continuous schizoparity, body segments are relocated to a progeny, which may itself produce additional fragments. As a result, some of these segments may be passed from generation to generation with little change and are potentially immortal. Akesson and Rice (1992) have carried out a series of experiments and their findings from three experiments are summarized: 1. In all the experiments, the trends run almost parallel to each other in those that are continuously fed and fed following a brief period of starvation. 2. In all the size classes, the number of segments/fragment is decreased totally beyond the 16th and 32nd segments in the mid-body and tail fragments, respectively. This clearly indicates that the fragment consisting of the first 15th to 31st segments determines the number and pattern of segmentation. 3. With advancing age from 0 to 15 months, the fragments holding 13 to 28 segments doubles the number of segments in the cephalic lineages but retains the same number in the non-cephalic lineages. Briefly, posterior fragments undergo senescence but not the mid-body segments and the fragmentation potency increases from the mid-body to the cephalic zone. In the marine paratomic oligochaete *Paranais litoralis* too, Martinez and Levinton (1992) have found age specific senescence. From the field study, Akesson and Rice (1992) have also noted that in the absence of adequate resource, *Parougia bermudensis* acquires the dispersal morphism. This migration strategy is more eloborately described in *P. litoralis* by Nilsson et al. (2000). On depletion of resource, the swimming morph foregoes clonal reproduction and grows by adding 40% more segments and also become thinner. With these morphogenetic changes, the migrant swims faster than the non-migrants.

Table 4.5 shows the clonally reproducing enchytraeids, lumbriculid and naidid oligochaetes also exhibit clonal features displayed by *P. bermudensis*. Of eight architomic clonal *Enchytraeus* species, more information is available for *E. bigeminus* and *E. variatus*. Life span of *E. variatus* is 81 days, grows to a length of 12 mm and consists of ~ 28 segments. Rearing clonal fragments of *E. variatus* under optimal conditions, Bouguenec and Giani (1989) have carried out three series of experiments. As they had not identified the fragments/worms, which underwent clonal and/or sexual reproduction, the results reported by them are confusing. An attempt is made to simplify and generalize their findings hereunder: when its fragments are reared under optimal condition in two series of experiments, some unidentified fragments generate immature and mature worms at the rate of 1.4 and 0.7 worms/d through sexual reproduction. Hence, clonally generated progenies are capable of restoring sexual reproduction (cf Ozpolat and Bely, 2015). Remarkably, other fragments have continued clonal reproduction and produced 5.1 fragments per day. The third series of experiment has shown that (a) clonal and asexual reproduction overlaps only for a short duration (Fig. 4.1) and (b) the number of fragments produced decreases from ~ 1.25/day during the initial 20-day period to 0.3/day during the period between 21st and 50th day, indicating the exhaustion of chloragogue to sustain clonal reproduction beyond 20 days. However, a sexually mature worm, which has not earlier undergone clonal reproduction, is able to produce ~ 0.13 cocoon or 1.5 eggs/d. Briefly, clonally produced progeny can restore sexual reproduction and clonal reproduction; clonal and sexual reproduction may overlap for a brief period (Fig. 4.1) and occurs at the cost of sexual reproduction.

E. bigeminus grows to a body length of up to 20 mm and consists of ~ 65 segments. At the age of 7 days, the worm obligately begins to clonally reproduce by dividing into seven fragments. Sexual maturity is attained by 2 and 3 weeks in anterior and other fragments, respectively. The worm is a polyploid. Hence, it is an obligate outbreeder. During the warmer season, the worm usually undergoes repeated clonal reproduction and attains high densities, when clonal reproduction is suppressed. As a result, sexual reproduction occurs mostly during colder months. On introduction of 21 fragments in a culture, the cumulative production of fragments by a single fragment is 276 immature worms. No sexual worm is produced, as clonal reproduction is suppressed only beyond the density of 400 worms in a culture. Interestingly, sexual reproduction is reduced in non-clonal *E. albidus* and *E. irregularis*, when they are cultured in combination with *E. variatus* (Christensen, 1973).

Naidids: Of 88 naidid species, in which clonal types are known, 19 and 69 are architomic and paratomic, respectively (Table 4.3). For tropical naidids, publications by Aiyer (1924, 1929) are mostly devoted to taxanomy. Through a series of publications, Hyman (1916, 1938) narrated the factors controlling clonal reproduction in architomic *Paranais paraguayensis*. Her major findings are listed below: 1. The worm grows to a length of > 35 mm and has > 50 segments. 2. The head fragment holds more number of segments than the trunk and caudal fragments, which consist of 10–25 large and 30–50 smaller segments, respectively. 3. Prior to fragmentation, the worm repeatedly twists its body and after the formation of a constriction, the daughter fragments pull apart. Irrespective of illumination or darkness, it fragments only during night. 4. It can fragment only at temperatures between 15°C and 27°C as well as between pH 5 and 8. Carbonate level plays a decisive role in controlling fragmentation. Fragmentation is retarded by potassium cyanide and 1/25,000 mol inhibits it. The fragments hold 15 and 18 segments at 26°C and 16°C, respectively. Clearly, elevation in temperature increases the fragmentation frequency and results in smaller daughter fragments. 5. Unlike the enchytraeids, in which smaller immature worms fragment more frequently, the naidid does not fragment until it attains a minimum size of 13 mm. At 15–20 mm and 30–35 mm sizes, it divides into two and three– eight fragments, respectively. 6. At low density, it fragments more frequently and produces smaller daughter fragments. With increasing density, clonal reproduction diminishes, as is the case in enchytraeids. 7. In fresh culture medium, the fragment length is shorter than that in old medium, indicating the inhibitory role of the accumulated excretory and putrefactive products.

Earthworms: Of ~ 500 species of earthworms (see Table 1.2), it is experimentally shown that *Perionyx excavatus* is capable of clonal reproduction (see Fig. 3.1, p 97).

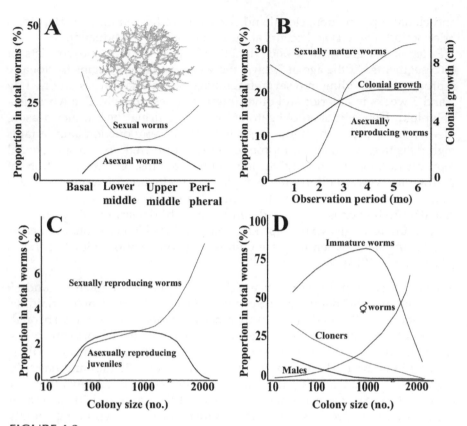

FIGURE 4.2

Salmacina dysteri (shown in the window): A. Shows the proportion (%) of total worms as function of position in the pseudocolony. B. Proportion of (%) sexually mature and asexually reproducing worms as function of the duration of observation period. Note growth of the colony is also shown as function of the duration of observation period. C. Proportion of (%) sexually and asexually reproducing worms as function of colony size. D. Proportion of (%) immature and asexually reproducing worms as well as percentage of males and hermaphrodites as function of colony size (compiled and modified from Nishi and Nishihira, 1994).

Pseudocolony: Clonal reproduction by budding, as a distinct form of architomy or fragmentation, occurs in serpulids (Faulkner, 1930) and syllids (Franke, 1999). In *Salmacina*, buds are always single, terminal and remain attached to the parent stock until all development, apart from size, is completed. Architomy leads to the formation of pseudocolony, in which the calcareous tubes of adjoining worms are attached to each other but without functional connection between them. From the report by Nishi and Nishihira (1994) on the florescent serpulid *S. dysteri* on the Okinawan coral reef, the following are summarized: 1. The worm grows up to 3 mm length and 0.2 mm width. 2. It consists of a head, three–eight segmented thorax and 10–25 segmented abdomen. 3. An immature worm is recognized by the presence of a bud at the posterior end of the abdomen and sexually mature one by eggs/embryos in the tube. 4. The approximate sex ratio is 0.3 for females, 0.1 for males and 0.6 for hermaphrodites, which seem to arise from males by addition of the ovarian component. 5. The experimental bisectioning of a pseudocolony reduces but not significantly the proportion of sexually reproducing worms. 6. Sexually reproducing worms occupy the basal and lower middle zones of the colony, while asexually reproducing worms are distributed in the upper middle and peripheral zones (Fig. 4.2A). 7. In a colony with a maximum of 2,000 members, the reverse trends are obtained for sexually and asexually reproducing worms with increasing colony size (Fig. 4.2C) and age (Fig. 4.2B). 8. With decreasing proportion of asexually reproducing worms, that of hermaphrodites increases (Fig. 4.2D). 9. The growth of the colony doubles and redoubles to a maximum of 9 cm (diameter) size during the observation period of 6 months (Fig. 4.2B).

4.5 Naidu's Monograph

In 1986, Brinkhurst had published a monograph on the North American tubificids, which are not known to reproduce asexually (Zattara and Bely, 2016). Thanks to the Zoological Survey of India, Naidu (2005) has brought out a monograph on Aquatic Oligochaetes in the Fauna of India series. Apart from taxonomy and distribution of these oligochaetes in the Indian waters, he has included very valuable information on asexual reproduction of eight aeolosomatid and 58 naidid species. His may prove to be the only literature source on asexual reproduction in the tropical aquatic oligochaetes. As an expert, Naidu has added to the paratomics 3 aeolosomatid and 32 naidid species to the list summarized for the oligochaetes by Zattara and Bely (2016) (Table 4.2). Assembling the scattered information from Naidu (2005), Table 4.7 summarizes clonal reproduction under two groups. According to him, clonal and sexual reproductions occurs concurrently in 18 naidid species and in the remaining 39 species either clonal or sexual reproduction occurs at a given body size or time. Naidu has noted the absence of ovary in all the 39 species but the presence of testes in a dozen species.

Group 1. *Clonal reproduction* alone is reported to occur in seven aeolosomatid and 39 naidid species. In the absence of sexual reproduction and production of the (diapausing) cocoon, clonal reproduction in naidids can be sustained only in such aquatic systems, where water is available round the year. India is characterized by the southwest monsoon sweeping the west coast, and West Bengal and adjoining the northeastern seven sister states. In Kerala, for example, the precipitation lasts for > 150 days and amounts to 400–600 cm/y. Conversely, a larger portion of Andhra Pradesh and Tamil Nadu receive

TABLE 4.7

Clonal reproduction in Aeolosomatidae and Naididae, as reported by Naidu (2005). All species undergo paratomy, except a few architomic species, indicated by arch, sex = sexual, clo = clonal, con = concurrent, bz = budding zone, fr = fragmentation, † indicates switching from clonal to sexual, * = gonads disappear, FW = freshwater, BW = brackish water, Mar = marine, Tin = tube inhabiting worm. Species in bold letters and bold letter[1] are already known for clonal reproduction from Zattara and Bely (2016) and Lohlein (1999), respectively

Species	Reproduction	Occurrence
Aeolosomatidae: Gonads absent (GA)		
Aeolosoma beddardi	GA	Kerala, Vizianagar, AP
A. headleyi	GA, 1–4 bz	Vehar lake, MH
A. hemprichi	GA, 2–4 bz	Kerala, Kashmir
A. niveum	Clonal, 3–4 bz	Kerala, Bihar lake, MH
A. ternarium	GA	Stagnant waters, Sri Lanka
A. travancorense	GA, bz	Kerala, Nagarkoil (TN)
A. viride	GA, bz	Chandigarh lake
Naididae: Gonads absent (GA)		
Allonais inaequalis	♀ A, ♂ +, arch	Kerala, West Bengal
A. paraguayensis	GA, fr, arch	Kerala
A. rayalaseemensis	GA, fr	Kerala, TN, AP
Aulophorus flabelliger	GA, bz	AP
A. gravelyi	GA, fr	Ennur (TN)
A. gwaliorensis	GA, ♂ +, fr	Kerala, Tanjavur (TN)
A. hymanae	GA, ♂ +, clo, fr	Mucous Tin, TN, AP
A. indicus	GA, fr, bz	AP ponds
A. michaelseni	GA, bz 1 + 1	Tin, Kerala
A. moghei	GA	Nagpur ponds, MP
*A. tonkinensis**	Clonal	Tin, Kerala, Bhim Tal, Varanasi
Branchiodrilus hortensis	GA, ♂ +, bz 1	Kerala, West Bengal, MP
B. semperi	GA	In burrows, cosmopolitan
Chaetogaster limnae bengalensis	GA, bz	West Bengal, Burma
Chaetogaster limnae limnae	GA, ♂ +, fr	Kashmir, Naini Tal
Dero cooperi	GA, ♂ +, bz 1	Mucous Tin, Kerala, Punjab
D. indica	GA, ♂ +, bz 1 + 1	Kerala, AP
D. nivea	GA, bz	Kerala, TN
D. palmata	GA, bz 2	Kerala
D. pectinata	GA, bz	Kerala
D. plumosa	GA, bz	Gelatinous Tin, AP
D. ravinensis	GA, bz	Ganga River belt
D. sawayai	GA, bz	Tin, AP, Nagpur, MP
D. zeylandica	GA, ♂ +, bz 1	Kerala
Nais andhrensis	GA, bz 1	Ooty (TN), permanent lakes
N. pardalis	GA, ♂ +, clo, bz	Afghanistan
N. variabilis	GA, bz 1	Kerala, Yercaud (TN)

Table 4.7 contd. ...

...Table 4.7 contd.

Species	Reproduction	Occurrence
Naididae: Gonads absent (GA)		
Pristinella acuminata	GA	North Indian ponds
P. jenkinae	GA, bz	FW, AP
P. menoni	GA, bz	Kerala, Yercaud (TN)
P. minuta	GA, bz	AP, Rabi river
Pristina aequiseta	GA, ♂ +, bz	In sponges, Kerala
P. breviseta	GA, ♂ +, bz	Kerala, Chennai ponds
P. evelinae	GA, ♂ +, bz	FW, Kerala, Ooty (TN)
P. foreli	GA, bz	FW, Dhaka
P. proboscidea	GA, bz 1	In sponges, coasts of Kerala, W. Bengal
P. sperberae	GA, bz 2	FW, Kolkata, AP ponds
P. synchites	GA, bz 1	FW, AP lakes
Slavina appendiculata	GA, ♂ +	Ganga, AP
Sexual and clonal		
Aeolosoma hyalinum	Sex, fr, 3 zooids	Vehar lake (MH)
Allonais pectinata	Sex, clo, arch	In sponge, Kerala, eggs don't develop
Aulophorus carteri	Sex, clo, bz	Tin, Kerala
A. furcatus	Sex, clo, bz	Mucous Tin, burrowers, Ooty (TN)
Chaetogaster cristallinus	Sex, clo, bz, con	Kerala, Ooty (TN), West Bengal
C. diaphanus†	Sex, clo, bz, con	UP, MP, BW in Sri Lanka
C. diastrophus	Sex, clo, fr, con	AP, Ganga
C. langi[1]	Sex, clo, bz, con	Cosmopolitan in India
Haemonais waldvozeli	Sex, clo, bz	Kerala, West Bengal
Nais andina†	Sex, clo, con	MP
N. barbata	Sex, clo, bz, con	Chandigarh, Lucknow (UP), W. Bengal
N. bretscheri	Sex, clo, fr, con	Afghanistan
N. communis	Sex, clo, fr, con	Cosmopolitan
N. elinguis	Sex, clo, fr, con	In sponges, Kerala, West Bengal
N. pseudobtusa	Sex, clo, bz	Afghanistan
N. simplex	Sex, clo	North Indian ponds
Pristina longiseta longiseta	Sex, clo	Mar, in sponges, Kerala
P. macrochaeta	Sex, clo, bz	FW, Afghanistan
Stephensoniana trivandrana	Sex, clo, con	Kerala
Stylaria fossularis†	Sex, clo, bz	Kerala, Ooty (TN)

AP = Andhra Pradesh, MH = Maharashtra, MP = Madhya Pradesh, UP = Uttar Pradesh, TN = Tamil Nadu

northwest monsoon of < 100 cm lasting for 30–40 d/y. Not surprisingly, Kerala happens to be the home for > 20 naidid and four aeolosomatid species that reproduce asexually alone. The others inhabit more or less permanent lakes in Kashmir and montane lakes like in Ooty and Yercaud of Tamil Nadu and aquatic bodies associated with live rivers like the Ganga. Notably,

Branchiodrilus semperi, as a burrowing worm, can survive within the moist soil, when water is not available.

Group 2. *Concurrent clonal and sexual reproduction*: Of 18 species in this group, nine species alone are recorded to undertake clonal and sexual reproduction concurrently. Even among these nine species, asexual reproduction is suspended with a commencement of sexual reproduction in *Chaetogaster diaphanus* and *Nais barbata*. For the remaining seven species, no information is provided. Typically, life span of tropical animals is shorter than their counterparts in temperate and arctic zones (e.g. crustaceans, Pandian, 2016). The life span of the tropical aquatic oligochaetes may not be an exception to this dictum. Hence, the life history strategy of the paratomic 18 naidid and one aeolosomatid (*Aeolosoma hyalinum*) species may fall into one of the following categories: 1. Within the short life span, a relatively longer duration of asexual reproduction is switched to a short duration of sexual reproduction with production of cocoons, which may hatch immediately, when water is present or switch to diapause, when water is not available. These worms may be characterized by semelparity (cf *Stylaria lacustris*, Schierwater and Hauenschild, 1990). 2. The third and more likely option is to have a very short period, during which clonal reproduction progressively diminishes and sexual reproduction gradually begins to dominate, as in *Enchytraeus variatus* (see Fig. 4.1). Research inputs are required to identify the option of 19 naidid and one aeolosomatid species, in which clonal and sexual reproduction are reported to co-occur.

According to Naidu (2005), India is endowed with seven clonal aeolosomatid species (out of 27) and 58 clonal naidid species (out of 175), although all the three aquatic enchytraeid species listed by him are not cloners. Indeed, these clonal worms are an important resource. Research input in some of these worms can promote aquaculture of these worms and provide gainful employment for the large number of Indians.

4.6 Paratomy

The incidence of paratomy is limited to two errant and two sedentary polychaete families as well as two oligochaete families (Tables 4.2, 4.3). In the errant syllids like *Procerastea*, *Autolytus*, *Myrianida*, *Pionosyllis* and *Trypanosyllis*, fission and budding are related to epitoky and are discussed in Chapter 5, although many authors have considered them with paratomy. It may be misleading to consider stolon formation as asexual reproduction. Stolonization is intimately associated with sexual reproduction and "stolons are little more than locomotive vessels for gametes" (Franke, 1999). Table 4.8 lists the number of progenies produced by some paratomic species. However, these numbers may be correct for the tubiculous spionids, in which space

TABLE 4.8

Progeny production and clonally reproducing paratomic polychaetes and oligochaetes

Family	Progeny (no.)	Species, Reference
Spionidae	1–5	*Polydorella kamakamai* (Williams, 2004)
	2	*P. tetrabranchia* (Campbell, 1955)
Naididae	2	*Nais communis* (Kharin et al., 2006)
	2	*Stylaria lacustris* (Schierwater and Hauenschild, 1990)
	3	*Pristina longiseta* (Kharin et al., 2006)
	7	*Dero digitata* (Drewes and Fourtner, 1991)

perhaps limits the formation of a chain, but may not be correct for the errant oligochaetes, which form chains of daughter and grand-daughter progenies. Secondly, these values have not taken into consideration for the frequency of paratomic fission in them. In *Enchytraeus japonensis*, each fragment regenerates into a small but complete worm in 4 days, which grows rapidly to divide again in another 10 days (Myohara et al., 1999). *Stylaria lacustris* splits once every fifth day from April to August/September, i.e. an individual may split > 30 times during this period. A theoretical estimate indicates that a single worm may give rise to a population of 3.4 billion worms (Schierwater and Hauenschild, 1990). Some polychaetes like *Polydorella tetrabranchia* produce only two offspring/fission once every fortnight (Campbell, 1955).

Paratomy involves the division of the body into two distinctive halves followed by regeneration of the missing body parts. Typical of paratomy, the anterior stock and posterior stolon remain attached together, the secondary and tertiary divisions proceed, resulting in the formation of a chain of zooids. Available information indicates the occurrence of division in the stock of sedentary tubiculous spionids but in the stolon of errant oligochaetes like aeolosomatids. In the spionid, *Polydorella dawydoffi*, which grows to 2 mm length and consists of > 15 segments, the first split occurs on the 13th segment. Hence, the stock inherits the head + 12 segments, while the stolon the pygidium + 4 parental segments (Fig. 4.3A). Preparing for the ensuing second split, the stock has regenerated 12 small segments. The second split occurs on the same 13th segment and the process is repeated during the tertiary split too. Contrastingly, the ambiguously named 'pygidial budding' (Herlant-Meewis, 1951) occurs in the errant oligochaete *Aeolosoma viride*. Falconi et al. (2015) have described the process correctly by stating that the secondary zooids (i.e. the daughter fragments of the first split) in *A. viride* are positioned posteriorly in inverse order with respect to their age and growth level. The older and more advanced ones are at the posterior end, while the younger and less developed are located immediately behind the parental zooid (Fig. 4.3B).

Both of these paratomic types occur in other errant oligochaete taxa, the naidids. In the naidian type, the second fission occurs at exactly on the same

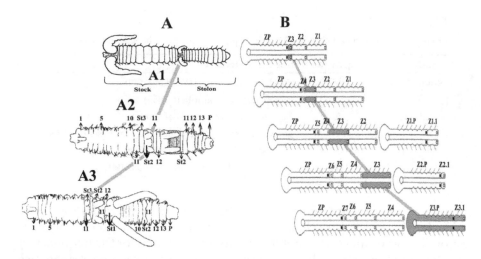

FIGURE 4.3

Freehand drawings to show the process of paratomic clonal reproduction in A. *Polydorella kamakamai*. A2 development of stolon 2 (St2) growth zone, A2 development of first 12 segments of stolon 2 and stolon 2 (St2) growth zone; secondary stolon is also shown by an arrow, A3 paratomic chain of 3 developing individuals prior to paratomic division between stolon 2 and stolon 1. Sites of paratomic divisions are indicated by darker arrows (modified from Williams, 2004). B. Schematic diagram to show paratomic clonal reproduction in *Aeolosoma viride*. Z3 shows the growth of secondary zooids and its origin in the budding area (filled square) of the main zooid (ZP). The Z3 grows forming new chaetigers produced in anterior direction by the growth area and moves the chain in the posterior direction by the inter position of new secondary zooids. Z3 zooids complete its morphogenesis and separate from the parental chain. ZP is the main zooid; Z1, Z2, Z3, Z4, Z5, Z6, Z7 are all secondary zooids. Z1.P + Z1.1, Z2.P + Z2.1 and Z3.P + Z3.1 represent the first, second and third filial chains, respectively (freehand drawing from Falconi et al., 2015).

segment number, as the first one has been. But, it occurs one segment behind the first one, designated as n-1 in the stylarian type. As a result, successive fissions are named as n-2, n-3 and so on, and each new individual contains one segment of the original parent. In *Stylaria*, the limit is n-7, after that the anterior stock elongates, as in normal posterior growth until the total number of segments exceeds 40, when a new fission zone is intercalated in the middle region (Berrill, 1952). In *Pristina*, the limit is n-12 (Hempelmann, 1923).

Nervous system and/or neurosecretion are known to play a dominant role in regeneration of clonal reproduction. The publications by Drewes and Fourtner (1990, 1991) on reorganization of segments and transformation of the giant nerve fiber sensory fields in the context of escape reflexes indicate a series of changes in the segment potential of *Lumbriculus variegatus* and *Dero digitata*. Following fission(s), a small and constant number of head segments are regenerated by the body segments, irrespective of their original axial origin. Thus, the original segment immediately behind the head becomes re-specified to match their altered position. The following morphallactic

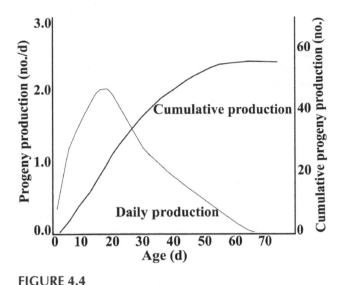

FIGURE 4.4

Clonal reproduction in *Aeolosoma viride*. Cumulative and daily production (thin line) of filial chains during the life time (modified and redrawn from Falconi et al., 2015).

changes include the rearrangement of segmental gradients in the giant nerve fibers also and progresses antero-posteriorly. Eventually, the medial and lateral giant nerve fiber sensory fields and conduction properties are also re-established.

Aeolosoma viride may serve as an example for the reproductive potential of asexual reproduction. *The Products*: 1. According to Falconi et al. (2015), the pygidial budding results in the formation of a chain composed of two–five zooids, designated as zA, zB, zC, zD and zE, based on their position in the antero-posterior axis of the body (Fig. 4.3B). 2. The worm produces 51 progenies during its Life Span (LS) of 64–66 days. *The process*: 3A. The proportion of a parental zooid is ll, 45, 40 and 5% for the 5th, 4th, 3rd and 2nd zooids, respectively. 3B. The number of chaetigers/zooid is decreased from 8 in zA to 5 in zD after deep reductions to 1 and 2 in zB and zC, respectively. 4. In this worm, a single budding area is located in the sub-terminal part of the posterior end of the parental zooid. In this area, a histochemically recognizable cell mass of the coelom produces the chaetiger. It is from this chaetiger, the secondary zooids arise at the rate of one every 24 hours. In the secondary zooids also, the same fragmentation process is repeated from the same chaetiger. 5. The maturation time, i.e. the interval between origin of a zooid and its separation from the chain is 80 hours. 6. The last zooid in a chain that has completed growth and cephalic differentiation separate itself from the chain at the rate of 1 at every 22 hours. *The dynamics*: 7. The production of cumulative number of zooids produced as a function of time shows a logistic trend with a raising slope lasting for 20 days or 30% of LS and a gradual asymptotic duration of 45 days or 70% of LS (Fig. 4.4). *Senescence*: 8. Correspondingly, the number of progenies produced is dropped to zero at the age of 60th day after an initial peak to 2.2 offspring/d. 9. From their observations on 1,559 fissions in the formation of G_1 zooids, Falconi et al. (2006) have found that the worm attains senescence at ~ the age of the 60th day of its life. Notably, the negative effect is imprinted on the progenies produced during the senescent age.

Not only age but also density and the consequent non-availability of food can induce irreversible effect on reproduction in sedimentivorous worms like *Paranais leidyi*. The worms that have been cultured at high densities (16–18 worms/10 cm^2) without renewal of fresh sediments crashed its asexual reproductive potency to zero level. The F_1 progenies produced during the crashing period, even when optimally fed, suffer 100% mortality and reduction in the proportion of clonal worms to 11%, in comparison to 25 and 58%, respectively in optimally fed worms. Hence, the effect of non-availability of food in the parental worm is also imprinted in F_1 generation of the worm.

4.7 Restoration of Sexual Reproduction

Germ cells constitute a key cell type. Being the sources of gametes, they are required to manifest sexual reproduction. In some polychaetes and oligochaetes, clonal reproduction has evolved independently at multiple numbers of times. Typically, a few of these species undergo many rounds of clonal generations (e.g. *Pristina leidyi*, Ozpolat and Bely, 2015) but cloning is terminated in semelparous oligochaetes (e.g. *Stylaria lacustris*, Schierwater and Hauenschild, 1990) or interspersed by short bouts of sexual reproduction (e.g. *Nais barbata*). These clonal annelids re-establish by transmission of the germ line post-embryonically. Homologs of the germline (*piwi, vasa* and *nanos*) and other multi-potent somatic stem cell genes are reported to express in the gonads as well as in proliferative tissues like Segment Addition Zone (SAZ) (e.g. *Platynereis dumerilii*, Gazave et al., 2013), regenerative blastema (e.g. *Nereis virens*, Kozin and Kostyuchenko, 2015) and fission zone (e.g. *Enchytraeus japonensis*, Tadokoro et al., 2006). Hence, both Primordial Germ Cells (PGCs) and somatic stem cells are characterized by the expression of the similar set of genes. Besides, PGCs and Mesodermal Posterior Growth Zone (MPGZ) cells or somatic stem cells are indistinguishable in morphology and expressed the germline markers *vaso, nanos* and *piwi*. Recently, the PGCs of *P. dumerilii* have been found to incorporate the proliferation marker 5-ethyl-2'deoxyuridine (EdU), a marker for proliferation shortly before gastrulation, which coincides with the emergence of four small blastomeres from the mid-

FIGURE 4.5

Distribution of PRlle-piwi1 positive ventral cells (PPVCs) along the body of *Pristina leidyi*. Asterisk indicates the axial position of individuals PPVCs in the worm. Arabic numbers indicate the mean number of PPVCs per segment in the six body regions along the body (modified and redrawn from Ozpolat and Bely, 2015).

blast lineage. Hence, the so called 'secondary mesoblast cells' constitute the definitive PGCs in *P. dumerilii*. In contrast, the cells of the MPGZ incorporate EdU only from the pre-trochophore stage onwards (Rebscher et al., 2012). Using the EdU pulse labeling technique, it has been possible to trace the independent emergence of PGCs from the *vasa*, *piwi* and *PL10* expressing MPGZ even earlier than that of somatic stem cells. Interestingly, the Oogonial Stem Cells (OSCs) only express *piwi* and *vasa* in embryos of the leech *Helobdella robusta*, while Spermatogonial Stem Cells (SSCs) only express *nanos* (Cho et al., 2014).

Regarding the restoration of germline lineage, two publications are considered. In *Enchytraeus japonensis*, the gonad can regenerate from any body segment produced by fission during clonal reproduction. Using homolog *piwi* gene (*Ej-piwi*) as a marker, Tadokoro et al. (2006) have found that *Ej-piwi* are distributed widely in the body as single cells. These cells serve as a reservoir of germ cell precursors and migrate into the regenerating tissue, where they eventually form the gonadal primordium and give rise to germ cells upon sexualization. These germline stem cells are distinct and differ from the somatic lineage arising from the neoblasts. In *P. leidyi*, PRlle-piwi1 is expressed in isolated spindle-shaped cells on the dorsal surface of the ventral nerve cord (Fig. 4.5). Following fission, the number and configuration of *Piwi*-Positive Ventral Cells (PPVC) are transmitted across asexual generation through migration along the ventral nerve cord. As a result, asexually reproducing *P. leidyi* expresses *PRlle-piwi1*. Amazingly, the clonal strain of *P. leidyi*, in which Ozpolat and Bely (2015) have accomplished their findings, has been reproducing asexually alone for over 60 years and has undergone clonal reproductive cycle for 1,000–3,000 generations.

4.8 Clonal Stem Cells

Using available information in the representative species belonging to 10 families of polychaetes and oligochaetes, an attempt has been made to trace the possible direction of clonal reproduction. The direction may provide a clue for the anlagen of the multipotent stem cells that manifest clonal reproduction (Table 4.9). In oligochaetes, it is the multipotent neoblasts that manifest clonal reproduction. At this juncture, two facts have to be noted. 1. Fragmentation can occur between any two post-cephalic segments in *Enchytraeus japonensis* (Myohara, 2012). However, the loss of neoblasts by some posterior segments by *Lumbriculus variegatus* limits the fragmentation from the trunk to the anterior abdominal segments (Morgulis, 1907). The extensive loss of neoblasts limits the fragmentation to the mid-body segments alone in *Stylaria lacustris* (Chu and Pai, 1944). 2. Paratomy may involve either fragmentation or budding (see Table 4.9). The number of the budding zone is limited to only one in many oligochaetes (e.g. *Aeolosoma viride*, Falconi et al., 2015) or rarely to two–four in a few naidids (Table 4.9). Consequent to these differences in the distribution pattern of neoblasts in segments, the direction of clonal reproduction is altered. With the budding zone limited to a single chaetiger at the posterior end of the parental zooid, the direction is from posterior to anterior in *A. viride*. Conversely, it is antero-posterior in *S. lacustris*, as the site of fragmentation is located in the stock.

In polychaetes, the clonal direction varies and is also complicated. Firstly, there are no equivalents of neoblasts. Secondly, some polychaetes are sedentary and tubiculous, while others are errants. Thirdly, within tubiculous polychaetes, cloning may be by architomic fragmentation, as in sabellids or paratomic budding, as in serpulids. Surprisingly, the direction originates from the posterior end in all the thus far investigated sabellids (Table 4.10). In *Sabella pavonina*, though all the eight fragments commence clonal regeneration, their survival increases from the posterior end. The posterior origin of clonal direction is also true for the paratomic serpulids (Faulkner, 1930), as represented by *Salmacina dysteri*. Understandably, the tubiculous sabellids and serpulids hold the clonal stem cells at the posterior segments in the depth of the tube, as the crown and thorax are subjected to more intense predation. Incidentally, the neurosecretion required for regeneration also arises from the ventral ganglia of posterior segments in the sedentary polychaetes (see p 111). With fission limited to the stock in the paratomic tubiculous *Polydorella dawydoffi*, the clonal direction is unusually antero-posterior (Table 4.9).

Among the errant polychaetes too, the direction varies. In the dorvilleid *Parougia bermudensis*, the direction originates also from posterior but more frequent from the mid-body segments (Table 4.9). Understandably, it is bidirectional in the cirratulid *Dodecaceria concharum* and the syllid *Procerastea halleziana*, as they possess one or two seminal segments in the mid-body, from which anterior and posterior segments are developed. In *Pygospio elegans*, the mid-body fragment develops both anterior and posterior; however, the anterior develops the posterior but the posterior may not. Neither the

TABLE 4.9

Clonal direction in polychaetes and oligochaetes. Arrows indicate the clonal direction and Arabic numbers indicate the position of amputation

Family	Habit/Type	Direction
Sabellidae *Bispira brunnea*	Tubiculous/architomy	 100 %
Pseudobranchiomma schizogenica	Tubiculous/architomy	
Pseudopotamilla reniformis	Tubiculous/architomy	 100 %
Sabellastarte sp	Tubiculous/architomy	
Sabella pavonina	Tubiculous/architomy	
Potamilla torelli	Tubiculous/architomy	
Serpulidae *Salmacina dysteri*	Tubiculous/paratomy	
Aeolosomatidae *Aeolosoma viride*	Errant/paratomy	
Dorvilleidae *Parougia bermudensis*	Errant/architomy	
Spionidae *Polydora colonia*	Tubiculous/architomy	
Cirratulidae *Dodecaceria concharum*	Errant/architomy	
Syllidae *Procerastea halleziana*	Errant/architomy	
Spionidae *Pygospio elegans*	Tubiculous/architomy	
Enchytraeidae *Enchytraeus japonensis*	Errant/paratomy	
Lumbriculidae *Lumbriculus variegatus*	Errant/paratomy	
Naididae *Stylaria lacustris*	Errant/paratomy	
Spionidae *Polydorella dawydoffi*	Tubiculous/paratomy	

TABLE 4.10

Clonal direction and potential anlage of clonal stem cells (vertical bars) in errant and sedentary polychaetes (for details see Table 4.9)

Family	Clonal Type	Clonal Direction
	Errant polychaetes	
Dorvilleidae	Architomy	←————————▪
Syllidae	Architomy	←————————▪
Cirratulidae	Architomy	←———▪———→
	Sedentary/Tubiculous polychaetes	
Sabellidae	Architomy	←————————▪
Serpulidae	Paratomy	←————————▪
Spionidae	Architomy	←————————▪
	Architomy	←▪←▪→▪→
	Paratomy	▪————————→

errant and sedentary habit nor architomy and paratomy in spionids seem to play a determining role in the clonal direction. In the absence of neoblasts, more than one type of clonal stem cells may be present at different anlage of polychaetes.

5

Epitoky

Introduction

Epitoky is a spectacular phenomenon, unique to a few errant polychaete species and is not known to occur in any other aquatic invertebrate taxa. Since 1868, this phenomenon has attracted the attention of a large number of zoologists. The transformation (metamorphosis) from benthic atokous form to a brief epitokous pelagic existence devoted to mating involves many morphological, physiological and behavioral modifications. It involves the following structural modifications: A. Morphology (i) body length and segment number (e.g. *Raricirrus variegatus*, Dean, 1995), (ii) enlarged eyes, both in cell number and volume (e.g. *Odontosyllis polycera*, Daly, 1975), (iii) broadened vascularized biramous parapodia with formation of spatulate natatory chaetae (e.g. *O. polycera*, Daly, 1975) and B. Anatomy (iv) atrophy of the gut, (v) histolysis of body wall to provide resources for gametogenesis (e.g. *O. polycera*, Daly, 1975), structural changes in musculature involving reduction in longitudinal muscle (cf Samuel et al., 2012, p 108) and reorganization of peripheral muscle cells in each segment and elaboration of structures for the release of gametes (*O. polycera*, Daly, 1975); these changes are radical in *Autolytus* but moderate in *Odontosyllis*. C. The physiological changes include adjustments in metabolic pathways facilitating higher muscular activity to sustain vertical swimming capacity (e.g. *Nereis virens*, Chatelain et al., 2008). D. Behavioral changes comprise of response to luminescence (e.g. *Odontosyllis enopla*, Fischer and Fischer, 1995) and pheromones (e.g. *Perinereis dumerilii*, Hardege et al., 1998) to synchronize swarming and spawning with precise timing guided by the lunar cycle (e.g. *Odontosyllis luminosa*, Gaston and Hall, 2000). As epigamic epitokes die soon after spawning, they lose 75–79% of available energy allocated for gametogenesis (e.g. *Nereis pelagica*, Olive et al., 1984, *Perinereis cultrifera*, Cassai and Prevedelli, 1998a). Not surprisingly, these epitokes release a few million eggs (e.g. *Glycera dibranchiata*, Creaser, 1973).

5.1 Types and Characteristics

Essentially, there are two major types of epitoky: epigamy and schizogamy. In epigamy, the reckless mating act commences with a burst of sustained vertical swimming activity, lifting the worm to the pelagic zone, where at the climax, a huge cloud of gametes is discharged and is terminated by death due to total exhaustion. For instance, it takes an entire life span of 4 years in *Glycera dibranchiata* to prepare for the event (Fig. 5.1). In fact, irreversible transition from somatic growth to reproductive development occurs several months prior to breeding. Briefly, all the epigamic polychaetes are semelparous.

In schizogamy, only a part (usually the stolon) of the body is transformed into an epitokous sexual stage (Fig. 5.1). Amassed with a load of gametes, equipped with a stolonial head and other sensory structures like the eyes, the stolon breaks off from the atokous benthic worm and migrates vertically for a brief existence in pelagic zone. An epitokous stolon lacks a mouth and a pharynx and its independent life exclusively is devoted to mating, followed by eventual death, i.e. the stolons are semelparous. The unchanged benthic parental stock, however, survives, continues to feed, regenerates the lost segments and then reproduces again (Franke, 1999). In these worms, schizogamy has restored iteroparity from the reckless semelparity.

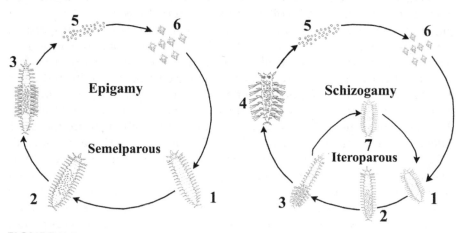

FIGURE 5.1

Schematic representation of semelparous epigamic epitoky in non-syllid errant polychaetes and iteroparous schizogamous epitoky in errant syllid polychaetes. 1. Atokous form, 2. Sexual maturation, 3. Epitokous form, 4. Stolonic epitokous form, 5. Spawned eggs by swarmed epitokous females, 6. Trochophore larvae, 7. Regenerated parental atokous stock (modified from Franke, 1999).

5.2 Epigamy

With taxonomy being in a fluid status and description of life history limited to < 3% of polychaetes (Giangrande, 1997) and the number of description increasing at three species/y (see p 41), it is to be expected that reviewers have also limited the description of epigamy citing a few examples alone. Secondly, a few reports on the incidence of epitoky are not reliable. Despite their admirable effort to assemble information on dispersal of polychaetes, Carson and Hentschel (2006) have wrongly indicated the incidence of epitoky in seven glycerid species. However, when checked with Strathmann (1987) and Shanks (2001), the cited information has not been found. With these constraints, the first attempt has been made to list the incidence of epigamy and schizogamy in Table 5.1 and Table 5.3, respectively. Incidence of epigamy is limited to 61 + species in a dozen errant polychaete families. Nereididae includes the maximum incidence in 19 species. Rightly, the epitokous form is named as 'heteronereis'. Epitoky occurs in these polychaetes from the Arctic to Antarctic (e.g. *Capitella capitata*, Table 5.1). Notably, half a dozen nereidid species are reported from estuaries (e.g. *Hediste japonica*) and rivers (e.g. *Dendronereis aestuarina*). In epigamics, especially in the nereidid, sex ratio is skewed in favor of males, except in *Perinereis vancaurica tetradentata* and *Eunice siciliensis*. Understandably, the former broods embryos and larvae in gelatinous mass (Arias et al., 2012). In nereidids, the ratio ranges from 1 ♀ : 1 ♂ in the pair-forming *Nereis limbata* to 1 ♀ : 3 ♂ in the riverine *H. diadroma* and *D. aestuarina* (Table 5.2). Hence, it is likely that the excess number of males ensures the highest Fertilization Success (FS). Regrettably, no information on FS is yet available for any epigamic epitokous species or for that matter, any broadcast spawning non-epitokous polychaetes. Simple plankton collection of eggs prior to the completion of swarming can readily provide the desired information (cf Pandian, 2010). Incidentally, 53% polychaetes are broadcast spawners (p 38) and, external fertilization occurs in 53% of polychaetes (p 44). Hence, it is difficult to justify whether epitoky has achieved a higher FS than non-epitokous broadcast spawning polychaetes or epitoky is simply a reckless mating act.

Incidentally, some hesionids like *Kefersteinia cirrata* swarm and spawn for 2 days during June–July in the UK waters (Olive and Pillai, 1983). In fact, Fage and Legendre (1927) have listed *K. cirrata*, *Oxydromus propinquius*, *Ophiodromus flexuosus* and *Podarke pallida* as epitokes. The epitokes of a few nereidids like *Hediste* spp and *Odontosyllis enopla* display neither the striking structural epigamous modifications nor look like heteronereis. However, these hesionids are not listed in Table 5.1, as they do not display epigamous modifications and are not semelparous. Likewise, the nephtyid

TABLE 5.1

Incidence of epigamic epitoky in errant polychaetes

Aerocirridae
Flabelligela macrochaeta (Mexican Pacific, Salazar-Vellajo and Londano-Mesa, 2004)
Amphinomidae
Eurythoe complanata (Bay of Bengal, Fauvell, 2010)
E. parvecarunculata (Bay of Bengal, Fauvell, 2010)
Aphroditidae
Drieschia pelagica (Bay of Bengal, Fauvell, 2010)
Capitellidae
Capitella capitata (Arctic, Antarctic, Mediterranean, Lopez-Jamar et al., 1986)
Cirratulidae
Aphelochaeta glandularia (USA, see Petersen, 1999)
A. monilaris (USA, see Petersen, 1999)
Caulleriella viridis (French coast, see Petersen, 1999)
Cirratulus cirratus (Gibson, 1981)
C. incertus (UK, see Petersen, 1999)
Dodecaceria caulleryi (Berrill, 1952)
D. concharum (Danish waters, see Petersen, 1999)
D. fimbriata (Northeast USA, Gibson, 1979)
Ctenodrilidae
Ctenodrilus serratus (Indo-Pacific, Atlantic, Harms, 1993)
Monticellina heterochaeta (Mediterranean, Martin and Gil, 2010)
Raricirrus variabilis (Virgin Islands, 17° N, Dean, 1995)
Tharyx perbranchiata (Mexican, Pacific, Salazar-Vellajo and Londano-Mesa, 2004)
Eunicidae
Eunice afra (Tropical, Bisby et al., 2005)
E. fucata (Atlantic, Mayer, 1902)
E. schizobranchia (Subtropical, Martin and Gil, 2010)
E. schemacephala (Red Sea, Mexico, Wehe and Fiege, 2002)
E. siciliensis (Mediterranean, Hofmann, 1974)
E. torquata (Tropical, Fishelson, 1971)
E. viridis (Pacific, Hauenschild et al., 1968)
Glyceridae
Glycera alba (Northeast Atlantic, Mediterranean, Gusso et al., 2001)
G. americana (Indo-Pacific, Western Atlantic, Fish Forum)
G. capitata (Belgium coast, Degraer et al., 2006)
G. dibranchiata (Maine, USA, Creaser, 1973)
G. macrobranchia
G. gigantea (Pacific, cosmopolitan, Gibbs, 1978)
G. oxycephala (Tropical, Salazar-Vallejo, 1996)
G. tenuis (Tropical, Pacific, Salazar-Vellajo and Londano-Mesa, 2004)

Table 5.1 contd. ...

...Table 5.1 contd.

Nereididae

Nereis succinea (Brazil, Aguiar and Santos, 2017)

N. virens (Canada, Chatelain et al., 2008) (Andaman Island, Muruganatham et al., 2015)

Dendronereis aestuarina (Freshwater in the southwest coast of India, Jayachandran et al., 2015)

Hediste diadroma (Japan, Hanafiah et al., 2006)

H. japonica (Japan, Sato, 1999)

H. osawai (Japan, Hanafiah et al., 2006)

H. oxypoda sensu (Japan, Hanafiah et al., 2006)

Marphysa disjuncta (New Zealand, Read, 2004)

N. falcaria (Australia, Read, 1974)

N. fucata (Atlantic, Gilpin-Brown, 1959)

N. grubei (Schroeder, 1967)

N. japonica (Japan, see Fischer, 1999)

N. limbata (Hauenschild and Hauenschild, 1951)

N. pelagica (Tropical, Salazar-Vallejo, 1996)

Perinereis cultrifera (Cassai and Prevedelli, 1998)

P. nuntia brevicirrus (see Gilpin-Brown, 1959)

P. vancaurica tetradentata (Mediterranean, Arias et al., 2012)

Platynereis dumerilii (Garcia-Alonso et al., 2013)

Tylorhynchus heterochaetus (Japan, Koya et al., 2003)

Ophelidae

Euzonus flabelligerus (Baltic, Rower, 2010)

Phyllodocidae

Mystides caeca (Mediterrean, Martin and Gil, 2010)

Nereiphylla castanea (Indo-Pacific, Salazar-Vellajo and Londano-Mesa, 2004)

Paranaitis polynoides (Western central Atlantic, Salazar-Vellajo and Londano-Mesa, 2004)

Phyllodoce cuspidata (Northeast Pacific, Macdonald et al., 2010)

P. groenlandica (East Pacific, Altantic, Salazar-Vellajo and Londano-Mesa, 2004)

P. hartmanae (Pacific, Macdonald et al., 2010)

P. longipes (Tropical, Vittor, 2002)

Protomystides confusa (Tropical, Western Atlantic, Salazar-Vallajo, 1996)

Pterocirrus foliosus (Central American Sea, Salazar-Vallajo, 1996)

Syllidae

Autolytus alexandri (Franke, 1999)

Nephtys caeca with a straightforward annual reproductive cycle is also not included, as it is polytelic (Olive, 1978). But both *Hediste* spp and *O. enopla* are listed, as the former is semelparous and the latter displays luminescence, a typical characteristic of epitokous genus *Odontosyllis*. Notably, many authors have reported either a fraction or an entire population undergoing reversible epigamy. For example, Creaser (1973) has observed a few partially milted males of *Glycera dibranchiata* returning to the bottom and burrows. In

TABLE 5.2

Sex ratio of some epigamic epitokes

Species	Sex Ratio ♀ : ♂	Reference
Nereidae		
Nereis limbata	1.0 : 1.0	Hauenschild and Hauenschild (1951)
N. fucata	1.0 : 1.0	Gilpin-Brown (1959)
N. brevicornis	1.0 : 1.25	See Gilpin-Brown (1959)
N. japonica	1.0 : 2.10	Hanafiah et al. (2006)
N. diadroma	1.0 : 3.0	Hanafiah et al. (2006)
Dendronereis aestuarina	1.0 : 3.0	Jayachandran et al. (2015)
Perinereis vancaurica tetradentata (brooder)	1.0 : 0.1	Arias et al. (2012)
Glyceridae		
Glycera dibranchiata	1.0 : 1.24	Creaser (1973)
Eunicidae		
Eunice siciliensis	1.8 : 1.0	Hofmann (1974)

O. enopla, a reversible epigamic epitoky is displayed. This Bermudan fireworm does not undergo structural modifications except for the large eyes and swimming chaetae; all the post-swarming worms return to a benthic life. In laboratory reared worms, the post-swarming females are filled with a new set of oocytes 4 months after the swarming (Fischer and Fischer, 1995).

Rearing the anterior fragment of the epitokous *Eunice siciliensis*, Hofmann (1974) has observed the following sequence of development: (i) within 10–14 days, all the anterior fragments begin caudal regeneration. (ii) more than 50 segments with parapodia are developed by the 30th day. (iii) between 10th and 16th month, almost all of them have developed gonadal sacs and germ cells. Briefly, the anterior part of the epitoky—when reared under optimal conditions—can survive, completely regenerate genital segments and produce another epitoky. In the recipient mature (non-epitokous) *N. diversicolor*, implantation of cerebral ganglia from the immature donor is reported to induce repeated gametogenic cycles but not spawning. Still, the publication by Golding and Yuwono (1994) has indicated that the trait of iteroparous gametogenic cycling is retained in the epitokous nereidids. Hence, the regenerative potency of the anterior fragment is inherent but for its initiation, promotion of gametogenesis and epitokous development, a gonadotrophic neurosecretion originating from the prostomium is required. Apparently, the genes responsible for iteroparity in these epigamics are indeed retained but are not expressed and results in an expensive semelparity.

TABLE 5.3

Incidence of schizogamic epitoky in syllid polychaetes

Autolytus brachycephalus (Schiedges, 1980)
A. charcoti
A. edwardsii (Okada, 1935)
A. magnus
A. prolifer (see Franke, 1999)
A. purpureimaculata (Okada, 1933)
Brania pusilla (Temperate, Mediterranean, Harms, 1993)
Eusyllis blomstrandi (Indo-Pacific, Atlantic, Arctic, Harms, 1993)
Exogone hebes (NewFoundland, Pocklington and Hutcheson, 1983)
Exogone naidina (Indian ocean, Atlantic, Arctic, Harms, 1993)
Haplosyllis spongicola (Indo-Pacific, Mediterranean, Bisby et al., 2000)
Haplosyllides floridana (see Franke, 1999)
Grubeosyllis clavata (Tropical, Circumglobal, MarineSpecies.org)
Myrianida pinnigera (Subtropical, Mediterranean, Martin and Gil, 2010)
Odontosyllis enopla (Bermuda, Fischer and Fischer, 1995)
O. hyalina (Indonesia, Lummel, 1932)
O. luminosa (Belize, 16°.5′ N, Gaston and Hall, 2000)
O. phosphorea (Bermuda, Tsuji and Hill, 1983)
O. polycera (New Zealand, Daly, 1975)
O. undecimdonta (Japan, Inoue et al., 1993)
Pionosyllis lamelligera (see Franke, 1999)
P. neapolitana (see Franke, 1999)
P. pulligera (see Franke, 1999)
P. procera (see Franke, 1999)
Proceraea cornuta (North Atlantic, North Pacific, Nygren, 1999)
P. okadai (North America, Calif Acad Sci, 2015)
P. picta (Subtropical, Martin and Gil, 2010)
Sphaerosyllis hystrix (Tropical, Mediterranean, Wehe and Fiege, 2002)
Streptosyllis verrilli (see Franke, 1999)
S. websteri (see Franke, 1999)
Syllides japonica (see Franke, 1999)
Syllis amica (Ruppert et al., 2004)
S. gracilis (Indo-Pacific, Mediterranean, Harms, 1993)
S. ramosa (Phillipines, Wikipedia)
S. vittata (Central American Mediterranean, Salazar-Vellajo, 1996)
Trypanosyllis asterobia (Okada, 1933, 1937)
T. coeliaca (Tropical, Salazar-Vallajo, 1996)
T. crosslandi (Bay of Bengal, Fauvel, 2010)
T. gemmipara (Tropical, Salazar-Vellajo and Londano-Mesa, 2004)
T. ingens (Northeast Pacific, Canada, Lamb et al., 2011)
T. zebra (Mediterranean, Western Pacific, Salazar-Vellajo and Londano-Mesa, 2004)
Typosyllis prolifera (Atlantic, Mediterranean, Wikipedia)
T. hyalina (see Franke, 1999)
T. pulchra (Tropical, Mexican Pacific, Salazar-Vellajo and Londano-Mesa, 2004)
T. variegata (Tropical, Mexican Pacific, Salazar-Vellajo and Londano-Mesa, 2004)

5.3 Schizogamy

The speciose (700 species) syllids display an incredible diversity in the reproductive phenomenon. Almost all the reported incidences of schizogamy occur in 45 + syllid species (Table 5.3); hence schizogamics make up only 6.4% of Syllidae. The schizogamic syllids reproduce by means of sexual stolons. The stock can generate one (e.g. *Streptosyllis verrilli*, iteroparous?) to 15 successive stolons (e.g. *Typosyllis prolifera*, see Franke, 1999). In *Autolytus prolifer* a single female can produce 1.4 (sacconereis) stolons/mo (Hauenschild, 1953). Indeed, the syllids have discovered stolonization as the most successful reproductive mode and have elaborated them into fascinating diverse types. Accordingly, schizogamy is divided into two subtypes: scissiparity, where the existing segments in the stock are transformed into a stolon and is followed by fragmentation. The fission plane is a fixed one, for instance, the 14th segment in *Proceraea picta*. It is further divided into paratomic stolonization, (e.g. *T. prolifera*) and architomic stolonization (e.g. *T. hyalina* and *T. variegata*, see Franke, 1999, *Polycera cornuta* [Fig. 5.2A], *Myrianida pachycera* [Fig. 5.2B]). The gemmiparous subtype is characterized by budding, in which the segments, destined from the very beginning to form a stolon, rapidly proliferate. Successive terminal buddings lead to the formation of stolon chains. For example, *Trypanosyllis asterobia* forms multiple collateral budding with a large number of successive posterior stolons (Fig. 5.2G). In *T. gemmipara*, a bundle of stolons are simultaneously formed within a limited proliferation area near the posterior end of the stock (Fig. 5.2H). A single stolon is formed/mo in *T. prolifera* and the number of regenerated stock segments is in a sort of temperature-dependent monthly rhythm (Franke, 1985). In *Syllis ramosa*, collateral budding results in a complex stock consisting of numerous branches extending through the canal system of the host sponge. Special branches bear a single stolon at their tip (Fig. 5.2E, F). Within the syllids, the subfamily Autolinae is the most complex. On the basis of sexual dimorphism, especially cephalic appendages, the female and male stolons are named as sacconereis and polybostrichos, respectively.

Procerastea halleziana undergoes rapid multiplication by a process of repeated fragmentations followed by regeneration of anterior and posterior ends in each fragment. The first fragmentation occurs between the 7th and 8th chaetigerous segments, resulting in the head and seven segments forming a fragment. This is followed by 3 fragmentations, resulting in 3 ramets each of with 2 segments, and another 3 ramets each with 3 segments, behind which 4 or 5 fragmentations *per se* with 4 fragments each followed by divisions *per se* again 2 fragments each with 3 segments up to pygidium. The mode of fragmentation is expressed in the following formula in species belonging to five different genera.

Procerastea: Hd 7 + 2 + 2 + 2 + 3 + 3 + 3 + 4 + 4 + 4 + 4 (+ 4) + 3 + 3.....Py

Autolytus: (Hd 7 + 2) + 2 + 2 + 3 + 3 + 3 + 4 + 4 + 4 + 4.................Py

Pionosyllis: (Hd 7 + 2) + 2 + 2 + 3 + 3 + 3 + 4 + 4 + 4 + 4.........Py

Myrianida: (Hd 7 + 2 + 2 + 2) + 3 + 3 + 3 + 4 + 4 + 4 + 4 + 3 + 3 + 4 + 4 + 4 + 4 + 4 + 4 + 3 + 3..Py

Trypanosyllis: Hd (7 + 2 + 2 + 2 + 3 + 3 + 3 + 4) + 4 + 4 + 4 + 3 + 3 + 4 + 4 + 4 + 3 + 3 + 4 + 4 + 4 + 4 + 3 + 3 + 3....Py

Hd = 7 = Head segments, Py = Pygidial segment, Head segments + Anterior region within brackets are usually not fragmented (Allen, 1923, Berrill, 1952).

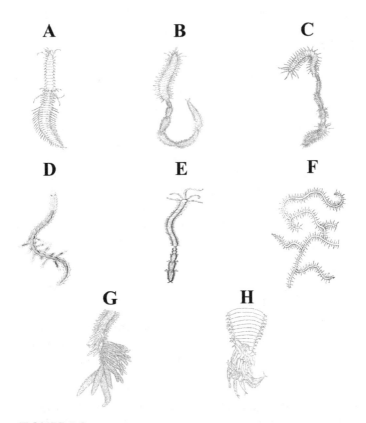

FIGURE 5.2

Some patterns of schizogamic epitoky in syllids. A. Scissiparous vertical *Polycera cornuta*, B. Gemmiparously budding *Myrianida pachycera*, C. *Autolytus prolifer*, D. Scissiparous laterally budding *Exogone rubescens*, E and F. Show individual and highly branching *Syllis ramosa*, G. Multiple collateral budding *Trypanosyllis asterobia* and H. Simultaneously bundled stolons at the posterior end of *T. gemmipara* (all are freehand drawings from Franke, 1999 and others).

Many hermaphroditic schizogamic syllids reproduce through stolons. As sex change occurs in the stolons of the hermaphrodites like *Syllis vittata* (Durchon, 1975), it is possible that germ cells are also transmitted through successive stolons. However, it is not known whether this sort of transmission also occurs in gonochoric epitokous syllids. In *Autolytus edwardsii*, fertile eggs produced in the terminal stock segments are passed on to the stolons, where they undergo further development with a support of nurse eggs produced in abortive stolonial ovaries (see Franke, 1999). As egg development is restricted to vitellogenesis alone in the stolon, research inputs are required to know whether the stolons of gonochoric have also received germ cells from the stock.

Schiedges (1979) has successfully hybridized *Autolytus prolifer* and *A. brachycephalus* by feeding them on laboratory-reared hydrozoid *Eirene viridis*. The life span of the hybrid is ~ 550 days. A detached sacconereis holds 2, 8–16 and 0–2 segments in its anterior, swimming mid-body and caudal zones, respectively, in comparison to 2, 4–20 and 0–1 segments in polybostrichos. A swarming polybostrichos commences the encircling dance around the sacconereis and begins to milt. It can inseminate eggs of several sacconereis prior to death. The stolonization potential increases from five numbers on the 8th day to 45 on the 180th day and is followed by gradual decrease to 5–10 stolons up to 540th day. Apparently, the stolonization process also undergoes age specific senescence (cf Martinez and Levinton, 1992, see also p 138), perhaps due to the reduction of juvenile hormone, the nereidine (see Golding, 1967). Besides questioning Hamond's (1969) idea of monophagy, Schiedges (1979) has also found that the hybrid produces scissiparous stolons initially but switches to gemmiparous stolons subsequently.

5.4 Vertical Migration

Understandably, the spectacular epitoky has attracted the attention of a large number of zoologists over long years. But only a few of them have cared to report the depth, from which vertical migration is undertaken by the epitokes to reach the surface waters, where swarming and mating occur. For the first time, Table 5.4 lists almost all available information on the depth, from which the migration is undertaken. Of 14 epigamic species, 11 of them undertake the migration over a distance of < 175 m. Similarly, of 13 schizogamics, the migration for 14 species is also limited to < 320 m. Only a capitellid (*Capitella capitata*), a glycerid (*Glycera oxycephala*) two ctenodrillid species and a syllid (*Eusyllis blomstrandi*) migrate over distances ranging from 1,310 to 4,000 m. In preparation of this migration, the epitokous form, for example, the epigamic *Raricirrus variabilis* grows to a larger size in length (14.6 mm), width (1.3 mm) and in chaetigerous segments (32), in comparison to the atokous form with

TABLE 5.4

Reported depths of collected epitokous species. For reference, see Tables 5.1 and 5.2

Species	Depth (m)	Species	Depth (m)
Epigamy		**Schizogamy**	
Nereididae		**Syllidae**	
Nereis falcaria	6	*Grubeosyllis clavata*	2
N. fucata	52	*Odontosyllis enopla*	2
Capitellidae		*O. phosphorea*	3
Capitella capitata	1500	*O. luminosa*	6
Eunicidae		*Typosyllis variegata*	10
Eunice schemacephala	150	*Trypanosyllis coeliaca*	43
E. siciliensis	33	*Odontosyllis polycera*	70
Glyceridae		*Syllis gracilis*	75
Glycera capitata	17	*Haplosyllis spongicola*	189
G. americana	37	*Exogone naidina*	210
G. macrobranchia	55	*Syllis ramosa*	250
G. oxycephala	2951	*Trypanosyllis zebra*	320
Phyllodocidae		*Eusyllis blomstrandi*	4000
Phyllodes groenlandica	38		
Nereiphylla castanea	45		
Paranaitis polynoides	82		
Phyllodoce longipes	175		
Ctenodrillidae			
Monticellina heterochaeta	1310		
Raricirrus variabilis	4000		

5.8 and 0.5 mm with 27 chaetigers (Dean, 1995). In the schizogamics too, the number of swimming segments is increased to 50% of the total number of segments and their chaetiger length is also increased to 1.5 mm length in *Odontosyllis polycera* (Daly, 1975). Swarming males of *O. enopla* have less number of segments (~ 85) than females (~ 116) (Fischer and Fischer, 1995). Except for these bit and pieces of information, no other relevant information is yet available on morphological changes in the epitokes. The duration of swarming lasts for 25 minutes in *O. luminosa*, which undertakes a vertical migration from 6 m depth (Gaston and Hall, 2000). *O. enopla* climbs a distance of just 2 m to reach the surface waters. Incidentally, the worms seem to swim fast. High speed swimming has been documented in film shots of *Platynereis dumerilii* (Fischer, 1985) and epitokous *Autolytus prolifer* (Fischer et al., 1992). The climb involves a concave bent on the dorsal side of the trunk followed by lateral waves passing from posterior to anterior (Fischer and Fischer, 1995). It is difficult to imagine how the little (15 mm length) *R. variabilis* can sustain its climb to cover a vertical distance of 4,000 m. Certainly, these small worms may adopt a different strategy like reduction in buoyancy to climb over 4,000 m distance. Till this day, this area remains a virgin field of research.

An attempt has been made to correlate vertical distance climbed by the epitokous worms with their body length and/or number of segments and/or body weight. Even with a computer search only limited information is available on body length and segment of epitokes (e.g. *Glycera dibranchiata*: Maryland 7–26 cm, Nova Scotia, 13–36 cm, Wiscasset 18–51 cm [Creaser, 1973], *Eunice viridis*, 20 cm body length [Caspers, 1984], 80 cm body length, 300–700 parapodial segments in *Eunice siciliensis* [Hofmann, 1974], 13 mm length, 60 segments in *Monticellina heterochaeta* [Martin and Gil, 2010], *Syllis vittata*, 25 mm length [Salazar-Vellajo, 1996]). Almost parallel trends are obtained for the relationship between body size and vertical distance traveled by epigamics and schizogamics (Fig. 5.3). Surprisingly but certainly, the ability

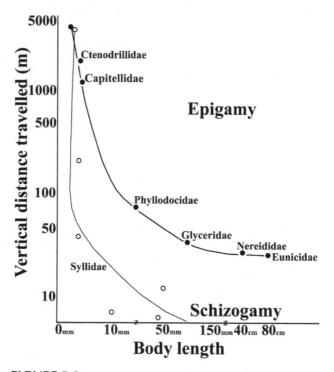

FIGURE 5.3

Effect of body size on vertical distance climbed by epigamic and schizogamic polychaetes. Data for vertical distribution are taken from Table 5.4. Due to non-availability of information on body length of many of these worms, only approximate trends are shown.

to travel vertical distance decreases with increasing body size in both epigamic and schizogamic epitokes. Of course, with more inputs of data, the level may change but not the trends. Among the epigamics, the ability for vertical climb decreases in the following order: Ctenodrillidae < Capitellidae < Phyllodocidae < Glyceridae < Nereididae < Eunicidae. Besides body length, departures from the stream-line body shape by curves, bends, projections and other structures may considerably decelerate the vertical migration, especially in the schizogamics (see Fig. 5.2).

5.5 Swarming Phenomenon

In epitokes, the swarming phenomenon is timed by a complex hierarchy of endogenous and environmental factors (e.g. temperature, photoperiod, luminescence). Long day length and elevation in temperature have timed the warmer spring-summer (in south the austral spring-summer) as a favorable season for swarming in temperate and tropical epigamics (Table 5.5). In semelparous epigamics, swarming occurs in the terminal end of their life during the warmer season, during July in the Mediterranean *Eunice siciliensis* and August in the Indian riverine *Dendronereis aestuarina* (Fig. 5.4A, B). In the epigamics, seasonal rhythms seem to play a role in timing the swarming phenomenon. In schizogamics also, long daylength and elevated temperature stimulate stolonization and short daylength and low temperature inhibit it (Franke, 1985).

Swarming rhythms: On the other hand, swarming in schizogamics is timed by a combination of annual, lunar and diel rhythms. It repeatedly occurs during a specific warmer season May–June in *Odontosyllis luminosa*. In *Typosyllis prolifera*, each cycle lasts for 31 days, which includes 17 days of regenerative phase and 14 days of stolonization phase (Franke, 1986a). Hence, the schizogamics can undertake repetitive swarmings. For example, *O. phosphorea* undertakes as many as nine swarming cycles within a season between June 30th and October 30th (Fig. 5.4C). In it, the timing of swarming is progressively postponed from ~ 30 minutes after sunset in July to ~ 60 minutes after sunset in October (Tsuji and Hill, 1983). Tuned to lunar rhythm, swarming occurs during a specific lunar phase (Table 5.5) and is reported to occur 3–5 days around that specific lunar phase. Amazingly, tuned to lunar cycle, swarming is precisely timed by the diel rhythm. Notably, it occurs during dusk in most species and dawn in *T. prolifera*. Dusk and dawn are selected to minimize their visibility and predation. Nocturnal high tides (e.g. *Syllis amica*) and neap tides (*O. phosphorea*) are also selected to ensure the maximum dispersal.

TABLE 5.5

Swarming timed by annual, lunar and diel rhythms in some epitokes

	Epigamics
Nereis fucata	April, Atlantic (Gilpin-Brown, 1959)
N. falcaria	January, austral summer, Australia (Read, 1974)
N. virens	July, Andaman Islands, India (Muruganantham et al., 2015)
Dendronereis aestuarina	August, Southwest Indian river (Jayachandran et al., 2015)
Glycera dibranchiata	July, USA (Creaser, 1973)
Eunice siciliensis	July, Mediterranean (Hofmann, 1974)
E. viridis	October–November, austral spring, Samoan Islands (Caspers, 1984)
Odontosyllis enopla	May, Bermuda (Markert et al., 1961, Fischer and Fischer, 1995)
	Schizogamics
	Annual rhythm
O. enopla	May, Bermuda (Markert et al., 1961, Fischer and Fischer, 1995)
O. luminosa	May–June, Caribbean Belize (Gaston and Hall, 2000)
O. phosphorea	June–October, South California (Tsuji and Hill, 1983)
O. polycera	October, austral apring, New Zealand (Daly, 1975)
Syllis amica	Mid-July–mid-August, France (Herpin, 1925)
Typosyllis prolifera	March–October, Adriatic Sea (Franke, 1985)
	Lunar rhythm
S. amica	A few days after the first quarter phase of the moon at nocturnal high tide
O. enopla	A few days immediately after the full moon
O. phosphorea	First and third quarter phases of the moon during neap tides
O. polycera	Last quarter phase of the moon
E. viridis	Third quarter phase of the moon
	Diel rhythm
O. polycera	Precisely 30-minutes after sunset
O. enopla	Between 15 and 55-minutes after astronomical sunset
O. phosphorea	60-minutes after sunset
O. luminosa	Between 55 and 75-minutes after astronomical sunset
T. prolifera	Between 30-minutes before and 20-minutes after sunrise

Besides temperature and photoperiod, some epitokes also engage luminescence to attract the opposite mating partners. Incidence of luminescence has been described in the epigamic *Odontosyllis enopla* (Fischer and Fischer, 1995) and schizogamics *O. hyalina* (Lummel, 1932), *O. undecimdonta* (Inoue et al., 1990), *Pionosyllis pulligera* (see Franke, 1999) and *O. luminosa* (Gaston and Hall, 2000). It is a secretory process and

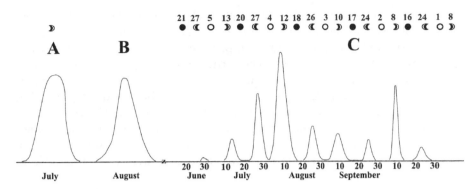

FIGURE 5.4

Swarming season and lunar periodicity in epitokes A. *Eunice siciliensis*, B. *Dendronereis aestuarina* and C. *Odontosyllis phosphorea*. Note a single event in the epigamics but repeated events in the schizogamics (Figure is compiled and redrawn from Hofmann, 1974, Jayachandran et al., 2015, Tsuji and Hill, 1983).

the luminescence slime is found even in one-month old *O. phosphorea*. Bright flashes of it are also observed in post-swarming *O. enopla* that has returned to benthic life. Hence, the role played by luminescence in swarming in *O. enopla* and *O. phosphorea* is not clear. However, swarming luminesced female *O. enopla* is reported to attract males, which also luminesce. *P. pulligera* displays vivid luminescence during swarming. The males are attracted by the luminescence; with their large eyes, they can precisely locate the luminescent female (see Franke, 1999). In *O. luminosa*, which is distributed throughout the tropical waters of the Western Hemisphere, the glow is bright and visible upto 30–50 m distance. Interestingly, it is the luminescent glow of *O. luminosa* in the Caribbean that is reported to have provided hope for the nearby land mass to the totally exhausted Christopher Columbus in 1492.

5.6 Pheromones and Spawning

In the pelagic realm, swarming leads to high density (e.g. $14,800/m^2$ *Dendronereis aestuarina*, Jayachandran et al., 2015) of mating partners and it lasts for 30 minutes in *Odontosyllis luminosa* (Gaston and Hall, 2000). During this brief pelagic stay, the pairing of mating partners is followed by species-specific and the sex-specific nuptial dance and other behavioral signals that eventually lead to the shedding of gametes. A series of publications by Boilly-Marer (1974, 1984), Boilly-Marer and Lassalle (1980), Boilly-Marer and

Lhomme (1986), Zeeck et al. (1988, 1996, 1998) and Hardege (1999) on the non-epitokous *Platynereis dumerilii* and *Arenicola marina* (Hardege et al., 1996) have contributed to our knowledge on pheromones and their role to induce spawning in these polychaetes. Their findings, though not immediately relevant to epitokes, are briefly summarized. Essentially, the following is the sequence of mating and spawning: 1. Male discharges an egg-releasing pheromone, 2. The Cysteine-Glutathione-Disulfide (CGD) pheromone stimulates a female to swim at high velocity in narrow circles surrounded by swarming males (Ram et al., 1999), 3. After an induction period of 30–40 seconds, the female spawns, 4. Subsequently, the male emits a cloud of sperm.

P. *dumerilii* detects a pheromone signals by their modified cirri. The bipolar neurons innervating the chemoreceptors transmit the electrophysiologically measureable excitation along the ventral nerve cord to the brain ganglia. From the coelomic fluid of males, a few low molecular and low polarity fractions have been extracted. These fractions induce the female to spawn. Of > 50 volatile substances, 5 methyl-3-heptanone, 3, 5-octadien-2-one and possibly 3-methyl-uric acid and xanthane are important. The heptanone is a Mate Recognition Pheromone (MRP), 3, 5-octadien-2-one is an Egg Releasing Pheromone (ERP) and uric acid acts as Sperm Release Pheromone (SRP) (see Andries, 2001). The actual concentration of the heptanone produced by the worm is 2.5-fold higher the threshold required to induce spawning in female. However, the 5 methyl-3-heptanone successfully induces 20% of females only. To induce 96% females, coelomic fluid has to be added to the heptanone. In *Nereis succinea*, injection of L-glutamic acid, inosine and quanosine induces egg release alone. Hence, *P. dumerilii* is reported to use a complex pheromone bouquet to co-ordinate the nuptial dance and gamete release.

Spawning act: With mating and spawning limited to a single event in semelparous epigamics, the mature worm begins to twists its body in a peculiar screw-like manner, breaks off from the infertile anterior body part and releases the gametes through ruptures of the body wall (e.g. *Eunice siciliensis*, Hofmann, 1974). In schizogamics, once a pair is formed, the female begins to quiver rapidly with periodical spinning circles and releases a cloud of gametes (e.g. *O. luminosa*, Gaston and Hall, 2000) and coelomic fluid (see Andries, 2001). These gametes stimulate the encircling males to milt. Sometimes, several males may encircle a single female (polyandry, cf polygyny in the hybrid male of *Autolytus* spp inseminating eggs of many females, Schiedges, 1979). The sequence of events from formation of heteronereis/stolon to spawning is briefly summarized in Fig. 5.5.

In polychaetes, especially the epitokous syllids, extirpation of the proventriculus induces stolonization in non-reproductive worms (Franke,

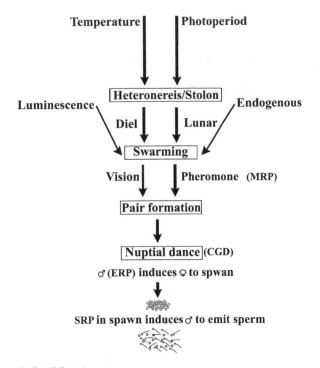

FIGURE 5.5

Sequence inductions that lead to the formation of heteronereis/stolon until gamete shedding in epitokes. MRP = Mate-Recognition Pheromone, ERP = Egg-Release Pheromone, CGD = Cysteine-Glutathione Disulfide, SRP = Sperm-Release Pheromone.

1980). Amputation of the prostomium causes rapid disintegration of growing oocytes followed by spermiogenesis (Kahmann and Franke, 1984). These contributions are elaborated in Chapter 7.

6

Sex Determination

Introduction

In annelids, sexuality includes gonochorism, parthenogenesis and hermaphroditism, which comprises of simultaneous, protandric, protogynic and serial hermaphrodites. More than 76% of annelids are gonochores. Hermaphroditic oligochaetes (18.7% of annelids), hirudineans (4.0%) and polychaetes (0.6%) make up the remaining 24% (see p 64). In gonochoric polychaetes, either male or female heterogametic chromosomal mechanisms of sex determination may be in operation. Sex ratio represents the cumulative end product of sex determination and sex differentiation processes. Table 2.2 shows that only in a few polychaetes, the expected ratio of 0.5 ♀ : 0.5 ♂ is reported. In many others, the ratio is, however, significantly deviated. These deviations indicate (i) amenability of sex chromosomes to overriding autosomal genes and alteration of the sex differentiation process exactly toward the opposite sex, and (ii) environment-dependent polygenic autosomal sex determination (Premoli et al., 1996). However, it must be indicated that our knowledge on sex determination in annelids is elementary, despite the presence of many freshwater oligochaetes with a short life span (e.g. naidids). With the presence of about 190 clonal species, the scope for their restoration to sexuality even after 1,000–3,000 rounds of clonal cycles may make the subject a fascinating one.

6.1 Karyotype and Heterogamety

Spermatogonia (male) and regenerating tail (female) of polychaetes have been used as tissues for karyotyping. In *Neanthes japonica*, the chromosome size ranges from 2.3 to 5.8 μm and most of them are metacentric (Sato and Ikeda, 1992). The number of 2n chromosome ranges from 12 in *Pristina aequiseta* to 54 in *Aporrectodea rosea* (Table 6.1). In *Serpula vermicularis*, the reported

TABLE 6.1

Chromosome number in annelids

Species	Number	Reference
Polychaetes		
Ditrupa arietina	20	see Kupriyanova et al. (2001)
Ficopomatus enigmaticus	26	see Kupriyanova et al. (2001)
Filograna implexa	20–44	see Kupriyanova et al. (2001)
Hydroides elegans	26	see Kupriyanova et al. (2001)
H. norvegica	22	see Kupriyanova et al. (2001)
Placostegus tridentatus	20	see Kupriyanova et al. (2001)
Pomatoceros triqueter	24–26	see Kupriyanova et al. (2001)
Serpula vermicularis	14–28	see Kupriyanova et al. (2001)
Spirorbis spirorbis	20	see Kupriyanova et al. (2001)
S. corallinae	20	see Kupriyanova et al. (2001)
S. tridentatus	20	see Kupriyanova et al. (2001)
Janua pagenstecheri	20	see Kupriyanova et al. (2001)
Circeis spirillum	20	see Kupriyanova et al. (2001)
Dinophilus gyrociliatus	XX-XO	Prevedelli and Vandini (1999)
Capitella capitata	ZW-ZZ	Petraitis (1985a, b)
Neanthes japonica	28 XX-XY	Sato and Ikeda (1992)
Hediste atoka	28 XX-XY	Tosuji et al. (2004)
H. diadroma	28 XX-XY	Tosuji et al. (2004)
H. japonica	28 XX-XY	Tosuji et al. (2004)
Nereis diversicolor	28	Christensen (1980)
N. acuminata	18, 22, 24, 28	Pesch and Pesch (1980)
Polydora curiosa	XX-XY	Korablev et al. (1999)
Aquatic oligochaetes		
Aeolosoma hemprichi	56	Naidu (2005)
Branchiodrilus sowerbyi	24	Naidu (2005)
Dero indica	36	Naidu (2005)
Frederiova bubosa	32	Naidu (2005)
Henlea ventriclosus	34	Naidu (2005)
Limnodrilus claparedianus	25	Christensen (1980)
L. hoffmeisteri	24	Naidu (2005)
L. undekemianus	25	Christensen (1980)
Pristina aequiseta	12	Naidu (2005)
Tubifex blanchardi	50	Marotta et al. (2014)
T. costatus	25	Christensen (1980)
T. tubifex	50, 100, 150	Marotta et al. (2014)
Lumbricillus lineatus	26, 39, 52, 65	Coates (1995)
Earthworms		
Allolopophora caliginosa	36	Muldal (1952)
A. chlorotica	32	Muldal (1952)
A. iterica	32	Muldal (1952)
A. nocturna	36	Muldal (1952)
A. terrestris	36	Muldal (1952)

Table 6.1 contd. ...

...Table 6.1 contd.

Species	Number	Reference
Earthworms		
Aporrectodea trapezoides	36, 54	Jaenike and Selander (1979)
A. rosea	54, 90, 108	Jaenike and Selander (1979)
Bimastos eiseni	32	Muldal (1952)
B. tenuis	48	Muldal (1952)
Eisenia foetida	22	Muldal (1952)
E. rosea f.typica	54	Muldal (1952)
E. rosea f.mut	33	Muldal (1952)
E. venata	36	Muldal (1952)
Eiseniella tetraedra	72	Muldal (1952)
Dendrobaena mammalis	34	Muldal (1952)
D. rubida	68	Muldal (1952)
D. subrucunda	68	Muldal (1952)
Dendrodrilus rubidus	34, 68, 102	Jaenike and Selander (1979)
D. octaedra	108, 124	Jaenike and Selander (1979)
Octolasion tytaeum	38, 54, 72	Jaenike and Selander (1979)
O. cyaneum	190	Muldal (1952)
O. lacteum	38	Muldal (1952)
Lumbricus castaneus	36	Muldal (1952)
L. festivas	36	Muldal (1952)
L. friendi	36	Muldal (1952)
L. rubellus	36	Muldal (1952)
L. terrestris	36	Muldal (1952)
Leeches		
Glossiphonia complanata	28	Davies and Singhal (1987)
Glossiphonia complanata concolor	28	see Davies and Singhal (1987)
Glossiphonia heteroclita	16	see Davies and Singhal (1987)
Glossiphonia heteroclita papillosa	16	see Davies and Singhal (1987)
Theromyzon tessulatum	16	Davies and Singhal (1987)
Branchellion torpedinis	12	see Davies and Singhal (1987)
Hemiclepsis marginata	32	see Davies and Singhal (1987)
Pontobdella muricata	20	see Davies and Singhal (1987)
Dina lineata	18	see Davies and Singhal (1987)
Erpobdella octoculata	16	see Davies and Singhal (1987)
Erpobdella testacea	16, 22	see Davies and Singhal (1987)
Nephelopsis obscura	22	see Davies and Singhal (1987)
Trocheta subviridis	22	see Davies and Singhal (1987)
Trocheta bykowskii	22	see Davies and Singhal (1987)
Haemopis sanguisuga	26	see Davies and Singhal (1987)
Hirudo nipponia	16	Yang et al. (1997)
Placobdella papillifera	24	Davies and Singhal (1987)
Theromyzon rude	14	Davies and Singhal (1987)
Hirudo medicinalis	28	Utevsky et al. (2009)
H. vertana	26	Utevsky et al. (2009)
H. orientalis	24	Utevsky et al. (2009)

numbers are 14 and 28 for specimens collected from the Mediterranean and Norwegian fjord, respectively. Similarly, they are 20 for *Filograna implexa* collected from the UK waters but 44 for that from Norway. It is not clear whether the Norwegians specimens are tetraploids. The 2n number of *Nereis acuminata* is 18, 22, 24 or 28 (Pesch and Pesch, 1980); it is also not clear whether the one with 28 chromosomes is a triploid.

Karyotyping has revealed the presence of male (XX-XY) heterogametic chromosomes in *Neanthes japonica* (Sato and Ikeda, 1992), *Hediste atoka*, *H. diadroma* and *H. japonica* (Tosuji et al., 2004). All these nereidids have 14 pairs of chromosomes including an unusually large sized Y chromosome. The spionid *Polydora curiosa* is also male heterogametic (Korablev et al., 1999), in which the Y chromosome is also larger than the X chromosome (Jablonka and Lamb, 1990). Possibly, the Y chromosome of polychaetes accumulates multiple transposable elements. Experimental crosses have also revealed the presence of male heterogamety in *Dinophilus gyrociliatus* (XX-XO, Prevedelli and Vandini, 1999) and female heterogamety in *Capitella capitata* (ZW-ZZ, Petraitis, 1985a, b).

6.2 Gametic Compatibility

In gonochores, reproduction is characterized by homospecific compatibility between egg and sperm and heterospecific incompatibility between them. Nevertheless, there are exceptions for each of heterospecific compatibility and homospecific incompatibility in annelids. Successful hybridization between the syllid *Autolytus prolifer* and *A. brachycephalus* as well as survival and reproduction of their hybrids have been described earlier (p 162). Another example for successful hybridization is between the nereidids *Hediste diadroma* and *H. japonica*. These two epitokes are sympatric only in the Omuta-gawa River, Japan. Their gametes are released without involving pairing between male and female partners and the nuptial dance. Their gametes are reciprocally compatible and hybrids between them are viable until 23rd day after fertilization. The differences in the developmental stages between these two nereids and their hybrids are listed in Table 6.2. The hybrid larvae are intermediate phenotypes but with a greater maternal influence in characteristics like the relative length of chaetae and lecithotrophic larval duration. Neither the pre-mating gametic incompatibility nor hybrid viability appears to ensure reproductive isolation between these nereids (Tosuji and Sato, 2006).

However, two points may have to be noted: 1. The hybrids between *Autolytus* spp sexually mature and successfully generate stolons. But the hybrids between *Hediste* spp are not reported to sexually mature and

TABLE 6.2

Time course of important developmental stages in *Hediste diadroma* and *H. japonica* and their hybrid offspring (condensed and compiled from Tosuji and Sato, 2006)

Development Stage	H. diadroma	H. japonica	H. diadroma egg × H. japonica sperm	H. japonica egg × H. diadroma sperm
Formation of I polar body	100 minutes	90 minutes	120 minutes	120 minutes
II Polar body formation	130 minutes	115 minutes	180 minutes	180 minutes
Trochophore hatching	38 hours	44 hours	38 hours	44 hours
Metatrochophore	72 hours	68 hours	72 hours	72 hours
Notochaeta				
4th chaetiger	16–30 days	7–8 days	12–23 days	9–10 days
5th chaetiger	> 30 days	10–20 days	~ 31 days	~ 30 days

reproduce, *albeit* both the species have the same number of 2n chromosomes (28), which morphologically look alike (Tosuji et al., 2004). 2. Fertilization success (FS) is ~ 95% in *Hediste* spp, indicating the near 100% compatibility between their respective eggs and sperm. Notably, the heterospecific compatibility between *H. diadroma* egg and *H. japonica* sperm, as indicated by FS, is reduced to ~ 80%, in comparison to that (~ 85%) between *H. japonica* egg and *H. diadroma* sperm. This minor level of heterospecific incompatibility can reach a very high level between two morphotypes within a species and surprisingly within populations of a single species.

In *Spirobranchus polycerus*, two morphotypes are recognized. Of them, the seven-(opercular) horned morph is hermaphrodite and two-horned morph gonochore. Gametic compatibility and hence FS for the homo-morphotypes is ~ 98%. Marsden (1992) has attempted to fertilize eggs of the gonochoric morph with sperms from hermaphroditic morph. None of the eggs have cleaved, a marker of egg-sperm compatibility. Hence, the gametic incompatibility between these two morphs within the species *S. polycerus* has been confirmed. Another serpulid *Galeolaria caespitosa* is widely distributed around Australia from the eastern coast at Sydney to the southern coast of Adelaide, covering a coastal distance of 2,200 km. In it, gametic compatibility and cross-fertilization occur among populations across temperate Australia. However, fertilization assays have revealed the pre-mating asymmetrical differences between very distantly located populations with near complete incompatibility between eggs from Sydney with sperm from Adelaide. However, the reverse cross is reasonably compatible (Styan et al., 2008).

The low level of gametic incompatibility observed between *Hediste* spp and high level of incompatibility reported between the two morphotypes within *S. polycerus* and that reported for the distantly located populations within *G. caespitosa* can be explained from our knowledge on echinoderms.

In them, the presence of bindin in the sperm acrosome serve as an adhesive responsible for sperm attachment to the vitelline layer of homospecific eggs. Hence, bindin plays an important role in determining whether the gametes of two species are compatible and fertilize each other. However, the role of bindin is not a simple lock- and key-mechanism, in which molecular change(s) in sperm need not necessarily result in analogous change(s) in the egg also and vice versa. As a consequence, heterospecific gametic incompatibility may be complete or incomplete. Gametic compatibility is defined as the ratio of mean percentage of eggs fertilized in homospecific crosses at a sperm concentration required to fertilize ~ 90% of eggs. In such species, analogous molecular change(s) have either occurred or not occurred in eggs and sperms, the ratio of reciprocal heterospecific compatibility is 1.00 (see Pandian, 2018). However, the ratio begins to decrease with slight changes in sperm of *H. japonica* but not in the egg of *H. diadroma*. The ratio is also significantly decreased to almost zero with major changes in eggs of *G. caespitosa* from Sydney but not in its sperm from Adelaide.

Off from these, allopatric isolation may lead to post-mating isolation due to karyotypic changes. Low dispersal and sexual selection are characteristics of coastal polychaetes like the monogamous *N. acuminata* with male parental care. From a karyotypic study, Weinberg et al. (1990) have reported that the *N. acuminata* populations from the North Atlantic have karyotypes with 11 pairs of small acrocentric chromosomes (2n = 22), while their Pacific counterparts have nine pairs of large metacentric or submetacentric 2n = 18 chromosomes. The Atlantic and Pacific North American populations have remained allopatric for a long time. They are now almost different species and post-mating reproductive isolation has evolved as an incidental byproduct of allopatric divergence between populations.

6.3 Sex Ratio and Variations

Male heterogamety has been recognized in the gonochoric *Dinophilus gyrociliatus*. This has been described earlier (p 90). However, a couple of sentences are required for the sake of continuity. In *D. gyrociliatus*, a single ovary simultaneously produces many small (XO) male eggs and one larger (XX) female egg. This unique mechanism allows the mother to overcome the chromosomal mechanism of sex determination by selective fertilization of larger eggs by X bearing sperms and small egg by sperm without sex chromosome. A number of environmental factors alter the ratio between small and larger eggs. Optimal temperature 28°C and salinity of 30‰ maintain the ratio at 1 ♀ : 3 ♂ by producing 1 large and 3 small eggs (Akesson and Costlow, 1991). Similarly, the optimal diet tetramin increases fecundity and thereby alters sex ratio (Prevedelli and Simonini, 2000). The presence of multiple females can also significantly skew sex ratio toward males (Minetti et al., 2013).

Through experimental crosses, heterogamety is identified in *Capitella capitata*. The 27 crosses between females and males of *C. capitata* have produced 2,110 offspring with sex ratio of 0.47 ♀ : 0.53 ♂. But 31 crosses between hermaphrodites and males have yielded 3,264 offspring; of them 96.7% are males. Apparently, females, males and hermaphrodites are characterized by ZW, ZZ and ZZ, respectively (Petraitis, 1985a, b). However, two subsequent series of experiments made the picture a little more complicated. In the first series, the effects of density have been tested on the expression of gender and hermaphroditism. The second one has tested the maternal effect on the incidence of hermaphroditism. From these experiments, Petraitis (1991) has drawn the following inferences: 1. Families determine the age of female maturity. The overall age of maturity of females is 39 days. 2. The first transition from male to hermaphroditism occurs at the age of 76 days. 3. Males are more prone to become hermaphrodites, especially in isolation or in the presence of excess number of males. 4. ZW mothers produce ZW females but hermaphroditic mothers (ZZ) only ZZ offsprings (Fig. 6.1). 5. ZZ females are more likely to become hermaphrodites. 6. Higher density induces sex transition but isolated ones switches to hermaphroditism. Briefly, isolated males and females are prone to become hermaphrodites. Clearly, ZW and ZZ genes are harbored on respective sex chromosomes. The environment-dependent differential expression of ZZ/ZW gene(s) stimulate simultaneous function of male and female and thereby hermaphroditism. Or the expression of Z gene(s) is inhibited to produce females. Incidentally, Akesson (1982) has

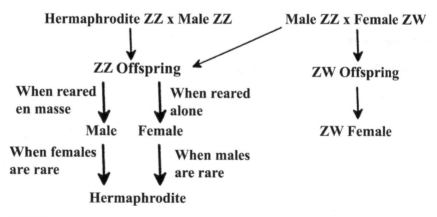

FIGURE 6.1

Types of sex change in *Capitella capitata* (modified and redrawn from Petraitis, 1991).

described the dominant yellow colored YY eggs and recessive (yy) eggs in *Ophryotrocha diadema* and effect of these genes on survivorship and spawning interval. This publication may not be of immediate relevance.

Unlike the widely distributed *G. caespitosa*, populations of *Ophryotrocha labronica* are found around the Italian coast. With differences in the biokinetic range for salinity tolerance, the Naple (Na) and Venice (Ve) populations are further divided into Na II, III and IV as well as Ve II and Ve IV. With gametic compatibility and successful hybridization between the representatives of these populations, male ratio among F_1 progenies ranges from 0.35 in a cross between ♂ Ve II and ♀ Ve II to 0.84 in another cross between ♂ Na III and ♀ Ve II (Table 6.3). When Na IV ♂ is crossed with Na IV, Na III and Ve II females, the male ratio is increased from 0.56 to 0.64 and 0.82. Hence, the female partner is responsible for the increased male ratio. It is also responsible for the increased ratio from 0.36 to 0.47 and 0.84 in combinations of ♂ Na III × ♀ Na III, ♂ Na III × ♀ Na IV and ♂ Na III × ♂ Ve II. Clearly, Ve II alters the ratio in favor of males. However, a cross between ♂ Ve II × ♂ Ve II has yielded the lowest male ratio, indicating the overriding impact of Ve II male on the increase of female ratio (Lanfranco and Rolando, 1981).

From their studies on heritable variation in sex ratio of *O. labronica*, Premoli et al. (1996) have proposed an updated version of an old hypothesis by Bull et al. (1982). Accordingly, sex is determined by a multilocus genetic system, facilitating the combined effects of a female major sex gene (as it is proposed for *Capitella capitata*) and masculanizing modifiers. Polygenic sex determination is a quantitative trait. It is rare in gonochoric animal species and evolutionary unstable (see Pandian, 2011). From their experiments, Premoli et al. (1996) have shown that *O. labronica* has a major sex determining gene with two alleles. Meiotic segregation imposes a sex ratio of 0.5 to the

TABLE 6.3

Effects of crossing Naple and Venice populations of *Ophryotrocha labronica* on male ratio (condensed and compiled from Lanfranco and Rolando, 1981)

Crosses Between Populations	Male Ratio
♂ Na IV × ♀ Na IV	0.56
♂ Na IV × ♀ Na III	0.64
♂ Na IV × ♀ Ve II	0.82
♂ Na III × ♀ Na III	0.36
♂ Na III × ♀ Na IV	0.47
♂ Na III × ♀ Ve II	0.84
♂ Ve II × ♀ Ve II	0.35
♂ Ve II × ♀ Na IV	0.51

oocytes of female. Variations from 0.2 to 0.8 in sex ratio are explained by the presence of polygenic system in the male parent, which may override the sex imposed by progamic maternal mechanism. Some evidences are brought for the heritable variations in sex ratio of *O. labronica*. Though attractive, the hypotheses of Premoli et al. (1996) remains to be tested in other annelid species, in which the factors responsible for variation in sex ratio are also inheritable.

7

Sex Differentiation

Introduction

In animals, sex determination and differentiation are successive but diverse processes that have evolved independently a number of times (Hodgkin, 1990). Hormones and neuroendocrines act as chemical messengers of genetic cascade that realizes sexualization and maintenance of sex as well as regulation of reproductive cycles in animals. In annelids, researches on these aspects have progressed from histological to surgical study to know the source and target of a specific endocrine as well as from injection (of a hormone) to *in vitro* and *in vivo* incubation studies to understand the action and biosynthesis of a hormone, respectively. Despite having a well-developed circulatory system, annelids do not possess a developed glandular endocrine system. No information is yet available on endocrines of hermaphroditic oligochaetes and hirudineans. Although most hirudineans are protandrics, they are not amenable to surgical study, as they do not have the ability to regenerate any tissue; most aquatic oligochaetes are too small (e.g. aeolosomatids measuring < 0.5 mm) for any surgical study. But the hermaphroditic earthworms are large in size (e.g. Australian giant earthworm *Megascolides australis* measuring 3 m) and have the ability to regenerate the 'head' including the prostomium harboring the 'brain' (see p 96). Our understanding of endocrine sexualization and regulation of reproductive cycles is based on findings from temperate polychaetes only; for their tropical counterparts, who have greater scope for aquaculture, no information is yet available. Using histology including immunohistochemistry (e.g. Weidhase et al., 2016), surgical (e.g. Pfannenstiel, 1978b) techniques, injection (e.g. Durchon, 1952) and incubation (e.g. Lawrence and Soame, 2009), subesophageal ganglia, prostomium and proventriculus have been identified as a source of endocrines in polychaetes. Expectedly, a fairly large number of publications is available; they have been reviewed from time to time; however, some of these reviews are restricted to nereidids alone (e.g. Andries, 2001). Unlike in crustaceans (Pandian, 2016) and echinoderms (Pandian, 2018), sex differentiation in polychaetes is a labile and protracted process.

7.1 Endocrine Regulation

Seasonal changes in temperature and photoperiod influence biological cycles like phytoplankton production, which, in turn, regulate reproductive cycle of temperate polychaetes (e.g. *Harmothoe imbricata*, Garwood and Olive, 1982). Conversely, the breeding season in cosmopolitan species like *Cirratulus cirratus* is not restricted; *C. cirratus* females spawn mostly two–three times/y and sometimes four–five times/y (Olive, 1970). This may also hold true for tropical polychaetes. Within temperate polychaetes, the breeding cycle is characterized by (i) semelparity, as in some errant nereidids, (ii) iteroparity (a) but with extended period of vitellogenesis, as in errant nephtyids, (b) with a short period of vitellogenesis, as in errant phyllodocids and (iii) sedentary polychaetes like *Arenicola marina*. Notably, tubiculous polychaetes have not yet received attention by endocrinologists. Table 7.1 summarizes available information on known endocrines and their action in polychaetes (see also Porchet et al., 1989).

7.1.1 Nereidids

Experimental grafting of the suboesophageal ganglion of immature *Nereis diversicolor* has led Golding (1967) to conclude that a single brain hormone is responsible for the promotion of somatic growth and inhibition of sexual maturation. Confirming his conclusion, Dhainaut (1970) has reported the accelerated growth of oocytes as small as 50 µm in *Nereis pelagica*. In *Perinereis dumerilii*, removal of the brain results in precocious sexual maturation but abnormality and degeneration of oocytes. However, these impairments are restored by replacing the brain (Hauenschild, 1966). Hence, the brain hormone also plays a gonadotrophic role by promoting progressive sexual development with gradual decrease in its titer. Briefly, the single brain hormone namely Juvenile Hormone (JS), the nereidin in nereidids performs a range of functions such as somatic growth and regeneration (see p 100), gonadotrophic support for oocyte growth and regulation of eleocyte activity as well as transition to sexual maturity and epitoky (see Andries, 2001, Lawrence and Soame, 2009). Incidentally, considerable asynchrony among smaller oocytes of neriedids is reported but the oocyte growth becomes synchronized with declining titer of JH and as the worm approaches epitoky (Fischer, 1974).

Through a series of experiments, in which oocytes of different sizes (47 or 50 µm) have been incubated with no ganglia or ganglia from juvenile or mature *P. dumerilii*, the following have been inferred: (i) growth is significantly faster in oocytes of 47 and 50 µm with no ganglia than that with ganglia from juvenile, (ii) it is significantly faster on incubation with fresh or boiled ganglia from mature *P. dumerilii*, (iii) it is also faster in

TABLE 7.1

Summary of known hormones, their source and action in some polychaetes

Family/Reference	Species	Source	Hormone	Effect
Nereidae, Golding (1967)	*Nereis, Platynereis*	Cerebral ganglia	Juvenile hormone (JH)	Gametogenesis inhibition
Durchon (1952)	*Perinereis cultrifera*	Coelomocytes, oocytes	Feedback substance	Inhibition of JH
Garcia-Alonso et al. (2006)	*N. virens*	Eleocytes	17-β Estradiol	Vitellogenin synthesis
Lawrence and Soame (2009)	*N. diversicolor*	–	Serotonin	+ ve effect on oocyte growth
Lawrence and Soame (2009)	*P. dumerilii*	–	Oxytocin	+ ve effect on oocyte growth
Lawrence and Soame (2009)	*N. succinea*	–	Dopamine	Switches off action of JH
Nephtyidae, Olive (1970)	*Nephtys hombergii*	Cerebral ganglia	Gonadotrophic hormone, Spawning hormone	Promotion of oogenesis Promotion of spawning
Phyllodocidae, Olive (1975)	*Eulalia viridis*	Prostomium	Vitellogenesis promotion hormone	Vitellogenesis promotion
Dorvilleidae, Pfannenstiel (1978b)	*Ophryotrocha* spp	Prostomium	Ootrophic hormone	Oogonial sex determination
Syllidae, Franke (1976, 1977)	*Syllis prolifera*	Proventriculus	Stolonization inhibition hormone	Inhibits stolonization
Arenicolidae, Howie (1963)	*Arenicola marina*	Prostomium	Prostomial maturation factor	Oocyte development
Watson and Bentley (1997)		Coelom	Coelomic maturation factor	Ripening and release of gametes
Bentley et al. (1990)		Coelom	8, 11, 14-eicosatrienoic acid	Spermatozoa are released from sperm cluster

P. dumerilii oocytes, when incubated with fresh ganglia from mature *N. virens* as well as (iv) oocytes (70 μm) of *N. succinea* with the ganglia from mature *P. dumerilii* female (at low concentration) (Olive and Lawrence, 1990, Lawrence and Olive, 1995). Arguably, these results have brought evidences that the JH is not species-specific within neriedids. The thermostable structure and action of JH are highly conserved among the neriedids (Lawrence and Soame, 2009).

Although the brain hormone from juvenile neriedids is reported to perform multiple functions, the incidence of a series of other hormones has also been reported. Estradiol-17β has been isolated from the coelomic fluid of *N. virens*, in which it promotes the secretion of vitellogenin by the eleocytes in mature females (Garcia-Alonso et al., 2006). The presence of an array of hormones has been reported such as serotonin (from immuno-positive staining) in *N. diversicolor* (Heuer and Loesel, 2007), oxytocin-like hormones in *Perinereis vancaurica* ganglia (Fewou and Dhainaut-Cortois, 1995, Matsushima et al., 2002) and vasopressin from *P. dumerilii* (expressed in developing forebrain, Tessmar-Raible et al., 2007). Melatonin is a key factor in the photoperiodic control of reproduction in vertebrates (e.g. fishes, see Pandian, 2013). As gonad maturation is controlled by photoperiod in many polychaetes, the possibility does exist for its presence in them also. Examining the effects of some of them, Lawrence and Soame (2009) have reported that dopamine and melatonin switch off the action of JH in *P. dumerilii* and *N. succinea* but serotonin and oxytocin have a positive effect on oocyte growth.

Feedback substance: In *Perinereis cultrifera*, the pioneering research by Durchon (1952) has initiated an idea of a feedback factor. Injection of oocytes of 90–100 μm from a submature worm into the coelomic cavity of juvenile eliminates the inhibition of the brain hormone. Increasing size of the injected oocytes accelerates sexual maturity (see also Pfannenstiel, 1978a). Confirming Durchon's findings, Porchet (1967) reported the down-regulation of juvenile hormone differs according to sex of the recipient. It results in precocious maturity and epitoky in males but abortive oocyte growth in female. Apparently, the feedback factor is sex specific. Hofmann (1974) suggested that the feedback substance emanates from coelomocytes and/or oocytes. However, its chemical identity is not yet known, though suspected to be a glycoprotein (Porchet and Cardon, 1976).

7.1.2 Nephtyids

Reproduction in the nephtyids are characterized by iteroparity, intra-ovarian pattern and discrete spawning. Many temperate nephtyids live up to 6–9 years and display an annual reproductive cycle (Olive and Bentley, 1980). In them, reproduction often fails due to (i) inability to initiate gametogenesis (e.g. *Nephtys caeca*, *N. cirrosa*), (ii) premature gametogenic degeneration and oosorption and (iii) non-release of viable gametes by gravid worm (= spawning failure) (e.g.

N. hombergii and *N. cirrosa*, Olive et al., 1985). The nephtyids sexually mature mostly at the age of 1+ year or latest by 2+ years. Unlike the nereidids, gametogenesis is well synchronized in nephtyids at all stages from meiotic prophase onward. Only fully grown oocytes are ovulated from the discrete ovary into the coelom. A long-term survey between 1978–1979 and 1983–1984 in the Tyne estuary, England has indicated successful spawning only during 1978–1979 and 1983–1984 (Olive et al., 1985). Our understanding of endocrine regulation of reproduction in nephtyids is based on the following findings: 1. Reproductive cycle is regulated by cyclic production of Gonadotrophic Hormone (GH) secreted and released from the subesophageal ganglia located in the prostomium and it supports the gametogenic development from October to March in *N. hombergii*. 2. Spawning Hormone (SH) is also released from the subesophageal ganglia during May. Injection of 10 µl of brain homogenate prepared from 40 ganglia/100 µl into gravid *N. hombergii* female induces spawning more successfully during May than April (Olive, 1970). Decerebration in the early gametogenic phase results in total failure of gametogenesis in *N. hombergii*. But the catastrophic effects of decerebration can be restored by implantation of whole alive subesophageal ganglia. This observation confirms that the cerebral subesophageal ganglia are the source of an Ootrophic Hormone (OH) (Olive and Bentley, 1980).

7.1.3 Phyllodocids

The non-epitokous phyllodocids like *Eulalia viridis* are characterized by release of oocytes from the dispersed ovary into the coelom (extra-ovarian pattern), where they accumulate from the previous autumn. Unlike the nereidids and nephytids, which require a prolonged duration for oocyte development, vitellogenesis is rapidly completed during April–May in *E. viridis* (Olive, 1975), when Vitellogenesis-Promoting-Hormone (VPH) is released from the prostomium. The response of an oocyte to VPH is dependent on its size. Smaller oocytes (> 40 µm) are refractory and do not commence vitellogenesis in a decerebrated worm. However, vitellogenesis is restored with an implantation of the ganglia into the coelom. The medium sized oocytes (> 40 but < 100 µm) degenerate, when VPH is deprived by decerebration. But the large oocytes (> 100 µm) survive in the absence of VPH, suggesting that a normal feature of oocyte maturation is by the withdrawal of VPH during the final stages of oocyte maturation (Oliver, 1976). The role of VPH has been elucidated by Lawrence and Olive (1995) using bioassays and incorporation of ^{3}H-leucine. Unlike in the nephytids, VPH promotes the uptake of external protein into the developing oocytes.

7.1.4 Arenicolids

Oocyte development in *Arenicola marina* requires not only Prostominal Maturation Hormone (PMH) but also coelomic fluid (Watson and Bentley,

1997). Hence, the oocyte development is controlled by a two-steps process namely PMH and Coelomic Maturation Factor (CMF). The CMF is thermolabile, trypsin-sensitive, has a molecular mass of > 30 kDa and suggests a proteinaceous nature. Clusters of spermatogonia are released from the testis into the coelom. In the presence of Sperm Maturation Factor (SMF) 8, 11, 14-eicosatrienoic acid, the breakdown of sperm clusters occurs (Bentley et al., 1990). Ripening of gametes and their release are inhibited by decerebration and are restored by administration of prostominal homogenate, *albeit* only during the breeding season (Howie, 1963).

7.2 Differentiation and Lability

Our understanding of endocrine sex differentiation is based on the findings from the dorvilleid protandric *Ophryotrocha* spp and gonochoric syllids.

7.2.1 Dorvilleids

Pfannenstiel (1971, 1973, 1974, 1975, 1976, 1977a, b, c, 1978a, b) made a series of homospecific and heterospecific prostominal transplantations. In 1980s, when Pfannenstiel made excellent publications (mostly in German language), when our understanding of Primordial Germ Cells (PGCs) in invertebrates has been at its nascent stage. Hence, his findings are briefly summarized but with insertions (in italics) of observations in the present day context. 1. At 6–8 and 16-segment stages, *O. puerilis puerilis* matures as male and switches to female phase, respectively. 2. (a) Decerebrated juveniles attain male phase but not female phase, even after passing the critical 16-segment stage. (b) On transplantation of prostominal graft from a female donor, the decerebrated recipient differentiates into a female but that from a male donor induces no change. Clearly, a factor originating from the prostomium, i.e. Ootrophic Hormone (OH) (cf Olive and Bentley, 1980) is obligately required for differentiation into a female. In males, factors like dopamine and/or melatonin may switch off the action of juvenile hormone (JH/OH). (c) In females, a portion of the germ cells, i.e. *PGCs retain their bisexual potency* (it also occurs in fishes, see Pandian, 2013), *while the other portion of PGCs is differentiated into oogonia under the influence of OH*. 3. Due to Paarkultureffekt (pair rearing effect), one individual in a pair becomes a female within 3–5 days under the influence of OH but the other is pheromonally (?) induced to differentiate into a male around the 8th day. Clearly, the *sex differentiation process in the protandric O. puerilis puerilis is labile with retention of a portion of PGCs with bisexual potency.* 4. In *O. labronica*, the process of sex differentiation is more labile than in *O. puerilis puerilis*. In this protandric, juveniles differentiate into 73% of (phenotypic) primary males and 27% of (genetic) stable females (Fig. 7.1A). Following decerebration, 48% (i.e. 67% of the total) of the phenotypic males remain as (genetic) stable males but the remaining 25% phenotypic males differentiate into secondary (phenotypic) females (Fig. 7.1B). Clearly, a half (24%) of primary males (not amenable to OH) are genetic males and the other half (24%) are amenable to OH and differentiate into phenotypic females (Fig. 7.1C). 5. Even with amputation of prostomium, the stable genetic females and genetic males differentiate into females and males, respectively (Fig. 7.1D). Therefore, *other factors like serotonin and dopamine*

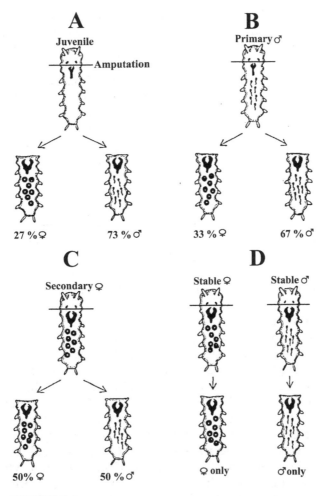

FIGURE 7.1

Decerebration and resulting sex ratio in A. juvenile, B. primary male and C. secondary female as well as D. stable female and male on sex ratio of *Ophryotrocha labronica* (modified freehand drawing from Pfannenstiel, 1978a).

may also be involved in sex differentiation. 6. All heterospecific transplantations involving the prostominal grafts from females or males of *O. labronica* into the decerebrated recipient *O. puerilis puerilis* females have induced differentiation females. Hence, the OH produced by the prostominal graft of *O. labronica* is compatible and completely compensates the OH required by the decerebrated females and males of *O. puerilis puerilis*. *Like JH within nereidids, OH is also not a species specific hormone within these dorvilleids.*

7.2.2 Syllids

Stolonization and stolon morphology: In syllids, stolonization is the common mode of reproduction. The proventriculus (PV) produces a hormone that at high titer attenuates stolonization but promotes regeneration of the posterior end, from which a stolon is detached. Weidhase et al. (2016) have described the stolonization cycle and stolon morphology in a representative syllid *Typosyllis antoni*. Accordingly, female stolons are full of oocytes and males harbor two packages of sperms per segment. The syllid undergoes many successive stolon cycles. The stolons are not generated by segment addition but arise from a transformation of posterior segments. Following detachment of a stolon, the posterior end is regenerated prior to development of a new one. Hence, the syllids have retained the regenerative potency of the posterior even after the commencement of reproduction. Briefly, regeneration is alternated with stolonization. The number of transformed segments in stolons varies between individuals and may also vary between successive stolonization events in an individual from 5 to 18 segments in *T. antoni*. The anterior most stolon segment bearing eyes and antennae undergoes extensive morphological changes to become the stolon's 'head'. Notably, having originated from the posterior tip of the ventral nerve cord, the new structure represents the stolon's 'brain' (Weidhase et al., 2016).

Our understanding on sexualization and sex reversal in stolons is based on the researches of Durchon (1975), Hauenschild (1975) and Franke (1976, 1977). The PV is a muscular structure with radially arranged striated muscle cells surrounding the gut (see Weidhase et al., 2016). These cells consist of usually only one or two sacromeres of up to 100 μm length, being the longest known sacromeres within Metazoa (Smith et al., 1973). A histological study of the PV has revealed no sign of glandular or secretory structure. The ventrical and caeca are composed of glandular tissues but they are not involved in the reproductive and regenerative processes (Weidhase et al., 2016). The PV is shown to release Stolonization Inhibition Hormone (SIH), which inhibits stolonization during the breeding season. 1. Sensing the absence of incoming food during starvation, the PV inhibits stolonization in *Syllis prolifera*, which has just released a stolon. 2. Implantation of PV from a juvenile donor into a recipient without its own PV also inhibits stolonization. 3. However, that from a sexualized donor does not inhibit stolonization in the recipient.

Apparently, *SIH promotes somatic growth but with its declining titer, sexualization of the stolon is commenced. Hence, SIH of the syllids functions like JH of nereidids.*

Sex reversal: In syllids, male differentiation is a stable process. Natural sex change from male to female is not yet reported. In fact, spontaneous sex change is rare in syllids. However, extirpation of PV or removal of SIH not only induces precocious stolonization but also leads to sex reversal (Fig. 7.2A). To test sex differentiation in stolons, homosexual transplantation of PV from an undifferentiated male stock of *S. prolifera* donor was grafted into a recipient male stock. More than 70% of juvenile recipient stocks became sterile and the remaining 30% of them produced male stolons only. However, receiving PV graft from a sexualized donor, the recipient continued to release male stolons only. Heterosexual implantation of PV from juvenile female stock was grafted into male stock that has already released male stolon. Of 27 implantations, 52% of the recipient stocks became sterile. *Obviously, heterosexual grafts are incompatible at least in half of the stocks.* The remaining stocks produced both female and male stolons at the ratio of 0.37 ♀ : 0.11 ♂ stolons. Hence, sex ratio of the stock was noted as 0.52 sterile: 0.48 fertile (Fig. 7.2B). On transplantation from sexualized donors, the ratio is 0.21 sterile stocks: 0.79 ♂ stolons. It seems that *the stocks of S. prolifera retain PGCs with bisexual potency and display a tendency toward the production of male stolon.*

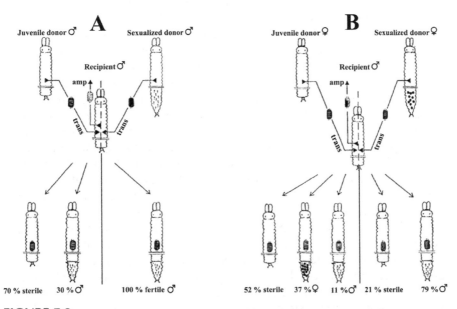

FIGURE 7.2

Stolonization and sexualization in *Syllis prolifera*. Effect of transplantations of proventriculus A. from juvenile and sexualized donor males on recipient's sex ratio and B. from juvenile and sexualized female donors to male recipient's sex ratio (modified freehand drawings from Franke, 1976, 1977).

To study the role played by cerebral and PV hormones in *Typosyllis pulchra*, Heacox and Schroeder (1982) removed either simultaneously prostomium and PV or one of them. In *T. pulchra*, the stolonization cycle lasts for 30 days and the stolon remains sexually undifferentiated until the 3rd day of the cycle. If prostomium and PV are removed on the 5th day, the stolon continues to undergo oogenesis. This observation is also confirmed by the occurrence of oogenesis in a recipient stock implanted with PV from an undifferentiated donor that has not yet stolonized. However, the removal of prostomium and PV prior to the 5th day induces sex reversal and the former female stock begins to produce stolon, in which spermiogenesis occurs. Further, the production of PV hormone is inhibited, when prostomium is removed during the initial days of stolon cycle. Hence, prostomium has an upper hand in regulation of the PV inhibitory hormone.

Weidhase et al. (2016) have further elaborated to know the exact source in and around the PV by amputating at the (i) pharynx, (ii) pharyngeal tube and (iii) middle of the PV in *T. antoni*. Their findings are listed below: (i) Incomplete anterior regeneration on removal of anterior end of the PV, (ii) Deviation from the usual reproductive pattern, (iii) Accelerated and masculanized stolonization and (iv) Limitation of regeneration of posterior segment.

Besides endocrines, density is an important factor in regulation of male ratio of stolons. *Typosyllis prolifera* flourishes on the thalli of *Halopteris scoporia*. Of 7,846 stolons surveyed by Franke (1986b), 25 and 27 stolons have been developing gametes of either sex in Porec and Pula populations of Yugoslovia, respectively. Irrespective of increasing density from 5 to 20 stolons/10 g thalli, male ratio of stolons remains around 0.5 in Rovinj population but decreases from 1.0 to 0.5 in Pula population. However, when stocks of these three populations are reared in singles, the ratio remains around 0.5. Hence, density plays an important role in sexually labile *T. prolifera*. But the level of lability is a population trait. When reared individually also, spontaneous sex change can occur in 2.5 and ~ 5.0% stolons in Rovinj and Pula stocks, respectively. At the 4th stolonization event, male ratio is 0.5 and 1.0 in Rovinj and Pula populations. Hence, Rovinj population is the least labile but that of Pula is the most labile. The spontaneous sex change begins to slowly but steadily increase in successive stolonization events. Consequently, the ratio increases from 20% in the 2nd stolonization event to 100% at the 5th event. Hence, masculanization is rapid and completes at the 6th event.

In another two series of experiments, Franke (1986b) investigated the effects of density and combination of partners on the stolon's male ratio of *T. prolifera*. The 1st series tested the effect of volume of water and presence of males, when female stolons were reared at 1 ♀/5 ml, 50 ♀♀/150 ml and 50 ♀♀ + 50 ♂♂/150 ml. The results indicated that (i) > 89% of female stolons reared in isolation in limited volume of water sex reversed into males; (ii) the sex reversal was reduced to 20%, when groups of female stolons were reared in larger volume of water; (iii) the presence of primary male stolons further reduced the reversal to 17%. Clearly, the co-presence of female and male stolons reduces spontaneous sex reversal. To study the effects of density and female-male combinations on male ratio during the 2nd, 3rd and 4th stolonization events in the Pula stock, Franke (1986b) has chosen the following three groups of combinations and three subgroups in each of them namely Group A: 2 ♀♀ : 0 ♂, 0 ♀ : 2 ♂♂, 1 ♀ : 1 ♂, Group B: 10 ♀♀ : 0 ♂, 0 ♀ : 10 ♂♂, 5 ♀♀ : 5 ♂♂ and Group C: 20 ♀♀ : 0 ♂, 0 ♀ : 20 ♂♂ and 10 ♀♀ : 10 ♂♂. The recalculated male ratios are plotted against the chosen female-male combinations at increasing density in Group A, B and C (Fig. 7.3A) and successive stolonization events (Fig. 7.3B). Following an initial increase of the ratio in the combinations of 0 ♀ : 2 ♂♂, the ratio progressively decreases, irrespective of the differences in combinations like 0 ♀ : 10 ♂♂ and 0 ♀ : 20 ♂♂. On the other hand, the ratio progressively increases with successive stolonization events but at different levels. The decrease in male ratio is in the following order for the chosen female : male combinations: 0 ♀ : 2 ♂♂ < 0 ♀ : 10 ♂♂ < 0 ♀ : 20 ♂♂. The trends for others in each of the groups fall below the respective ones. From these observations, the following may be inferred: 1. Male ratio increases with successive stolonization events 2. The ratio decreases with increasing density and 3. However, the presence of males alone at any chosen density increases male ratio to a higher level than that in which females alone is present. Hence, the presence of females reduces the male ratio and the reverse is true for the males.

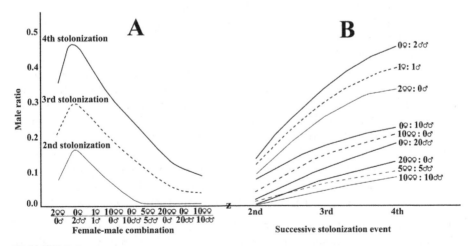

FIGURE 7.3

Male ratio in stolons of *Typosyllis prolifera*. A. Effect of different female-male combinations on male ratio during the 2nd, 3rd and 4th stolonization events. B. Effect of successive stolonization events on male ratio in different female-male combinations (drawn using data reported by Franke [1986b] in his Table 5).

7.3 Pollutants and Reproduction

Annelids play an important role in decomposition and bioturbation. In this process, they accumulate the pollutants; in its tissues, *Tubifex tubifex* may concentrate cadmium by ~ 35 times (Gillis et al., 2002). Not surprisingly, aquatic oligochaetes are reported to have been used in bioassays since the times of Aristotle (Lobo and Espindola, 2014). A fairly large volume of literature is available on this subject. However, this account elucidates selected mosaic representative examples on gametogenesis and reproductive output, regenerative potency and brooding in acidic water as well as inherent recovery ability from cadmium and pCO_2 pollutants.

For annelids, information related to the negative effects of pollutants is available on survival (e.g. *Enchytraeus crypticus*, Castro-Ferreira et al., 2012), body size (e.g. *T. tubifex*, Gillis et al., 2002) and reproductive output (e.g. *Neanthes arenaceodentata*, Oshida et al., 1981). However, information is not yet available on metabolic pathways through which endocrine disruption occurs. Many insecticides and metallic pollutants are reported to induce (i) spermatogonial damages in *Eisenia foetida*, (ii) incomplete sperm maturation in *Branchiura sowerbyi* and (iii) reduce sperm count in *Pheretima guillelmii* (Table 7.2). In a review, Yasmin and D'Souza (2010) listed the negative effects on growth and reproduction in *Eisenia foetida* exposed to copper oxychloride, malathion, achetochlor, chloropyrifos, cypernethrin or benomyl.

For aquatic annelids, more information is available on cadmium (cd) pollution. Exposure to sedimental cd reduces body weight of *T. tubifex* from 2.7 mg to 0.9 mg and reproductive output from 15 offspring/adult to 5/adult (Table 7.2). Undertaking a long term (440 days) chronic exposure to trivalent cadmium covering three generations of *Neanthes arenaceodentata* (Table 7.3), Oshida et al. (1981) have reported the negative effects in the first generation and the possible recovery during the third generation. With increasing dose, the following negative effects are recorded in the parental generation: (i) decrease in the number of spawning pairs (6 to 5), (ii) extended inter-spawning interval (112 to 123 days), (iii) reduced brood size (255 to 78) and (iv) total offspring production (1,628 to 391). However, the worm recovers during the second regeneration by (i) an increase in spawning pairs (5 to 10), (ii) reduction in inter-spawning interval (123 to 118 days), (iii) increases in brood size (78 to 111) and (iv) total offspring production (391 to 1,112).

Ocean acidification: Oceans cover 70% of the earth's surface and hold 97% of its water and serve to buffer CO_2. Thankfully, the daily uptake of atmospheric CO_2 by the oceans is 22 million metric tons. Since the advent of the industrial era, oceans have absorbed 127 billion metric tons of carbon as CO_2 from atmosphere. Carbon dioxide combines with water chemically. Hydrolysis of CO_2 increases the hydrogen ion (H^+) concentration with

TABLE 7.2

Effect of pollutants on spermatogenesis and fecundity of annelids

Species/References	Reported Observation
Branchiura sowerbyi Casellato et al. (2011)	Treated at (50–100 times above the normal) 0.1–0.3 mg fluoride/1 for 5 months inhibits the completion of sperm maturation
Eudrilus eugeniae Yesudhason et al. (2012)	X-rays exposure fragments acrosome in the head and breaks the zig-zag tail
Lumbricus terrestris Cikutova et al. (1993)	Exposure to cadmium reduces sperm count
Eisenia foetida Corpas et al. (1995)	Exposure to lead damages spermatogonia and spermatocytes
Pheretima guillelmii Reinecke and Reinecke (1997)	Chronic exposure to heavy metals reduces sperm production
Hirudinaria manillensis Singhal and Davies (1996)	Exposure to organo-phosphorous insecticide Temphos at > 0.1 mg/1 for 22 hours reduces fecundity
Tubifex tubifex Gillis et al. (2002)	Exposure to cadmium reduces body weight from 2.7 mg at sediment concentration of 3.8 μmol/g x 10^{-3} to 0.9 mg at 5 μmol/g x 10^{-3} and reproductive output from 15 offspring/adult in control to 5/adult at 5 μmol/g sediment

concomitant reductions in pH and carbonate ion (CO_3^{2-}) concentration. The progressive reduction in availability of carbonate (CO_3^{2-}) renders the acquisition of biogenic calcium carbonate ($CaCO_3$) by calcareous tubiculous worm more difficult and energetically costlier (cf Pandian, 2017). As a result, marine annelids may encounter a series of significant changes in oceanic pCO_2 and pH levels in the future. Increases in oceanic pCO_2 are predicted to occur at an unprecedented rate, leading to a significant decrease in pH, a phenomenon commonly known as 'ocean acidification'. However, evolutionary adaptation occurs, when selection on existing genetic variation shifts the average phenotype of a population toward the fitness peak that matches with changing environmental conditions (Rodriquez-Romero et al., 2015) like elevation in pCO_2 and declining pH. With a short life span, many polychaetes have served as an ideal model for multi-generational studies indicating initial negative effects and subsequent recovery.

In a rare study, Pires et al. (2015) have investigated the effects of increase in temperature and decrease in pH on regenerative potency of *Diopatra neapolitana* (Table 7.4). Increase in temperature below 24°C accelerates the regenerative process. But decreases in pH from alkaline to neutral level cause a drastic increase in regeneration duration from 52 to 63 days and decreases segment regenerative potency from 81 at pH 7.8 to 68 at pH 7.1.

Rodriquez-Romero et al. (2015) have carried out two series of experiments in the errant *Ophryotrocha labronica*. In the first one, reproductive traits have

TABLE 7.3

Effect of different doses of trivalent cadmium chronic exposure on breeding characteristics in parental generation and recovery in second generation of *Neanthes arenaceodentata* (condensed and simplified from Oshida et al., 1981)

Breeding Characteristics	Control	Cadmium Dose (mg/l)	
		0.0125	0.05
Parental generation			
Spawning pairs (no.)	6	9	5
Inter-spawning interval (d)	112	100	123
Brood size (no.)	255	133	78
Total offspring (no.)	1628	1199	391
First generation			
Spawning pairs (no.)	9	10	7
Inter-spawning interval (days)	153	130	111
Brood size (no.)	292	258	50
Total offspring (no.)	2628	2580	415
Second generation			
Spawning pairs (no.)	8	10	10
Inter-spawning interval (days)	129	132	118
Brood size (no.)	273	190	111
Total offspring (no.)	2186	1395	1112

TABLE 7.4

Effects of temperature on the duration and level of regeneration in *Diopatra neapolitana* (condensed and simplified from Pires et al., 2015)

Temperature (°C)	Regeneration Completed		pH	Regeneration Completed	
	(d)	Chaetiger (no.)		(d)	Chaetiger (no.)
17	61	75	7.8	52	81
20	48	82	7.5	54	77
24	43	89	7.3	57	72
			7.1	63	68

been estimated in the worms exposed to elevated ($pCO_2 = 1,000$ µatom) level for seven generations. The second one has involved the transplantation from elevated level to low ($pCO_2 = 400$ µatom) level and vice versa. The biological assay has shown that all the tested traits are recovered and even improved, except the egg size, which has remained constant at 0.6×10^{-3} mm³. The increase for growth is from 0.9 to 1.4 chaetigers per day and for fecundity from 5.13 to 8.55 eggs per chaetiger in the first and seventh generation, respectively (Table 7.5), i.e. the increases represent improvement of 55% for growth and 67% for fecundity. With transplantations, growth rate (1.40 chaetigers per d), adult size (~ 15 chaetigers) and egg size (0.6 × 10⁻³ mm³) have remained constant. However, there are decreases in survival (from 82% at elevated-elevated transfer to 79% in low to elevated transfer) and fecundity (from 22.3 to 18.7 eggs per chaetiger) (Table 7.5). Studies on calcareous tube-building sedentary polychaetes are desirable, as they may find it difficult and costlier to acquire the required calcium carbonate with increasing pCO_2 and decreasing pH.

In the field of ocean acidification research, our knowledge is based on experimental studies involving a short or long term exposure to different levels of predicted future pH. In a model investigation, Lucey et al. (2015) have circumvented the limitations related to experimental studies by relating the dominance and distribution of the known polychaete species inhabiting a range of natural acidic waters from extremely low pH to low pH and ambient pH. They have found that on exposure to extreme low pH and low pH, the proportion of brooders within a species progressively increases, especially in interstitial species (Table 7.6). However, a small portion (6–17%) spawns in 'chemical islands' characterized by low pH. Broadcasters (e.g. *Platynereis dumerilii*) and others with short non-feeding pelagic phase (e.g. *Pileolaris* spp) choose to spawn in open waters with normal ambient pH.

TABLE 7.5

Multi-generational responses of life history traits of *Ophryotrocha labronica* on exposure to pCO_2. Also the effects of transplantation from elevated (pCO_2 = 1,000 µatom) to low (pCO_2 = 400 µatom) level and vice versa (compiled and simplified from Rodriquez-Romero et al., 2015)

Multi-generational responses							
Trait	F_1	F_2	F_3	F_4	F_5	F_6	F_7
Survival (%)	80	83	80	87	90	89	83
Growth rate (chaetigers/d)	0.9	1.3	1.3	1.4	1.5	1.3	1.4
Adult size (no. of chaetigers)	14	15	15	16	16	–	15
Fecundity (no./chaetiger)	5	7	10	10	9	–	9
Egg volume (× 10⁻³ mm³)	0.6	0.6	0.6	0.6	0.6	–	0.6

Transplant assay			
Trait	Elevated to Elevated	Elevated to Low	Low to Low	Low to Elevated
Survival (%)	82	78	85	79
Growth rate (chaetigers/d)	1.4	1.4	1.4	1.4
Adult size (no. of chaetigers)	15.4	15.1	15.4	15.0
Fecundity (no./chaetiger)	22.3	19.5	21.4	18.7
Egg volume (x 10⁻³ mm³)	0.6	0.6	0.6	0.6

TABLE 7.6

Adaptive strategies in response to ocean acidification and lowering pH in vent-inhabiting polychaetes (simplified from Lucey et al., 2015)

Species/Family	Adaptive Strategy	Abundance (%) at		
		Extreme Low pH	Low pH	Ambient pH
Positive increase in pelagic				
Platynereis dumerilii Nereididae	Broadcaster, External fertilization, PLK larvae	6	–	94
Pileolaris spp Serpulidae	Brooder, LEC non-feeding larva at pelagic	19	38	43
Spio decoratus Spionidae	Brooder, pelagic/benthic juvenile development	17	17	67
Positive increase in brooding				
Platynereis massiliensis Neredidae	Brooder, Mucus tube brooding Direct development	91	–	9
Exogone naidina Syllidae	Brooder, Direct development, Interstitial	27	35	38
E. meridionalis Syllidae	Brooder, Direct development, Interstitial	44	39	17
Syllis prolifera Syllidae	Stolonization, Swarming, Benthic fertilized eggs	48	21	21
Rubifabriciola tonerella Fabricidae	Intra-tubular brooding, Direct development	67	21	31
Parafabricia mazzellae Fabricidae	Intra-tubular brooding, Direct development	85	6	9
Brifacia araiargonensis Fabricidae	Intra-tubular brooding, Direct development	74	19	7

7.4 Vectors and Borers

In crustaceans (see Pandian, 2016) and molluscs (see Pandian, 2017), parasites are known to disrupt the endocrine sex differentiation process. However, no parasite is reported to disrupt the endocrine sex differentiation in annelids. But many annelids, especially tubificids serve as vectors to transmit diseases. For example, *Tubifex tubifex* serve as an obligate vector of the sporidean *Myxobolus cerebralis*, the causative agent of salmonid whirling disease (Stevens et al., 2001).

The cosmopolitan spionids *Polydora* and *Boccardia* compose a large number of burrowing species of 'mudworms'. For example, *P. websteri* damages oysters by burrowing into the shell through boring aided by a viscous fluid, which dissolves the shell. Burrowing on the shell in and around the adductor muscle area by *P. ciliata* renders the shell to open readily and thereby increases predation of mussels like *Mytilus edulis* (see Pandian, 2017).

8

Vermiculture

Introduction

One objective of this book is to elevate annelids from their academic interest to economic importance. For a long time, the usefulness of some annelids was known but is not adequately recognized. Due to page limitation, this chapter may not elaborate nascent technical details related to (i) rack culture system for *Tubifex tubifex* (Marian et al., 1989), (ii) culture of sedentary polychaetes (e.g. *Sabella spallanzanii*, Giangrande et al., 2014b), (iii) regeneration as a novel method for ornamental sabellids (Murray et al., 2013), (iv) compacted culture system (Garcia-Alonso et al., 2013) and (v) ideal method for changing of water from simulated beach set-up (Serebiah, 2015). Rather, it attempts to assemble widely scattered basic information to (i) identify the fast growing candidate species, (ii) assess biomass production in the contexts of (a) nutrients; (b) temperature and (c) rearing density; (iii) distinguish 'layers' from 'brooders'; (iv) estimate offspring production in the candidate species characterized by (a) sexual, (b) parthenogenic and (c) clonal reproduction; and (v) explore the scope for increased biomass production in elevated ploids. Information on this fundamental biological inputs may direct future research on more profitable vermiculture utilizing waste as a resource, and discourage odd researches, in which costlier shrimp meat (e.g. Nielsen et al., 1995) and trout pellet (e.g. Memis et al., 2004) are used as feed in vermiculture. Briefly, 'wealth from waste' must be the theme of vermiculture.

8.1 Characteristics and Features

Earthworms are known to (i) improve aeration, (ii) drainage, (iii) availability of nutrients to plants and (iv) integrate soil organic and mineral elements (Butt, 1993). For example, more than 80% of litter-fall is decomposed and its nutrients are made available to *Salix gigantea* plantation (Curry and Bolger,

1984). The vermicomposts increase the yield of commercial crops (e.g. *Lablab purpureus*, Karmegam and Daniel, 2009, *Brassica compensis*, Guerrero III, 2009). In recent years, the dramatic increase in demand for polychaetes as baits by anglers is so high that even developed countries like UK and South Korea have begun to aquaculture them in a big way (Olive, 1999, E Costa et al., 2006). The role played by the enchytraeids in composting municipal waste and production of live feed for salmon and sturgeon is known, especially from subantarctic countries like Russia (see Fairchild et al., 2017). In waste water treatment plants, polychaetes (e.g. *Marphysa sanguinea*, Parandavar et al., 2015) and aquatic oligochaetes (see Ratsak and Verkuijlen, 2006) serve as biological agents to minimize activated sludge. In percolating filter sewage system of UK, *Lumbricillus rivalis* (incidence frequency, 91%), *Enchytraeus buchholzi* (57%), *E. coronatus* (22%), *E. albidus* (9%) and *Fridericia* spp (4%) are present. Their density fluctuates but is mostly in the range of 200/l for *L. rivalis* and > 300/l for *E. coronatus* (Learner, 1972).

Paddy fields: In a very useful investigation, the International Institute of Rice Research has made a survey of the presence and dominance of aquatic oligochaetes in the Philippine ricefields (Simpson et al., 1993). The survey has recorded the following: (i) Nine species of aquatic oligochaetes are present: (a) tubificids: *Limnodrilus hoffmeisteri*, *Branchiura sowerbyi*, *Aulodrilus limnobius*, *A.* sp, (b) naidids: *Dero digitata*, *Aulophorus hymanae*, *Pristina* sp, (c) enchytraeids: *Mesenchytraeus* sp and (d) lumbriculid: *Lumbriculidae* spp, (ii) of them, *L. hoffmeisteri* (81%) and *B. sowerbyi* (13%) dominate, (iii) population increases with increasing soil carbon content up to 3.5% and application of nitrogen fertilizer up to 140 kg/ha, (iv) density frequency decreases from 45% at 5,000/m^2 to 1% at 35,000/m^2 and (v) release of soil organic substances possibly dependent on bacterial decomposition. The occurrence of *Aulophorus vagus*, *A. furcatus* and *Dero digitata* in Indian ricefields has been reported (Hegde and Sreepada, 2014). In sediments, where *L. hoffmeisteri* flourishes, the processes of bacterial de-nitrification and nitrification occur simultaneously. Aquatic oligochaetes accelerate the release of nitrate (NH_4^+N) and phosphate (PO_4^{3-}) from overlying water and soil. However, population density is increased to a greater extend by organic fertilizer than inorganic ones (see Yokota and Kaneko, 2002). The de-nitrification rates are 90 and 50 mg N/m^2/d in the presence and absence of *L. hoffmeisteri*, respectively; the corresponding values for nitrification are 69 and 29 mg N/m^2/d (Chatarpaul et al., 1980). Hence, the loss due to de-nitrification is almost completely compensated by nitrification. However, there are other reports indicating the stimulation of sediment nitrification at low density of *Tubifex tubifex* but inhibition of it at high density (Pelegri and Blackburn, 1995). Research inputs in this field are urgently required to know whether the harvest of *L. hoffmeisteri* and *T. tubifex* at appropriate intervals shall reduce the application of nitrogen fertilizer and provide live feed as a byproduct.

Size and Density: A notable feature is that with decreasing body size of annelids, cultivable density can be increased (Fig. 8.1). Individual size decreases from 7 g in *Marphysa sanguinea* (Parandavar et al., 2015) and *Aporrectodea longa* (Butt, 1993) to a few milligrams in the tubificids *Branchiura sowerbyi* (100–150 mg, Aston, 1968), *Tubifex tubifex* (7–17 mg, Marian and Pandian, 1984, Finogenova and Lobasheva, 1987) and *L. hoffmeisteri* (9.4 mg, Nasciomento and Alves, 2009) and to a few micrograms (recalculated highest live weight of aquatic enchytraeids: 900, 300, 240 and 180 µg for *Lumbricillus rivalis, Cognettia cognettia, Marionina southerni* and *Achaeta eiseni*, respectively, see Lindegaard et al., 1994). As aeolosomatids weigh < 100 µg, their growth is measured in number alone (e.g. *Aeolosoma viride*, Falconi et al., 2015). However, cultivable density increases in the following order: 50/m² in *Perionyx ceylanensis* (Karmegam and Daniel, 2009) > 2,000/m² in *M. sanguinea* > 20,000/m² in *T. tubifex* > 35,000/m² in smaller aquatic oligochaetes. Most remarkably, cultivable density of earthworms is far less than that of most aquatic worms. Differences in food availability and motility to acquire food from the substratum/culture medium may be responsible for the observed low density requirement of earthworms. The highest organic substance, in

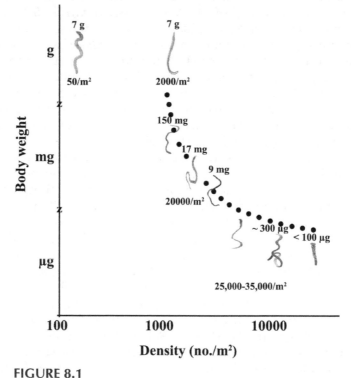

FIGURE 8.1

Effect of body weight on density of some oligochaetes (data assembled from different sources, see text).

which *Perinereis* spp (Palmer, 2010) and *Lumbricus terrestris* (Butt et al., 1995) have been reared, is in the range of 4.5% for the former and 3.9% for the latter. Hence, food availability may not be the reason for the low density requirement of earthworms. The reason seems to the denser substratum, through which earthworms have to move to acquire food. Both earthworms and nereidids move by alternate muscular contraction and relaxation. However, the resistance encountered by earthworms against movement in the denser soil substratum is enormous, in comparison to the 'lose' sediment submerged in aquatic medium. Consequently, the reported values for motility is 6.7 cm/d for earthworm (e.g. estimated distance of 200 cm/10 months in *Aporrectodea longa*, Butt et al., 1995) and 7,344 m/d for nereidid (e.g. estimated potential distance of 85 mm/s, see Pandian, 2016). Of course, the nereidid cannot sustain motility for all the 24 hours in a day and the earthworm motility may be altered by organic content and soil texture. Despite reduced motility, earthworms seem to require a larger volume of substratum to acquire food. In low nutrient sediments, polychaetes may switch to acquire dissolved organic substances from the surrounding medium (p 21–22) but earthworms can only increase the gut loading frequency up to 6 times (p 23). As a result, they require a larger volume of substratum. The calculated highest cultivable density values for earthworm range between 50/m² for *P. ceylanensis* weighing 0.8 g and 159/m² for *Aporrectodea caliginosa* weighing 0.8 g. It is possible to increase density with worms smaller than 0.2 g (Johnston et al., 2014).

Taxa and Reproduction: Reproductive modes impose a profound impact on individual growth and biomass production and are important in vermiculture. Of 13,000 and odd polychaete species, only 207 are hermaphrodites and three are parthenogens (Table 8.1). Among 79 clonal species, the majority of them are capable of bidirectional cloning, i.e. a single parent (genet) produces two offsprings (ramets). In the errant stolonizing syllids, a parent is capable of producing up to 18 stolons (ramets). In some tubiculous sabellids, a single genet is shown to produce three to eight ramets (e.g. *Potamilla torelli*, *Sabella pavonina*, *Sabellastarte* spp, see Fig. 3.7, Table 4.9). However, the syllids and

TABLE 8.1

Taxa (no.) and reproductive modes of annelids (compiled from Tables 1.2, 2.4, 2.5, 4.3, 6.1)

Taxa	Species	Parthenogens	Ploids	Parthenogens + Ploids	Cloners
Polychaetes	13,002	3	3	–	79
Oligochaetes	3,175				
Earthworms	432	5	5	6	0
Enchytraeids	670	6 + 3?	?	?	9
Tubificids	1,113	?	4	?	0
Naidids	175	?	?	?	88
Aeolosomatids	27	?	?	?	12

sabellids have not yet received much attention by aquaculturists. Notably, much efforts are being made to cultivate nereidids, which may or may not be cloners.

Among oligochaetes, earthworms are unidirectional cloners, i.e. on fission, only a single ramet is developed and the other dies. However, of 432 earthworm species, 56 of them are parthenogens and five of these parthenogens are polyploids (Table 2.7). These earthworms do not grow as fast as gonochorics. Of 70 enchytraeid species, 6 + 3(?) are parthenogens and 9 cloners. They are multidirectional cloners, i.e. each genet produces as many as eight ramets (e.g. *Enchytraeus japonensis*) and is smaller in size and weight, in comparison to sexually reproducing enchytraeids. In fact, the fast growing cloning naidids and aeolosomatids may prove to be great biomass producers only from wastewaters.

Collection and Harvest: Not many authors have reported methods for collection and harvest of annelids. With decreasing size, they become more and more important. Earthworms are collected by digging and hand picking (e.g. *Aporrectodea trapezoides*, Fernandez et al., 2010). Terrestrial enchytraeids can be extracted from soil cores up to 10 cm^2 area and 6 cm depth by wet funnel method (for details see O'Connor, 1957). Aquatic oligochaetes are collected by sieving through a hand net at depth of 0.5 cm (Ratsak and Verkuijlen, 2006). To collect *Dero digitata*, a net with 2000 μm copper mesh screen is first sieved to remove large debris and subsequently a net with 300 μm mesh screen is used (Mischke and Griffin, 2011). More worms of *Uncinais uncinata* are collected, when 100 μm mesh screen is used instead of a 300 μm screen (Lohlein, 1999). Briefly, the mesh size is progressively reduced with decreasing size of aquatic worms.

The need for a simple but effective method of harvesting is obvious. In this regard, Marian and Pandian (1984) have made a key finding. *T. tubifex* is too sensitive to day light. More than 90% of worms remain at the surface between 20 and 24 hours midnight, when they can be sieved and collected. However, an easier and more effective method is to employ the self-assemblage behavior of annelids. In many oligochaetes, self-assemblage is a behavioral response to stress. An exposure to hypoxic water containing 1.5 mg O_2/l, the self-assembled *T. tubifex* forms a ball, enabling the easiest method of harvest. Stressed by immersion into water, *Eudrilus eugeniae* self-assemble into a ball (Daisy et al., 2016, Fig. 3.6). Research is required to know whether the other aquatic annelids also self-assemble, when stressed by hypoxia.

Feeds and ingredients: In aquaculture, feed is a single most important item, as nearly 60% cost is associated with the feed. Many worms show great promise as live feed for fragile early stages and brooders, and ingredient in processed feeds, as well. As feeds, they are more advantageous than the pelleted and pelagic feeds. Their color (white worm, red worm, the tubifex) and wriggle attract predators. They do not impair water quality, when added to aquaculture system. Being benthic, they remain alive at the bottom and are not easily flushed out from the rearing tanks, as the traditional feeds like artemia nauplii. Some of them are highly productive. During the 1940s, Russian biologists have made an admirable contribution to mass culture of white and red worms. Stalked culture boxes can readily produce 30 kg worm/d (see Fairchild et al., 2017). However, their size, even at hatching, is two-times larger than the first instar artemia larva (Table 8.2). Hence, they may serve as feed more for the larvae of fishes than shrimp. Interestingly, night-crawling earthworm *E. eugeniae* has been used as bait since the 1940s.

Dietary fatty acids and some vitamins are obligately required to ensure optimal growth of finfish and shellfish in aquaculture farms. Of 14% fat present in *E. eugeniae*, nearly half of it is constituted by saturated and unsaturated fatty acids (Guerrero III, 2009). Available information on levels of fatty acids in *Enchytraeus albidus* and *Nereis virens* is listed in Table 8.2. The presence of 14% fat in *E. eugeniae* and 10–25% lipids in *E. albidus* are 2 to > 5 times more than that present in commercial pellet feeds (cf Leelatanawit et al., 2014). A comparative study on sperm production in the broodstock of tiger shrimp *Penaeus monodon* fed on commercial feed and polychaetes has indicated that the sperm cell production is 20×10^3/ml and 37×10^3/ml in the former and latter, respectively (Leelatanawit et al., 2014). At School of Biological Sciences, Madurai Kamaraj University, India, red tilapia, guppy, molly, fighter fish, barbs and tetras have been reared for nearly 15 years. To ensure quality eggs on expected dates, their brooder females obligately require to be fed at least for 7–10 days with tubifex (e.g. Kavumpuruth,

TABLE 8.2

Fatty acid contents (mg/g dry substance) of *Enchytraeus albidus* and *Nereis virens* (condensed and compiled from Evjemo and Olsen, 1997, Brown et al., 2011, Fairchild et al., 2017). * coffee ground waste, † stale bread waste

20 : 5n–3	22 : 6n–3	n–3	n–6	SFA	MUFA	PUFA
			Enchytraeus albidus (**1.0–1.5 mm** at hatch)			
5	1	11	126	52	34	154*
6	0	11	39	33	42	62†
			Artemia franciscana (**0.42–0.52 mm**, I instar)			
36	22	80	9	22	36	246
			Calanus finmarchicus (**0.22–0.79** mm)			
9	18	31	4	9	22	66
			Brachionus plicatilis (**0.13–0.34** mm)			
13	18	62	4	9	27	119
			Nereis virens			
16.0	16.1	18 : 2ω6	20 : 1ω6	20 : 4ω6	20 : 5ω3	22 : 1ω11	22 : 1ω3
22	8	12	10	2	12	12	6

1992). Incidentally, many annelids are capable of synthesizing essential fatty acids and vitamins or they engage symbiotic microbes to synthesize them. The presence of many microbes in the gut of many annelids is reported (e.g. Sruthy et al., 2013). Subramanian et al. (2017) have demonstrated that *E. eugeniae* engage *Bacillus endophyticus* to synthesize riboflavin to facilitate regeneration (see p 108, 110).

Protein content of annelids ranges between ~ 59% in *E. albidus* (Fairchild et al., 2017) and 65% in earthworms (Guerrero III, 2009). Vermimeal can replace fishmeal as an ingredient and as protein source for *Oreochromis niloticus* and *Macrobrachium idella* (Guerrero III, 2009). Studies on the vermicost of *E. eugeniae*, using biodegradable waste for culture, have indicated that the efficiency and cost-effectiveness in reducing chemical fertilizer for crop cultivation up to 100% (Guerrero III, 2009). Rearing Atlantic halibut in a recirculating culture system with *N. virens* in raceway water containing pellet waste, fecal waste or both, Brown et al. (2011) have estimated the cost-profit of the polyculture system. For every unit of feed input at US$1.04, the halibut fetches US$9.5 plus 0.3 kg sludge, which produces 0.11 kg *N. virens* valued at US$3.3.

8.2 Candidate Species

To identify the fast growing candidate species for vermiculture, information on growth-age relationship is required. Unfortunately, description of life history characteristics is limited to 3% of polychaetes (p 41). As a result, information available for polychaete is limited, while a reasonable amount of information is available for earthworms, enchytraeids and tubificids with relatively longer history of their cultivation. Growth and reproduction are important factors in biomass production and are considered separately.

8.2.1 Growth

Earthworms: In earthworms, there is a lime-lag between sexual maturity and cocoon laying (see also Fig. 8.9A). Their growth trends are more or less sigmoidal. At the end of log-phase of growth, cocoon production commences. Subsequently, their body weight may be retained, as in *Aporrectodea trapezoides* (Fig. 8.2A) or decreased, as in *Eisenia foetida* (Fig. 8.2C). Considering 150 days rearing period, growth achieved by the investigated earthworm decreases in the following order: *A. longa* > *Eudrilus eugeniae* > *Lumbricus terrestris* > *E. foetida* > *A. trapezoides* and *Octolasion cyaneum* > *Perionyx ceylanensis* > *Hyperiodrilus africanus*. Fed on paper pulp and reared singly at 20°C, *A. longa* grows to a size of 6.5 g within 150 days. Reared in group of two to four and fed on waste dung manure at 25°C, *E. eugeniae* grows to 4.4 g within the same period of

150 days. Both of them are 'layers', as they are sexually mature but have not yet commenced laying cocoons. At 30°C, *E. eugeniae* grows to 2.3 g on the 60th day (Dominguez et al., 2001) and is likely to reach ~ 4.2 g on the 150th day (Fig. 8.2B). Hence, *A. longa* may be more suited to temperate zone and *E. eugeniae* to tropics. Notably, the Spanish *E. eugeniae* grows significantly faster, even when reared in a group, but the South African grown individually achieves 3.5 g body weight at 25°C (Viljoen and Reincke, 1989). A search for an *E. eugeniae* that achieves super-fast growth is required for tropical vermiculture.

It is known that temperate animals grow larger than their tropical counterparts. Not surprisingly, the tropical earthworms like *E. foetida*, *P. ceylanensis* and *H. africanus* do not attain a large size, as temperate earthworms. However, it is not clear why the growth rate of tropical earthworms is at abysmal slow rate. As moisture plays an important role in sustaining growth (Fig. 8.2E), it is not clear whether the relatively lower moisture content prevailing in humid tropical soil (e.g. 10–40% in most soils of Tripura, India, Bhattacharjee and Chaudhuri, 2002) reduces growth rate. Expectedly, growth is accelerated up to 20°C in *L. terrestris* and 25°C in *E. eugeniae* (Fig. 8.2B).

Earthworms survive and grow on litter and/or waste manure. *E. foetida* grows nearly two-times faster, when fed on cow dung manure than on dry leaves (Fig. 8.2C). Hitherto, semi-moist, half-digested dung of cow, horse or pig have been found to sustain growth of earthworms. However, fecal pellets of goats and sheep do not support cocoon laying in them (Siddique et al., 2005). Raw soil, which supports poor growth, can be enriched with saw dust and coconut husk. The 'skin' of coconut pod contains fiber (used for rope manufacture) and husk. The husk is not burnable but retains moisture for a long duration. In coconut-cultivating Asian countries, the coconut husk is a waste and remains as environmental hazard. Hence, research is required to estimate the moisture retention capacity and enriching ability of raw soil by coconut husk.

Apart from these environmental factors, growth of earthworms can considerably be altered by internal factors like parthenogenesis and polyploidy. *O. cyaneum* is a parthenogen (see Table 2.5) and a pentaploid (see Table 6.1). Likewise, *A. trapezoides* is also a parthenogen and occurs as diploid or triploid (Table 6.1). Unfortunately, Fernandez et al. (2010) have not indicated the ploidy status of *A. trapezoides*. *Metaphire houlleti* is suspected as parthenogen (Kaushal et al., 1999, not shown in Fig. 8.2A). The growth trends of sexually reproducing diploid *A. longa* and *E. eugeniae* are considerably at higher levels than those of parthenogens. Hence, parthenogenic earthworms are not good candidate species for vermiculture.

Polychaetes: Within polychaetes, researchers have concentrated mostly on commercially valuable (as bait) nereidids. A few nereidids are semelparous, while others are iteroparous. The former allocate much of available resource

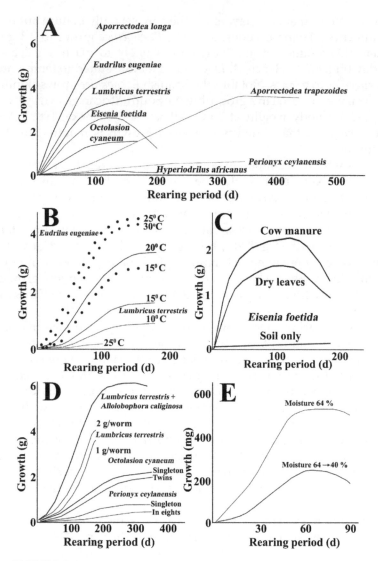

FIGURE 8.2

Growth of selected earthworms A. *Hyperiodrilus africanus* (Tondoh and Lavelle, 1997), *Perionyx ceylanensis* (Karmegam and Daniel, 2009), *Octolasion cyaneum* (Lowe and Butt, 2008), *Eisenia foetida* (Siddique et al., 2005), *Lumbricus terrestris* and *Aporrectodea longa* (Butt, 1993), *A. trapezoides* (Fernandez et al., 2010) and *Eudrilus eugeniae* (Dominguez et al., 2001) B. Effect of temperature on *L. terrestris* (Lowe and Butt, 2005) and *E. eugeniae* (Dominguez et al., 2001) C. soil and manure on *E. foetida* (Siddique et al., 2005), D. density on *P. ceylanensis* (Karmegam and Daniel, 2009), *O. cyaneum* and *L. terrestris + Allolobophora caliginosa* (Curry and Bolger, 1984) and E. soil moisture on *E. foetida* (Reinecke and Venter, 1987) (figures are all redrawn).

on gamete production, whereas the latter allocate bulk of resource on somatic growth (Cassai and Prevedelli, 1998b). Irrespective of any of them, harvest of 'layers' is to be made prior to sexual maturity, i.e. prior to the commencement of allocation to gametic production. Growth of polychaetes has been measured as body length in units of cm (e.g. *Glycera dibranchiata*, Fig. 8.3A) or number of segments (e.g. *Perinereis rullieri*, Fig. 8.3D) or weight (e.g. *Marphysa sanguinea*, Fig. 8.3B), making comparison difficult. In natural habitats, *Nereis virens* requires 2 years to attain sexual maturity at the size of 35 cm body length (Creaser and Clifford, 1982). Olive (1999) has suggested that optimal rearing conditions and feeding can reduce this period to 1 year. Hence, many attempts have been made to use costly enriched feeds. For example, the feeding regime of *Neanthes arenaceodentata* consists of initial inoculation of 15 mg dry powder of green alga *Enteromorpha* sp, subsequent addition of 15 mg powder on the 44th, 51st and 70th days and costly prawn flakes thrice weekly (Pesch et al., 1987). The others have gone for high protein fish feed pellets (hpffp) for *M. sanguinea* (Parandavar et al., 2015) and shrimp meat for *N. virens* (Nielsen et al., 1995). In fact, sedimented pellets and feces due to excess feeding are reported to reduce feeding in *Pseudopolydora kempi japonica* (Miller and Jumars, 1986).

Figure 8.3B shows that the fastest growing *M. sanguinea* attains a body weight of 7 g fed on hpffp at the density of $500/m^2$. Considering 150 days period in the investigated polychaetes, growth decreases in the following order: *N. virens* > *M. sanguinea* > *Perinereis nuntia* (Fig. 8.3A–C). These worms have been fed on uneaten pellet + halibut fecal waste, hpffp or waste water containing ~ 4% organic matter, respectively. As in earthworms, increase in temperature accelerates growth of polychaetes (Fig. 8.3D); in *P. rullieri*, a reduction in ration decreases growth (Fig. 8.3E) and animal (nauplii) or protein enriched diet (e.g. *N. aranaceodentata*) enhances growth better than algal tetramin (Fig. 8.3D). Clearly, nereidids require relatively protein-rich feed than earthworms. With substratum containing 3–4% organic substance, earthworms achieve 4–6 g body weight but polychaetes > 1 g (e.g. *Perinereis* spp, Fig. 8.3C). With increasing protein enrichment in diet, the production cost of nereidids shall be higher than that of earthworms. Briefly, the waste water recirculating system of Brown et al. (2011) with an area of $12.6 \ m^2$ containing 1,110 worms (i.e. 407 g/0.37 g initial size), i.e. 87 no./m^2 *N. virens* reared in waste water containing uneaten pellet and halibut feces seem to be the optimal system and best candidate species to produce 2 g sized worm within 80 days.

Tubificids: The live feed tubificids are potentially important in commercial vermiculture. *Branchiura sowerbyi*, *Tubifex tubifex*, *Limnodrilus* sp and *Lumbriculus* sp can produce ~ 15 kg/m^2 (Lietz, 1987).

B. sowerbyi: The relatively large sized *B. sowerbyi* is a deserving cultivable candidate. It grows to 150 mg in 180 days (Fig. 8.4A). However, it attains

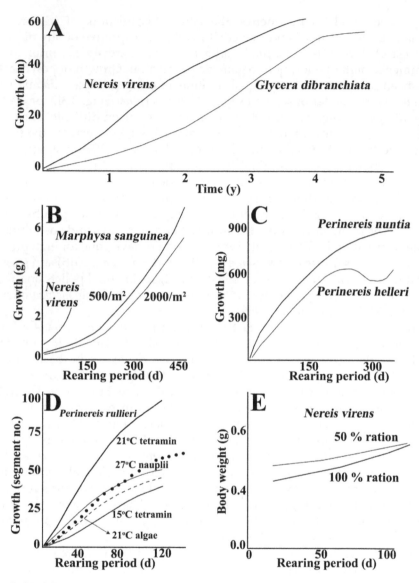

FIGURE 8.3

Growth in selected polychaetes. A. *Glycera dibranchiata* (drawn from data reported by Creaser, 1973) and *Nereis virens* (drawn considering about 35 cm growth in 2 y, Creaser and Clifford, 1982). B. *Marphysa sanguinea* (growth data pooled from 0.5, 1.0 and 2.0 g size reared at 500 or 2000 worms/m² , compiled and redrawn from Parandavar et al., 2015) and *Nereis virens* fed on waste in recirculating culture system (Brown et al., 2011). C. *Perinereis nuntia* and *P. helleri* in a sand filter system (redrawn from Palmer, 2010). D. Effect of temperature and food quality on growth of *Perinereis rullieri* (compiled and redrawn from Prevedelli, 1992) and *Neanthes arenaceodentata* (in dotted line, redrawn from Pesch et al., 1987). E. Effect of ration on growth of 400 mg weighing *Nereis virens*, redrawn from Nielsen et al., 1995).

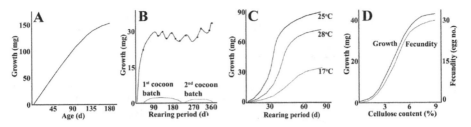

FIGURE 8.4

Growth and reproductive output in *Branchiura sowerbyi*. A. Growth as function of age (drawn from information reported by Aston, 1968). B. Growth and cocoon production as function of age (compiled and redrawn from Lobo and Alves, 2011). C. Effect of temperature on growth (redrawn from Aston et al., 1982). D. Effect of nutrients (cellulose content) on growth and egg production (compiled and redrawn from Aston, 1984).

maturity at the age of 30 days and subsequently lays cocoons almost once a week in two temporally separated batches (Fig. 8.4B); the first batch lasts from the 35th to the 190th day and the second from the 217th to the 360th day. More cocoons and eggs (see Fig. 2.8B) are laid in the first batch. Hence, harvest of 'layers' can be made on the 28th day at ~ 30 mg size and 'brooders' on the 190th day. As in other worms, elevation in temperature accelerates growth up to 25°C (Fig. 8.4C). Notable is that it grows to 30 mg size within 30 days, in comparison to 7–17 mg size of *T. tubifex* (Fig. 8.5A). Further, *B. sowerbyi* is more herbivorous than *T. tubifex*, as it can grow to 40 mg size within 21 days, when fed on filter paper containing 6–9% cellulose (Fig. 8.4C).

T. tubifex grows to a size of 7 and 17 mg, when reared on cow dung manure and activated sludge, respectively (Fig. 4.5A). Marian and Pandian (1985) have noted the entry of chironomus into the open continuous flow system. With increasing chironomus density from 0 to 7/cm², the number and production of tubifex are reduced from 146 to 10 no./cm² and 181 to 13 mg/cm². Improving the culture system with recirculating water system and rack culture tanks covered to prevent the entry of chironomus and mosquitoes, Marian et al. (1989) have reduced water consumption from 38,000 l to 193 l with a production of 5.6 kg tubifex/mo. In Bangladesh, Jewel et al. (2016) have also reared tubifex with cow dung supplemented with yeast and the like. Their production system has yielded less tubifex than that of Marian and Pandian (1984). Hossain et al. (2011) have also reared tubifex feeding on costlier combination of 30% soybean, 20% mustard oil, 20% wheat bran, 20% cow dung and 10% sand and have achieved tubifex production of 659 mg/cm². Clearly, protein requirement of tubifex is higher than that of *B. sowerbyi* (see Fig. 8.4D). Oplinger et al. (2011) have made a series of experiments to estimate the effects of feed type, ration, temperature and density on juvenile and biomass production. However, initial stocking level, which is known to have a profound effect on production, was not kept uniform; for example, it was 31 in cow dung manure, instead of 50 in feed type experiment. It also ranged

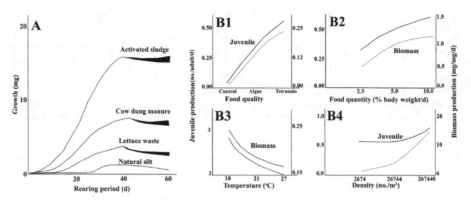

FIGURE 8.5

A. *Tubifex tubifex* growth as function of age in worms reared in activated sludge in Leningrad, Russia and natural silt (from Finogenova and Lobasheva, 1987), cow dung manure in Madurai, India (from Marian and Pandian, 1984) and lettuce waste (from Kosiorek, 1974). Dark shade indicates the cocoon biomass. B. Effects of (B1) food quality, (B2) food quantity, (B3) temperature and (B4) density on juvenile and biomass production (drawn from recalculated data reported by Oplinger et al., 2011).

from 166 to 397 worms, instead of 50 in temperature series experiment. Secondly, tubifex was reared in circular beakers and biomass production was reported in mg/mg stock, instead of reporting in mg/cm². Hence, the results are not comparable with those of others. Thirdly, with elevation of temperature from 18°C to 27°C, both juvenile and biomass production was reported to decrease (Fig. 8.5B3). *T. tubifex* is cosmopolitan and almost all the worms belonging to different annelid taxa are reported to accelerate growth with elevation of temperature. As far as possible, their data were recalculated and summarized in Fig. 8.4B. With increasing protein enriched feed type and ration, the number of juvenile production is increased but only up to 0.4–0.5/adult/d; however, the increase is upto 0.7/adult/d with increasing density and the decrease from 2.7/adult/d with increasing temperature. Biomass production is also increased in the following order: increasing density < increasing ration < decreasing temperature < feed type. Briefly, stocking density has an overriding role in biomass production of tubifex.

Enchytraeids: Of 670 species, nine are clonal and another nine may be parthenogens (Table 8.1). A few like *Enchytraeus bigeminus* are parthenogens and polyploids (Christensen, 1973). Most enchytraeids are terrestrial and more abundant in arctic and temperate zones (density: > 100,000/m²) than in humid tropics (< 10,000/m²) (Schmelz et al., 2013). Their density can also reach 21,500/m² for *Marionina southerni* in Lake Esrow, Denmark (Lindegaard et al., 1994) and as high as 300,000/m² in moorland (see Gonclaves et al., 2017). Their body weight ranges from 15–240 µg in *M. southerni* (Lindegaard et al., 1994) to 770 µg in *E. bigeminus* (Christensen, 1973). As they are too small, their growth is measured in body length or number of segments or mostly in numbers. As mentioned earlier, clonals are more of academic interest and gonochorics are of economic importance. Life span of cloners is half of that of gonochores. For example, it is 15 days in *E. japonensis* in cloners but 29 days in gonochores. When the worm grows to 60–80 segments size, it spontaneously autotomizes into 5–10 fragments and each one of them is regenerated into a complete worm in 4–5 days but the ramet's size is reduced to < 50% (Yoshida-Noro and Tochinai, 2010). Reared in 200 ml jar containing sand and abundant porridge, *E. bigeminus* grows to a maximum size of 770 µg within 2–3 weeks. Initially, it clonally reproduces until the density of 300–350 is reached, when it switches to sexual reproduction. In natural habitats, clonal reproduction occurs during favorable conditions but with a commencement of winter, it switches to sexual reproduction. In a 200 ml jar, the maximum biomass production during 8 weeks experiments can be 2.2 kg (Christensen, 1973). Incidentally, density also imposes profound effect on body size and juvenile production in F_0 and F_1 generations of *E. crypticus*. With increasing density, reductions are 50% in body size and 75% in juvenile production (Gonclaves et al., 2017).

Information on growth of tropical aquatic enchytraeids is not available. Available information pertains to temperate enchytraeids thriving in activated sludge arising from sewage waters. Considering a 50-day period, growth of the investigated worms decreases in the following order: *Lumbricillus rivalis* < *L. lineatus* < *E. albidus* < *E. crypticus* < *E. coronatus* (Fig. 8.6A). In almost all these worms, temperature accelerates growth up to 20°C (Fig. 8.6B). As no decline in the acceleration is observed in *Lumbricillus lineatus* even at the 20°C, it may grow even faster up to 30°C. Biomass production of *Cognettia sphagnetorum* is ~ 200 worms per day (Springett, 1970). In Lake Esrow, annual net production of *M. southerni* is in the range of 5.1 Kcal/m²/y (Lindegaard et al., 1994).

Doubling time (Dt): Gravimetric or linear measurements to estimate growth in smaller worms become increasingly difficult, time-consuming and requires sophisticated balance. Growth can also be measured in numbers or specific growth/intrinsic growth rate. But the latter suffers from the following: (i) it decreases with increasing body size (e.g. *Dero dorsalis*, Ratsak and Verkuijlen, 2006). It may not be possible to fix a rate for a species and compare it with another species, as their body lengths may differ considerably. (ii) it can not be used in clonal species, as they only regenerate the missing body parts. A better parameter is the doubling time. Dt represents the period from the day, when sexual or clonal reproduction is commenced to that day, when F_1 offspring begins to reproduce.

An analysis of available data suggests that body size and temperature impose profound effects on Dt. Expectedly, Dt increases with increasing size in enchytraeids and tubificids. For example, Dt is < 3.61 days in relatively larger (7–17 mg) tubifex than in the smallest enchytraeids

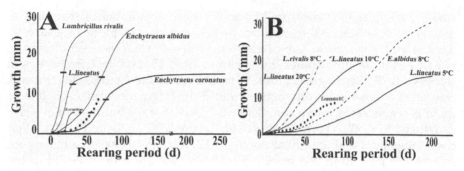

FIGURE 8.6

Growth of selected gonochoric enchytraeids. A. Growth as function of age at 20°C in gonochoric *Lumbricillus rivalis, Enchytraes albidus* (o represents data reported by Reynoldson, 1943), *E. coronatus* and *E. buchholzi* (in dotted line) in sewage percolating filters (redrawn from Learner, 1972), *L. lineatus* (redrawn from Reynoldson, 1943) and *E. crypticus* (redrawn from Bicho, 2015). Cross bars indicate the size and age at sexual maturity. B. Effect of temperature on growth of body length *L. lineatus, L. rivalis, E. coronatus* (redrawn from Learner, 1972).

(< 2 days). It requires the longest duration of 21 days in the largest (100–150 mg) *Branchiura sowerbyi* (Fig. 8.7A). It is known that 54% of naidids are cloners (p 127); hence, they are considered separately. In the naidids too, a linear direct relationship between body size and Dt becomes apparent (Fig. 8.7B); irrespective of sexually or clonally reproducing species within relatively shorter temperature ranges from 20°C to 30°C. Also, irrespective of clonal or sexual reproduction, naidids are capable of biomass production. In them, Dt increases from 2 days in the smallest gonochoric *Chaetogaster langi* at 20°C to 3.3 days in *Dero dorsalis* at 30°C and > 6 days in the largest cloner *Nais elinguis*. However, relatively a wider range of temperature does influence Dt

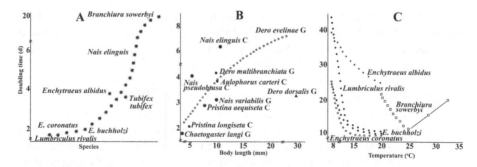

FIGURE 8.7

A. Doubling time in some aquatic oligochaetes. B. Doubling time in naidids, G = gonochore, C = cloner, ● = 20°C, ■ = 25°C and ▲ = 30°C (drawn from data reported by Ratsak and Verkuijlen, 2006). C. Doubling time as function of temperature in enchytraeids and *B. sowerbyi* (drawn from data reported by Aston et al., 1982).

TABLE 8.3

Effect of density on successful pairing, reproduction and fecundity in *Neanthes arenaceodentata* (condensed from Pesch et al., 1987)

Trait	Density (no./840 cm²)			
	Control	40	80	160
Survival (%)	97	90	81	74
Sex ratio	0.54	0.55	0.56	0.53
Age at spawning (d)	85	101	102	109
Paired reproducing female (%)	54	54	49	23
Fecundity (egg no./brood)	695	881	622	598
Paired but				
(a) not successful females (%)	5.3	0	0.8	0.5
(b) laid eggs (%)	5.1	5.5	13.0	11.0
(c) no egg laying (%)	35	41	37	66

in enchytraeids and tubificids. With increasing temperature, Dt decreases but at different levels. In *Lumbricillus rivalis*, it decreases from 45 days at 8°C to 11 days at 20°C; the corresponding values for *Enchytraeus coronatus* are 24 days at 8°C and 11 days at 20°C. For *B. sowerbyi*, it increases from 11 days at 25°C to 20 days at 33°C but decreases from 21 days at 21°C. Apparently, the optimum for Dt is 25°C (Fig. 8.7C). From this analysis too, *B. sowerbyi* and *T. tubifex* are suggested as more suitable candidate species.

8.2.2 Reproduction

In polychaetes, nereidids are cultivated. They are all gonochores and are not cloners. Some of them like *Nereis virens* are epitokous and semelparous (Table 5.1), while others like *Marphysa sanguinea* are atokous and iteroparous. Information on their reproduction, especially fecundity has been elaborated in Chapter 2. From the point of maintaining brooders, only one publication is available. A vast majority of oligochaetes are hermaphrodites. In them, a few are parthenogens (e.g. *Lumbricus terrestris*, Table 2.5) and a very few are parthenogens cum polyploids (e.g. *Dendrodrilus rubidus*, *Lumbricillus lineatus*, Tables 2.7, 6.1). For maintenance of brooders in oligochaetes, relatively more information is available.

Polychaetes: Pesch et al. (1987) have investigated the effect of density on pair formation and reproduction in *Neanthes arenaceodentata*. With increasing density from 40, 80, and 160/840 cm², i.e. 48, 95 and 190 no./m², survival, male ratio and age at spawning are marginally affected (Table 8.3). However, the number of successfully paired females is decreased from 54% to 23% and fecundity from 881 eggs/brood to 598 eggs/brood at the highest density of 160 no./840 cm². Further, the number of females, which are unable to pair (0 to 0.8%) and which are unable to lay eggs (41 to 66%) is significantly increased. Considering the control levels, maintaining brooders of *N. arenaceodentata* at ~ 50 no./m² is the recommended density. It must be noted that *N. arenaceodentata* is a fairly small nereidid. *Nereis virens*, the recommended candidate nereidids, grows to > 40 cm (Fig. 8.3A). Hence, the density at which larger brooding nereidids can be maintained may be in the range of only a few. However, research inputs are required.

Oligochaetes: During their life time, oligochaetes lay cocoons in two temporally separated batches. The second batch contains less number of eggs/cocoon than the first one (Fig. 8.8A). Further, fertility of eggs decreases at elevated

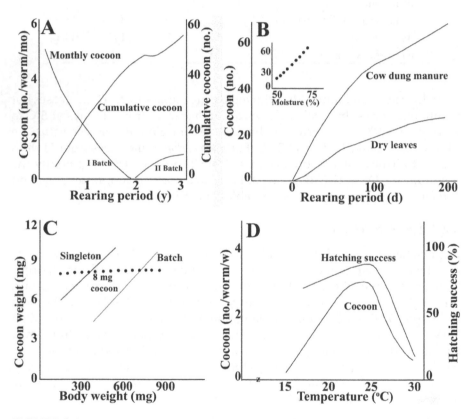

FIGURE 8.8

Reproduction in earthworms. A. Monthly and cumulative cocoon production in *Lumbricus terrestris* (simplified and redrawn from Butt et al., 1994). B. Effect of feeds on cocoon production in *Eisenia foetida* (simplified and redrawn from Siddique et al., 2005). Window shows the effect of moisture on cocoon production in *E. foetida* (redrawn from Reinecke and Venter, 1987). C. Effect of culture density on number and biomass of cocoon production in *Metaphire houlleti* (compiled and redrawn from Kaushal et al., 1999). D. Effect of temperature on cocoon production and hatching success in *Eudrilus eugeniae* (drawn from data reported by Dominguez et al., 2001).

temperature (Fig. 8.9D). Hence, the maintenance of brooders beyond 2 years in *Lumbricus terrestris* (Fig. 8.8A), 190 days in *Branchiura sowerbyi* (Fig. 8.4B) and 60 days in *Eisenia foetida* (Fig. 8.8B) is not recommended.

With advancing age, fecundity is considerably decreased and fertility of egg is also decreased. Considering the respective initial values and 50% decrease in fecundity with advancing age, it is suggested that brooders older than 1 year in *L. terrestris* (Fig. 8.8A) and 50 days in *Lumbricillus rivalis* (Fig. 8.9B) are not maintained.

Food quality and moisture play an important role in cocoon production. Cow dung manure and maintenance of 75% moisture facilitate production of more number of cocoons, especially in tropical earthworms, as in *Eisenia*

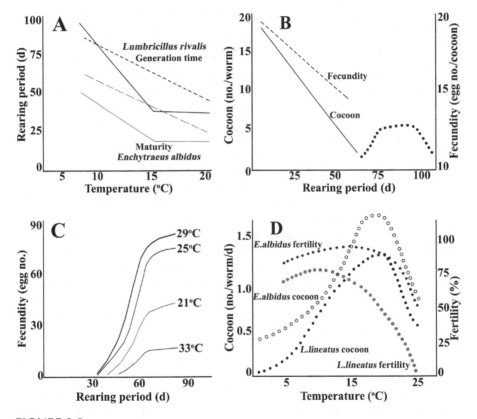

FIGURE 8.9

Reproduction in aquatic oligochaetes. A. Sexual maturity and generation time (discontinuous line) as a function of age at different temperatures in *Lumbricillus rivalis* (thick lines) and *Enchytraeus albidus* (drawn from data reported by Learner, 1972). B. Fecundity as function of age in *L. rivalis* (drawn from data reported by Kirk, 1971). C. Effect of temperature on fecundity of *Branchiura sowerbyi* (redrawn from Aston et al., 1982). D. Effect of temperature on cocoon production and egg fertility in *L. lineatus* and *E. albidus* (drawn from data reported by Reynoldson, 1943).

foetida (Fig. 8.8B). Incidentally, *E. foetida* brooders older than 80 days may not be as productive as the young ones. Irrespective of rearing in single or batch, the suspected parthenogen *Metaphire houlleti* invests 8 mg on cocoon biomass (Fig. 8.8C). As age advances from 75 days to 350 days in *Perionyx ceylanensis* too, the cocoon production rate (no./worm/d) also decreases from 1.15 to 1.00 in singles and 1.25 to 1.15 in batches of eight worms (Karmegam and Daniel, 2009). Hence, density seems not to affect cocoon production. However, more research inputs are required.

In general, oligochaetes are sensitive to temperature. The period required for sexual maturity decreases with increasing temperature in *Lumbricillus rivalis* and *E. albidus*. However, there is a time-lag between sexual maturity and the day, on which the first cocoon is laid, i.e. generation time. This time-lag between maturity and generation time is maintained with increasing temperature, though the trend is linear in *L. rivalis* but L-shaped in *E. albidus* (Fig. 8.9A). The thermal optima for brooders is 25°C for *Eudrilus eugeniae* (Fig. 8.8D), 29°C for *B. sowerbyi* (Fig. 8.9C) and 15°C for *L. lineatus* and *E. albidus* (Fig. 8.9D). Of course, these two enchytraeids are temperate and subarctic species. A feature that deserves to be noted is that the optimum temperature for somatic growth differs from that for cocoon production. For example, it is 25°C for growth in *B. sowerbyi* (Fig. 8.4C) but 29°C for cocoon production (Fig. 8.9C). However, it is 28°C for growth in *E. albidus* and *L. lineatus* (Fig. 8.6B) but 15°C for cocoon production (Fig. 8.9D). For *E. eugeniae*, 25°C is the optimum for both growth and reproduction (Fig. 8.2B, 8.8D).

Oligochaete embryos are enclosed in eggs, which, in turn, are enclosed in a cocoon. No information is available on the mechanism of hatching, and the time course and sequence, through which hatching occurs. Firstly, fertility of eggs is an important factor in determining hatching success. It decreases on either side of the optimum temperature in *L. lineatus* and *E. albidus* (Fig. 8.9D). Clearly, on either side of 15°C, fertilizability of eggs and sperm is reduced. The presence of infertile eggs within cocoons may be responsible for the reported decrease in hatching success of *Eudrilus eugeniae* on either side of 25°C, the optimum temperature (Fig. 8.8D). In view of wide differences in hatching success reported for oligochaetes, information on cocoon structure and hatching mechanism is required. Considering earthworms, for which relatively more information is available, hatching success ranges for 20% in *Eutyphoeus gammiei* to 100% in *Metaphire houlleti* fed on moist filter paper. In *E. foetida* too, it is 54 and 63% in worms fed on cow manure and dry leaves. It is not clear whether the water content of feed and/or desiccation plays a role in determining hatching success. For cocoons of *M. houlleti*, which have been fed on moist filter paper or incubated in distilled water, the success is 100%. It is then difficult to comprehend 70–79% hatching success reported for the cocoons of *Aporrectodea longa*, *L. terrestris* and *Octolasion cyaneum* incubated in water at 15°C, especially 47% success reported for *A. longa* cocoons incubated in water medium at 20°C. Hence, attempts have been made to relate cocoon

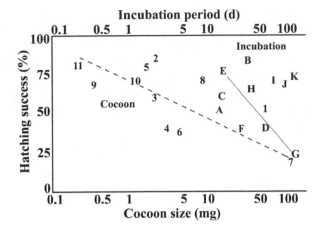

FIGURE 8.10

Effect of cocoon size (in numbers) and incubation period (in alphabets) on hatching success of earthworms. 1/A *Perionyx excavatus*, 2/B *Pentoscolex corethrurus*, 3/C *Lampito mauritii*, 4/D *Polypheretima elongata*, 5/E *Dichogaster modigilianii*, 6/F *Drawida nepalensis* (P), 7/G *Eutyphoeus gammiei* (data from Bhattacharjee and Chaudhuri, 2002), 8/H *Metaphire houlleti* (P?) (data from Kaushal et al., 1999), 9/I *Aporrectodea longa*, 10/J *Lumbricus terrestris* (P) and 11/K *Octolasion cyaneum* (data from Butt, 1993).

size and hatching success as well as incubation period and hatching success (Fig. 8.10). On plotting, all the available values are found scattered widely. Still, cocoon size and incubation period seem to determine the level of hatching success. The longer the incubation, the possibilities for desiccation and/or predation are greater. Hence, incubation of cocoon in safe aquatic medium is recommended.

8.2.3 Polyploids and Parthenogens

In animals, cell volume of triploids and polyploids triply and multiply increases; but, the maximum attainable body size by polyploids is regulated by reducing the number of cells (e.g. fishes, Pandian, 2011). However, both cell volume and cell number are increased in polyploid bivalves (Pandian, 2017). As eggs of oligochaetes are enclosed in a cocoon, ploidy induction is difficult to achieve. However, studies on cell number and cell volume in natural polyploid annelids are required to know whether or not the cell volume and number are increased. The ensuing information seems to demand research input in this area. *Bimastos eiseni* and *B. tenuis* exist only as diploid and triploid, respectively. So are *Eisenia venata* and *E. rosea*. *Dendrobaena rubida* exist only as diploid but its sister species *D. subrucunda* exists only as tetraploid. In *Octolasion, O. lacteum* exists only as diploid, while its sister species *O. cyaneum* as polyploid. Interesting information on their growth has been assembled (Table 8.4). Triploid *B. tenuis* grows to a larger length (52 mm), width (3 mm) and with more number of segments (105), in comparison to diploid *B. eiseni*. Body lengths of triploid *B. tenuis* (10%), *E. rosea* (56%), tetraploid *D. subrucunda* (40%) and polyploid *O. cyaneum* (26%) grow to larger sizes than their respective diploid sister species. It is not clear whether triploid *Bimastos, Eisenia*, tetraploid *Dendrobaena* and *Octolasion* grow both in cell volume and number. Notably, *E. rosea* triploid grows 1.6 times larger than its diploid counterpart *E. venata*. Hence, these two earthworms may be subjected to studies on cell volume and number

TABLE 8.4

Body size in some diploid and polyploidy earthworms (modified from Muldal, 1952)

Species	Ploidy	Length (mm)	Diameter (mm)	Segment (no.)
Bimastos eiseni	2n	47	2–5	93
B. tenuis	3n	52	3	105
Eisenia venata	2n	34	3–4	102
E. rosea	3n	55	3–4	135
Dendrobaena rubida	2n	43	3.5	75
D. subrucunda	4n	60	4	85
Octolasion lacteum	2n	97	4	126
O. cyaneum	9n	122	7.5	127

TABLE 8.5

Polyploidy, parthenogenesis and reproduction in *Tubifex tubifex* and *Limnodrilus hoffmeisteri* (inferences from Poddubnaya, 1984)

Traits	Semelparous	Iteroparous	
Tubifex tubifex			
Life span (days)	100	245	380
Age of maturity (days)	55	60	72
Reproductive cycle (no.)	1	2–3	2–6
Fecundity (egg no./cycle)	36	40	23
% in population	60	40	60
Limnodrilus hoffmeisteri			
Life span (days)	135	181	349
Age of maturity (days)	78	90	66
Reproductive cycle (no.)	1	2–5	2–4
Fecundity (egg no./cycle)	65	35	64
% in population	27	69	78

on a priority. If found that the triploid *E. rosea* grows by increasing both cell number and cell volume, as in bivalves, then it is a matter of great academic and economic importance.

Rearing parthenogenic *Tubifex tubifex* and *Limnodrilus hoffmeisteri* for two years, the former Soviet scientist Poddubnaya (1984) has briefly summarized the results in Table 1 of the publication. The results reported therein are confusing and difficult to understand. However, two groups in each species are recognizable with a short and long life span. Describing fecundity of the Italian *T. tubifex*, Pasteris et al. (1996) has recognized two cohorts with production of less and more number of eggs (see Fig. 2.8D, E). Investigating the ploidy status of the Italian *T. tubifex*, Marotta et al. (2014) have reported that the chromosome number of *T. tubifex* can be 2n = 50, 4n = 100 and 6n = 150. It is in this context, that an attempt has been made to understand and infer some information from Table 1 of Poddubnaya. The life span is 100, 245 and 380 days for *T. tubifex* and 135, 181 and 349 days for *L. hoffmeisteri* (Table 8.5). Those with the shortest life span have a single reproductive cycle. Hence, these diploids may be semelparous. The other two groups have two to five reproductive cycles and are iteroparous. Fecundity per reproductive cycle is reduced to ~ 50% in the long living *T. tubifex* and in *L. hoffmeisteri*. It is likely that they are tetraploids. However, more research on the effect of ploidy on growth and reproduction is urgently required, especially in the commercially valuable *T. tubifex*.

9

Summary and New Findings

As in other invertebrates, taxonomy of annelids is in a fluid but dynamic state. The number of described species, which has remained ~2,000–3,000 until the 19th century, has now grown to 17,000. Correspondingly, publications on polychaetes alone have also increased from a dozen/decade until 1960 to ~ 300/decade during 2,000–2,010 (Fig. 1.2). Yet, description of life history characteristics is limited to 3% of polychaetes and is increasing at the rate of three species/y. Some 81, 12 and 7% annelids are marine, freshwater and terrestrial inhabitants. Vertical distribution of the soft-bodied annelids ranges from 4,900 m depth (tubificid, *Abyssidrilus stilus*) to 2,000 m altitude (enchytraeid, *Buchholzia appendicularia*). They thrive in unusual habitats like hydrothermal vents (siboglinids) and sub-terranean aquatic system (stigobiont oligochaetes belonging to 42 genera in 27 families). For the first time, relevant information has been highlighted on gutless oligochaetes and polychaetes, osmotrophism and anaerobiosis in some annelids. For example, constantly migrating between the nutrient-rich anoxic and oxic zones in the sediments, the gutless tubificids have successfully colonized an ecological niche, so far unoccupied by any other interstitial fauna (Giere et al., 1984).

A vast majority of polychaetes are gonochores, whereas almost all clitellates are hermaphrodites characterized by internal fertilization and laying cocoon containing a few eggs. Polychaetes display fascinating and incredibly diverse reproductive modes. From his extensive survey, Wilson (1991) has estimated that 47% of polychaetes brood their eggs and remaining 53% are broadcast spawners. Another estimate has indicated 42% are brooders and 58% are broadcasters. In 45% of polychaetes, fertilization is external; it is also external in another 9% of brooders. With hermaphroditism in clitellates, gonochorism is limited to 76% of annelids. Hermaphroditism occurs also in 23 families of polychaetes. From their survey, Schroeder and Hermans (1975) have reported the incidence of hermaphroditism in 67 polychaete species. Without validating and/or updating the report of Schroeder and Hermans, many reviewers have virtually repeated the same number of 67 hermaphroditic species. From an intensive survey of literature up to 2017, the number is updated to 207 species (Table 2.4). For the first time, a directory has been documented for the incidence of parthenogenesis in 75 annelid species, of which 56 are earthworms (Table 2.5). In annelids, fecundity decreases with advancing age in *Lumbricus terrestris* (Fig. 2.8A), and in tube-dwelling *Streblospio benedicti* (Fig. 2.8H) or increases with advancing age and body weight but beyond a particular age/body weight, it decreases in *Tubifex tubifex* (Fig. 2.8D, E). It is not clear whether with increasing body length, the girth grows differently in these worms or they undergo reproductive senescence at different ages and sizes. The flattened body in *Ophryotrocha puerilis puerilis* and dorso-ventrally compressed body in hirudineans provide a relatively larger surface area than that in cylindrical annelids; in the hirudineans and *O. puerilis puerilis*, fecundity continues to increase with increasing body size (Fig. 2.7). Besides body size and egg size, sexuality and oogenic anlage are shown to affect fecundity. Poecilogony is a rare reproductive mode reported only from polychaetes and opisthobranch molluscs. Whereas it is limited to two alternative morphs in a half a dozen opisthobranchs, the boccardian polychaetes exhibit three different morphs. In a reciprocal mating system, each partner fertilizes the eggs of the other. Amazingly, the partners simultaneously change sex several times in their life time in *O. puerilis puerilis* to ensure that no partner cheats the other.

The account on regeneration has brought to light a whole range of new findings. Among annelids, hirudineans are not capable of either regeneration of missing body parts/organs or clonal reproduction; a few annelids are capable of regeneration of the entire anterior or posterior or anterior cum posterior fragment of the body. For example, the head of *Lumbriculus variegatus* can be regenerated 21 times and the tail 42 times and both together 20 times. Posterior regeneration is induced by neurosecretion from the brain in errant polychaetes. However, the ventral ganglia are capable of inducing posterior regeneration in sedentary polychaetes, whose projected 'heads' are more often subjected to sub-lethal predation. Notably, regeneration in annelids is a fairly complex but orderly process and proceeds from wound-healing to blastema formation and differentiation, segmental reorganization, and growth and elongation. It ranges from species that has no ability to regenerate even a single segment (e.g. *Streblospio benedicti*) to species that are capable of regenerating an entire worm from a single seminal (newly coined term) segment (e.g. *Chaetopterus variopedatus*). In polychaetes and oligochaetes, ectodermal, mesodermal and endodermal regeneration involves the injury-activated, dedifferentiated multipotent stem cells arising from the respective germ layers. Oligochaetes with chloragogue and polychaetes without it exhibit contrasting features. In chloragogue, reserves are inadequate to simultaneously meet the costs of both regeneration and reproduction and are temporally separated. But the sedentary polychaetes undertake them together at the cost of reduced reproduction. In annelids, the origin and loss of regenerative potency have occurred independently at multiple numbers of times (Bely, 2010). Hence, the seminal segments are positioned at different locations in different taxonomic groups *albeit* mostly between the cephalic region and anterior trunk.

The assemblage of incidences of regeneration in annelids by Zattara and Bely (2016) has paved the way for further analysis. The number of species capable of anterior, posterior and anterior cum posterior regeneration is 149, 206 and 144, respectively, i.e. only 0.88, 1.22 and 0.85% of annelids are capable of anterior, posterior and anterior cum posterior regeneration. The reasons for the less number of incidences of anterior regeneration are: (i) the anterior contains vital organs like the brain, heart and metanephridia and (ii) as new segments are regenerated from the pygidial zone, the regenerative potency diminishes faster in the 'older' anterior segments than in the 'young' posterior segments. When the respective number of these regeneratives is considered as fractions of 13,012 polychaete and 3,175 oligochaete species, the percentage values indicate that the potency is 1.5–2.0 times more prevalent in oligochaetes (1.57, 1.80, 1.42) than the respective ones (0.88, 1.22, 0.85) of polychaetes.

In annelids, clonal reproduction is characterized by the following general features: (i) Reared under optimal conditions, a dozen annelid species are sustained by clonal reproduction alone up to 30–60 years. However, the Primordial Germ Cells (PGCs) can be transmitted upto 1,000–3,000 generations in the clones (e.g. *Pristina leidyi*). When stressed or induced, the cloners switch to sexual reproduction. (ii) Abundant food supply, low density and favorable temperature may either singly or in combination trigger clonal reproduction. (iii) Except for a couple of species, the duration required for completion of the sensitive cloning ranges between 3 and 8 days (Table 4.5). (iv) Intense predation alone may not manifest clonal reproduction; for example, of 290 tubiculous sabellids species, only 17 are cloners. (v) Unlike in echinoderms (Pandian, 2018), hermaphroditism and brooding in annelids do not prevent clonal reproduction. However, larval cloning is not yet reported in annelids. (vi) In clonal species, sex ratio is altered and gamete production is reduced. (vii) Clonal reproduction is usually succeeded by sexual reproduction but both can occur simultaneously in a very few naidids. (viii) Clonal reproduction can be grouped into architomy and paratomy. In architomy, fission is followed by the completion of regeneration and formation of new progeny (ramet). But it occurs even before the ramets are fully formed in paratomy.

Assigning clonal species to their respective families in relation to architomy and paratomy has revealed the following new findings: (i) Of 100 and odd annelid families, clonal reproduction occurs only in 12 polychaete and five oligochaete families (Table 4.3). Architomy occurs exclusively in eight polychaete families but paratomy in aeolosomatids alone. (ii) In clonal reproduction, the incidence ranges from 2% of spionids to 54% of naidids. (iii) Further, it is limited to 79 polychaete species but to as many as 111 oligochaete species (Table 4.3), i.e. cloning occurs only in 0.61% of polychaetes but 2.14% of oligochaetes. Hence, clonal potency is ~ 4.6 times more prevalent in oligochaetes than that of polychaetes. Analyzing the data of clonal species, Zattara and Bely (2016) have viewed that cloning may have been derived from regeneration. This view may not be correct for following reasons: (a) Cloning in 111 oligochaete species do obligately requires the presence of neoblasts (Myohara, 2012). In the absence of neoblasts, almost all earthworms, many enchytraeids (e.g. *Enchytraeus buchholzi*, Myohara, 2012), naidids and tubificids are unable to clonally reproduce. (b) Without having the ability to regenerate anterior cum posterior regeneration, *Chaetogaster diaphanus* and *C. diastrophus* reproduce clonally. Clearly, the clonal potency of oligochaetes is vested with neoblasts and not from anterior cum posterior regenerative potency. (c) Even with anterior cum posterior regenerative potency, 31 out of 63 polychaete species do not reproduce clonally (Table 3.8). Hence, a search for the equivalent of neoblasts-like multipotent stem cells in polychaetes is required.

Available information on the possible anlage for the clonal stem cells is assembled in Table 4.9 and 4.10. In majority of sedentary/tubiculous sabellids and serpulids, these cells are located deep at the posterior end, irrespective of architomic or paratomic fission. This is also true of the errant dorvilleid *Parougia bermudensis* and aeolosomatid *Aeolosoma viride*, and possibly the spionid *Pygospio elegans*. With the presence of neoblasts in all the non-cephalic segments, the segments are capable of clonal reproduction in *E. japonensis* and *L. variegatus*. Due to progressive loss of neoblasts in *Stylaria lacustris* and their equivalents in polychaetes, the stem cells are restricted to the seminal segments or to the mid-body. However, investigations are required to discover the equivalent of neoblasts in polychaetes.

Epitoky is unique to errant polychaetes. The epitokes are divided into epigamics and schizogamics. In epigamy, the entire body of the worm undergoes epitokous transformation but only a fraction of it in schizogamics. As a result, the former is semelparous but the latter iteroparous. For the first time, a directory is documented listing the epigamic incidence in 62 species from 12 families (Table 5.1) and schizogamic incidence in 45 syllid species (Table 5.3). Hitherto, not a single author or reviewer has ever considered the energy demanding vertical migration over distances. Again for the first time, widely scattered information on vertical distance travelled by 15 epigamics and 13 schizogamics has been assembled (Table 5.4). Surprisingly, the vertical distance climbed by the epitokes decreases with increasing body size (Fig. 5.3). Obviously, the larger epitokes glycerids, nereidids and eunicids utilize muscular energy to climb up < 50 m distance. The smaller phyllodocids and ctenodrilids may engage reduction in buoyancy to migrate the vertical distance of > 100–4,000 m. This also holds true for schizogamics. In the epitokes swarming is timed by a combination of annual, lunar and diel rhythms. Temperature and photoperiod induce the formation of heteronereis in epigamics and stolon in schizogamics. The subsequent events are all induced by specific pheromones (Fig. 5.5).

In annelids, sex is determined by genes harbored on one or more chromosomes. Karyotypic heterogametism is described in five polychaete species. Selective breeding has led to the discovery of heterogametism in *Dinophilus gyrociliatus* (XX-XO) and *Capitella capitata* (ZW-ZZ) (Table 6.1). For the first time, a directory is assembled for chromosome numbers in annelids. By selective fertilization of large eggs by X-carrying sperm, *D. gyrociliatus* females have nullified the chromosomal mechanism of sex determination. In *C. capitata*, the expression of W gene(s) is relatively more stable and not amenable to environmental factors like density. But the gene(s) in Z chromosome are amenable. Consequently, phenotypic ZZ females and ZZ hermaphrodites are generated. The low level of gametic compatibility between *Hediste* spp and high level of incompatibility between two morphs of *Spirobranchus polycerus* and that for the distantly located populations in *Galeolaria caespitosa* can be explained by the role played by bindin present in the acrosome as an adhesive to attach the sperm to the vitelline layers of eggs.

Despite having a well-developed circulatory system, annelids do not possess a glandular hormonal system. Our understanding of endocrine sexualization and regulation of reproduction cycle is based on temperate polychaetes alone. About a dozen neuroendrocrines/hormones secreted mostly by the brain are responsible for regulation of the reproductive cycle. The role played by stolonization inhibiting hormone arising from the proventriculus is responsible for sexualization in syllids. But its action can be reversed by prostominal hormone. The negative effects of cadmium and elevated pCO_2 are neutralized in the third generation of polychaetes.

Elevation of annelids from an academic interest to economic importance and 'wealth from waste' in vermiculture has been one of the objectives of this book. Possibly engaging symbiotic microbes, the annelids produce valuable fatty acids and vitamins, which cannot be synthesized by finfishes and shellfish. Hence, the worms can serve as valuable live feed in aquaculture. In the ricefields, tubificids and naidids play a complex role in nitrification and denitrification processes. Harvesting them at appropriate intervals may reduce the application of nitrogen fertilizer. Research inputs are urgently required in this vitally important field of rice cultivation. The need for research inputs on basic information on growth and reproduction in cultivable worms is emphasized. An early harvest of 'layers' is recommended. With minimal genetic diversity, parthenogens and cloners are not good candidates for vermiculture. Secondly, cloners rapidly increase in numbers but may not in biomass production. The fast growing *Aporrectodea longa* and *Eudrilus eugeniae* are recommended earthworm species for vermiculture in temperate and tropical zones, respectively. They can be harvested as baits at 4.2 and 2.3 g size on the 60th day (Fig. 8.2A). Cultivation of *Nereis virens* in wastewater containing waste feed and fecal waste in aquaculture system produces 2 g sized worms within 80 days. In rack culture system with water containing activated sludge or aerated sewage or cow dung manure, cultivation of *Tubifex tubifex* is profitable. For the first time *Branchiura sowerbyi* is recognized as a candidate species for vermiculture, as it can be reared in water containing waste paper and harvested at 30 mg size on the 30th day at 25°C (Fig. 8.4D). Among enchytraeids, *Lumbriculus* spp and *Enchytraeus albidus* grow fastest at 20°C. A search for suitable tropical enchytraeids is urgently needed. Oligochaetes lay cocoons in two temporally separated batches (Fig. 8.8A). The second batch contains less number of eggs than the first one. Hence, maintenance of oligochaete brooders till the day, when laying of the first batch of cocoons, is more profitable.

10

References

Adiyodi, K.G. 1988. Annelida. In: *Reproductive Biology of Invertebrates*. (eds) Adiyodi, K.G. and Adiyodi, R.G. Oxford and IBH Publishing, New Delhi, 3: 189–250.

Aguado, M.T., San Martin, G. and Siddall, M.E. 2011. Systematics and evolution of syllids (Annelida: Syllidae). Cladistics, 28: 234–250.

Aguado, M.T., Capa, M., Oceguera-Figueroa, A. and Rouse, G.W. 2014. *The Tree of Life: Evolution and Classification of Living Organisms*. (eds) Vargas, P. and Jardoya, R. Saunder, pp 254–269.

Aguiar, T.M. and Santos, C.S.G. 2017. Reproudctive biology of *Alitta succinea* (Annelida: Nereididae) in a Brazilian tropical lagoon. Invert Biol, 137: 17–28.

Ahearn, G.A. and Gomme, J. 1975. Transport of exogenous D-glucose by the integument of a polychaete worm (*Nereis diversicolor* Muller). J Exp Biol, 62: 243–264.

Aiyer, K.S.P. 1924. Reproduction of the aquatic oligochaete *Nais paraguayensis* Michaelsen. Ann Mag Natl Hist, 14 : 615–616.

Aiyer, K.S.P. 1926. Notes on the aquatic Oligochæta of Travancore. J Natl Hist, 18: 131–142.

Aiyer, K.S.P. 1929. An account of oligochaetes of Travancore. Rec Ind Mus, 31: 13–76.

Akesson, B. 1973. Morphology and life history of *Ophryotrocha maculata* sp (Polychaeta, Dorvilleidae). Zool Scrip, 2: 141–144.

Akesson, B. 1982. A life table study on three genetic strains of *Ophryotrocha diadema* (Polychaeta, Dorvilleidae). Int J Invert Reprod, 5: 59–69.

Akesson, B. and Costlow, J.D. 1991. Effects of constant and cyclic temperatures at different salinity levels on survival and reproduction in *Dinophilus gyrociliatus* (Polychaeta, Dinophilidae). Bull Mar Sci, 48: 485–499.

Akesson, B. and Rice, S.A. 1992. Two new *Dorvillea* species (Polychaeta, Dorvilleidae) with obligate asexual reproduction. Zool Scrip, 21: 351–362.

Alikunhi, K.H. 1951. On the reproductive organs of *Pisione remota* (Southern) together with a review of the family Pisionidae. Proc Ind Acad Sci, 33b: 14–31.

Allen, E.J. 1923. Regeneration and reproduction of the syllid *Procerastea*. Phil Trans R Soc Lond, 211B: 131–177.

Andries, J.C. 2001. Endocrine and environmental control of reproduction in Polychaeta. Can J Zool, 79: 254–270.

Anger, V. 1984. Reproduction in *Pygospio elegans* (Spionidae) in relation to its geographical origin and to environmental conditions: a preliminary report. Fortschrit Zool, 29: 45–51.

Appleby, A.G. and Brinkhurst, R.O. 1970. Defecation rate of three tubificid oligochaetes found in the sediment of Toronto Harbour, Ontario. J Fish Res Bd Can, 27: 1971–1982.

Arias, A., Richter, A., Anadon, A. and Glasby, C.J. 2012. Reproductive biology of the alien Korean bait-worm, *Perinereis vancaurica tetradentata* (Annelida: Nereididae), from the Mar Menor Lagoon (Western Mediterranean). Ecol Impact Biol Invasion, Poster Session 5.

Arias, A., Richter, A., Anadon, A. and Paxton, H. 2013. Evidence of simultaneous hermaphroditism in the brooding *Diopatra marocensis* (Annelida: Onuphidae) from northern Spain. J Mar Biol Ass UK, 93: 1533–1542.

Arp, A.J. and Childress, J.J. 1981. Blood function in the hydrothermal vent vestimenifera tube worm. Science, 213: 342–344.

Aston, R.J. 1968. The effect of temperature on the life cycle, growth and fecundity of *Branchiura sowerbyi* (Oligochaeta: Tubificidae). J Zool, 154: 29–40.

Aston, R.J., Sadler, K. and Milner, A.G.P. 1982. The effects of temperature and the culture of *Branchiura sowerbyi* (Oligochaeta, Tubificidae) on activated sludge. Aquaculture, 29: 137–145.

Aston, R.J. 1984. The culture of *Branchiura sowerbyi* (Tubificidae, Oligochaeta) using cellulose substrate. Aquaculture, 40: 89–94.

Australian Museum Annual Report. 2015. Australian Museum Trust, Sydney, ISSN 2206–8473.

Bahl, K.N. 1947. Excretion in Oligochaeta. Biol Rev, 22: 109–147.

Balavione, G. 2014. Segment formation in annelids: patterns, processes and evolution. Int J Dev Biol, 58: 469–483.

Baldo, L. and Ferraguti, M. 2005. Mixed strategy in *Tubifex tubifex* (Oligochaeta, Tubificidae). J Exp Zool, 303A: 168–177.

Banse, K. 1979. On weight dependence of net growth efficiency and specific respiration rates among field populations of invertebrates. Oecologia, 38: 111–126.

Barnes, R. 1974. *Invertebrate Zoology*. Saunders, International Student Edition, p 870.

Bateman, A.J. 1948. Intra-sexual selection in *Drosophila*. Heredity, 2: 349–368.

Beatty, R.A. 1967. Parthenogenesis in vertebrates. In: *Fertilization*. (eds) Metz, C.B. and Monroy, A. Academic Press, New York, 1: 413–440.

Beckmann, M., Harder, T. and Qian, P.Y. 1999. Induction of larval attachment and metamorphosis in the serpulid polychaete *Hydroides elegans* by dissolved free amino acids: mode of action in laboratory bioassays. Mar Ecol Prog Ser, 190: 167–178.

Bely, A.E. 1999. Decoupling of fission and regenerative capabilities in an oligochaete. Hydrobiologia, 406: 243–251.

Bely, A.E. 2006. Distribution of segment regeneration ability in the Annelida. Integ Comp Biol, 46: 508–518.

Bely, A.E. 2010. Evolutionary loss of animal regeneration: Pattern and process. Integ Comp Biol, 50: 515–527.

Bely, A.E. and Sikes, J.M. 2010. Latent regeneration abilities persist following recent evolutiory loss in asexual annelids. Proc Natl Acad Sci USA, 107: 1464–1469.

Bely, A.E. 2014. Early events in annelid regeneration: A cellular perspective. Integ Comp Biol, 54: 688–699.

Bentley, M.G., Clark, S. and Pacey, A.A. 1990. The role of arachidonic acid and eicosatrienoic acids in the activation of spermatozoa in *Arenicola marina* L. (Annelida: Polychaeta). Biol Bull, 178: 1–9.

Berglund, A. 1986. Sex change by a polychaete: effects of social and reproductive costs. Ecology, 67: 837–845.

Berglund, A. 1990. Sequential hermaphroditism and size advantage hypothesis: an experimental test. Ani Behav, 39: 426–433.

Berglund, A. 1991. To change or not to change sex: a comparison between two *Ophryotrocha* species (Polychaeta). Evol Ecol, 5: 128–135.

Berkeley, E. and Berkeley, C. 1954. Notes on the life history of the polychaete *Dodecaceria fewkesi* (nom. n). J Fish Res Bd Can, 2: 326–334.

Berrill, N.J. 1928. Regeneration in the polychaete *Chaetopterus variopedatus*. J Mar Biol Ass UK, 15: 151–158.

Berrill, N.J. 1952. Regeneration and budding in worms. Biol Rev, 27: 401–438.

Bhattacharjee, G. and Chaudhuri, P.S. 2002. Cocoon production morphology, hatching pattern and fecundity in seven tropical earthworm species—a laboratory based investigation. J Biosci, 27: 283–294.

Bhaud, M.R. and Cazaux, C. 1990. Buoyancy characteristics of *Lanice conchilega* (Pallas) larvae (Terebellidae). Implications for settlement. J Exp Mar Biol Ecol, 141: 31–45.

Bicho, R.C., Santos, F.C.F., Gonclaves, M.F.M. et al. 2015. Enchytraeid reproduction test PLUS: hatching, growth and full life cycle test-an optimal multi-endpoint test with *Enchytraeus crypticus*. Ecotoxicology, 24: 1053–1063.

Biggers, W.J. and Laufer, H. 1992. Chemical induction of settlement and metamorphosis of *Capitella capitata* sp I (Polychaeta) larvae by juvenile hormone-active compounds. Invert Reprod Dev, 22: 39–46.

Bilello, A.A. and Potswald, H.E. 1974. A cytological and quantitative study of neoblasts in the naid *Ophidonais serpentina* (Oligochaeta). Wilhelm Roux Arch Entwicklungsmech Org, 174: 234–249.

Bisby, F.A., Ruggiero, M.A., Wilson, K.L. et al. 2005. Species 2000 & IT IS Catalogue of Life: 2005 Annual Checklist. CD-ROM; Species 2000: Reading, U.K.

Blake, J.A. 1983. Polychaetes of the family Spionidae from South America, Antarctica and adjacent seas and islands. Biology of Antarctic Seas XIV. Antarct Res Ser, 39: 205–288.

Blake, J.A. 1996. Family Spionidae. In: *Taxonomic Atlas of the Santa Maria Basin and Western Santa Barbara Channel*. (eds) Blake, J.A., Hilbig, B. and Scott, P.H. Mus Nat Hist, Santa Barbara, 6 Part 3: 81–223.

Blake, J.A. and Amofsky, P.L. 1999. Reproduction and larval development of the spioniform Polychaeta with application to systematics and phylogeny. Hydrobiologia, 402: 57–106.

Blakemore, J. 2003. Japanese earthworms (Annelida: Oligochaeta): a review and checklist of species. Org Divers Evol, 3/11: 1–43.

Bleidorn, C., Helm, C., Weigart, A. and Aquado, M.T. 2015. Annelida. In: *Evolutionary Development of Invertebrate Biology*. (ed) Wanninger, A. Springer Verlag, Vienna, pp 194–230.

Boi, S. and Ferraguti, M. 2001. Temporal pattern of the double sperm line production in *Tubifex tubifex* (Anellida, Oligochaeta). Hydrobiologia, 463: 103–106.

Boi, S., Fscio, U. and Ferraguti, M. 2001. Nuclear fragmentation characterizes paraspermiogenesis in *Tubifex tubifex* (Annelida: Oligochaeta). Mol Reprod Dev, 59: 442–450.

Boilly, B., Boilly-Marer, Y. and Bely, A.E. 2017. Regulation of dorso-ventral polarity by the nerve cord during annelid regeneration: A review of experimental evidence. Regeneration, 4: 54–68.

Boilly-Marer, Y. 1974. Etude experimentale du comportement nuptial de *Platynereis dumerilii* Aud. et M. Edw. (Annelida: Polychaeta): Chemoreception, emission des produits genitaux. Mar Biol, 24: 167–169.

Boilly-Marer, Y. and Lassalle, B. 1980. Electrophysiological responses of the central nervous system in the presence of homospecific and heterospecific sex pheromones in *Nereis* (Annelida Polychaeta). J Exp Zool, 213: 33–39.

Boilly-Marer, Y. 1984. Comportement nuptial et pheromones d'annklides marines. Oceanis, 10: 169–178.

Boilly-Marer, Y. and Lhomme, M.F. 1986. Sex pheromones in the marine annelid *Platynereis dumerilii* Aud. and M. Edw. In: *Advances in Invertebrate Reproduction*. (eds) Porchet, M., Andries, J-C. and Dhainaut, A. Elsevier, Amsterdam, 4: 1–494.

Bomfleur, B., Mors, T., Ferraguti, M. et al. 2017. Fossilized spermatozoa preserved I a 50-Myr-old annelid cocoon from Antarctica. Biol Let, 11: 20150431.

Bondarenko, N.A., Guselnikova, N.E., Logacheva, N.F. and Pomazkina, G.V. 1996. Spatial distribution of phytoplankton in Lake Baikal, Spring 1991. Freshwat Biol, 35: 517–523.

Bonifazi, A., Ventura, D. and Gravina, M.F. 2016. New records of old species: some pelagic polychaetes along the Indian coast. Italian J Zool, 83: 364–371.

Bouguenec, V. and Giani, N. 1989. Biological studies upon *Enchytraeus variatus* Bouguenec & Giani 1987 in breeding cultures. Hydrobiologia, 180: 151–165.

Bridges, T.S. and Heppell, S. 1996. Fitness consequences of maternal effects in *Streblospio benedictii* (Annelida: Polychaeta). Am Zool, 36: 132–146.

Brinkhurst, R.O. 1964. Observations on the biology of the marine oligochaete *Tubifex costatus*. J Mar Biol Ass UK, 44: 11–16.

Brinkhurst, R.O. 1970. Distribution and abundance of tubificid (Oligochaeta) species in Toronto Harbour, Lake Ontario. J Fish Res Bd Can, 27: 1961–1969.

Brinkhurst, R.O. 1986. Guide to the freshwater aquatic microdrile oligochaetes of North America. Dept Fisher Ocean, Canada, p 268.

Brinkhurst, R.O. and Gelder, S.R. 1991. Annelida: Oligochaeta and Branchiobdellida. In: *Ecology and Classification of North American Freshwater Invertebrates*. (eds) Thorp, J.H. and Covich, A.P. Academic Press, San Diego, pp 401–433.

Britayev, T.A. and Belov, V.V. 1994. Age determination of polynoidea polychaetes based on growth lines on the jaws. Hydrobiol J, 30: 55–60.

Brown, N., Eddy, S. and Plaud, S. 2011. Utilization of waste from a marine recirculating fish culture system as a feed source for the polychaete worm, *Nereis virens*. Aquaculture, 322-323: 177–183.

Bryan, J.P., Kreider, J.L. and Qian, P.Y. 1998. Settlement of the polychaete *Hydroides elegans* on surfaces of the cheilostome bryozoan *Buguia neritina*: Evidence for a chemically mediated relationship. J Exp Mar Biol Ecol, 220: 171–190.

Bull, J.J., Vogt, R.C. and Bulmer, M.G. 1982. Heritability of sex ratio in turtles with environmental sex determination. Evolution, 36: 333–341.

Bullock, T.H. and Horridge, G.A. 1965. *Structure and Function in the Nervous System of Invertebrates*. W.H. Freeman, San Francisco, Volume 2, p 1722.

Bunke, D. 1986. Ultrastructural investigations on the spermatozoon and its genesis in *Aeolosoma litorale* with considerations on the phylogenetic implications for the Aeolosomatidae (Annelida). J Ultr Mol Struct Res, 95: 113–130.

Butt, K.R. 1993. Reproduction and growth of three deep-burrowing earthworms (Lumbricidae) in laboratory culture in order to assess production for soil restoration. Biol Fert Soils, 6: 135–138.

Butt, K.R., Frederickson, J. and Morris, R.M. 1994. Effect of earthworm density on the growth and reproduction of *Lumbricus terrestris* L. (Oligochaeta: Lumbricidae) in culture. Pedobiologia, 38: 254–261.

Butt, K.R., Frederickson, J. and Morris, R.M. 1995. An earthworm cultivation and soil inoculation technique for land restoration. Ecol Eng, 4: 1–9.

Buzhinskaja, G.N. and Smirnov, R.V. 2017. A new species of *Raricirrus* (Polychaeta, Ctenodrilidae) from the continental slope of the Laptev Sea near the Gakkel Ridge. Proc Zool Inst, 321: 425–432.

Bybee, D.R., Bailey-Brock, J.H. and Tamaru, C.S. 2006. Evidence for sequential hermaphroditism in *Sabellastarte spectabilis* (Polychaeta: Sabellidae) in Hawai. Pacific Sci, 60: 541–547.

California Academy of Sciences 2015. *Proceraea okadai* (Imajima, 1966). http://researcharchive.calacademy.org/research/izg/SFBay2K/wormhead.htm.

Cammen, L.M. 1980a. A method for measuring ingestion rate of deposit feeders and its use with the polychaete *Nereis succinea*. Estuaries, 3: 55–60.

Cammen, L.M. 1980b. The significance of microbial carbon in the nutrition of the deposit feeding polychaete *Nereis succinea*. Mar Biol, 61: 9–20.

Cammen, L.M. 1987. Polychaeta. In: *Animal Energetics*. (eds) Pandian, T.J. and Vernberg, F.J. Academic Press, San Diego, 1: 217–260.

Campbell, M. 1955. Asexual reproduction and larval development in *Polydora tetrabranchia* Hartman. Ph.D. Thesis, Duke University, Durham, NC, United States, 67 p.

Cannarsa, E., Lorenzi, M.C. and Sella, G. 2015. Early social conditions affect female fecundity in hermaphrodites. Cur Zool, 61: 983–990.

Carson, H.S. and Hentschel, B. 2006. Estimating the dispersal potential of polychaete species in the Southern California Bight: implications for designing marine reserves. Mar Ecol Prog Ser, 316: 105–113 (with an appendix of 28 pages).

Casellato, S., Piero, S.D., Masiero, L. and Covre, V. 2011. Fluoride toxicity and its effects on gametogenesis in the aquatic oligochaete *Branchiura sowerbyi* Beddard. Res Rep Fluoride, 46: 7–18.

Caspers, H. 1984. Spawning periodicity and habitat of the palolo worm *Eunice viridis* (Polychaeta: Eunicidae) in the Samoan Islands. Mar Biol, 79: 229–236.

Cassai, C. and Prevedelli, D. 1998a. Reproductive effort, fecundity and energy allocation in two species of the genus *Perinereis* (Polychaeta: Nereididae). Invert Reprod Dev, 34: 125–131.

Cassai, C. and Prevedelli, D. 1998b. Reproductive effort, fecundity and energy allocation in *Marphysa sanguinea* (Polychaeta: Eunicidae). Invert Reprod Dev, 34: 133–138.

Castro-Ferreira, M.P., Roelofs, D., Cornelis, A.M. et al. 2012. *Enchytraeus crypticus* as model species in soil ecotoxicology. Chemosphere, 87: 1222–1227.

Cavanaugh, C.M., Gardiner, S.L., Jones, M.L. et al. 1981. Prokaryotic cells in the hydrothermal vent tube worm *Riftia pachyptila* Jones: Possible chemoautotrophic symbionts. Science, 213: 340–342.

Chapman, A.D. 2009. Number of living species in Australia and the world. Report for the Australian Biological Study. Department of Environment, Government of Australia, p 80.

Charnov, E.L. 1982. *The Theory of Sex Allocation*. Princeton, New Jersey: Princeton University Press.

Charnov, E.L. 1987. Local mate competition and sex ratio in the diploid worm *Dinophilus*. Int J Invert Reprod Dev, 12: 223–225.

Chatarpaul, L., Robinson, J.B. and Kaushik, N.K. 1980. Effects of tubificid worms on denitrification and nitrification in stream sediment. Can J Fish Aquat Sci, 37: 656–663.

Chatelain, E.H., Breton, S., Lemieux, H. and Blier, P.U. 2008. Epitoky in *Nereis* (Neanthes) *virens* (Polychaeta: Nereididae): A story about sex and death. Comp Biochem Physiol, 149B: 202–208.

Chatelliers, M.C., Juget, J., Lafont, M. and Martin, P. 2009. Subterranean aquatic Oligochaeta. Freshwat Biol, 54: 678–690.

Chevaldonne, P., Jollivet, D., Vangriesheim, A. and Desbruyeres, D. 1997. Hydrothermal vent alvinellid polychaete dispersal in the eastern Pacific. 1. Influence of vent site distribution, bottom currents and biological patterns. Limnol Oceanogr, 42: 67–80.

Cho, S.J., Valles, Y. and Weisblat, D.A. 2014. Differential expression of conserved germ line markers and delayed segregation of male and female primordial germ cells in a hermaphrodite, the leech *Helobdella*. Mol Biol Evol, 31: 341–354.

Christensen, B. 1959. Asexual reproduction in the Enchytraeidae (Oligochaeta). Nature, 184: 1159–1166.

Christensen, B. 1960. A comparative cytological investigation of the reproductive cycle of an amphimictic diploid and parthenogenetic triploid form of *Lumbricillus lineatus* (O.F.M.) (Oligochaeta: Enchytraeidae). Chromosoma, 11: 365–369.

Christensen, B. 1961. Studies on cyto-taxonomy and reproduction in the Enchytraeidae. With notes on parthenogenesis and polyploidy in the animal kingdom. Hereditas, 47: 387–450.

Christensen, B. 1973. Density dependence of sexual reproduction in *Enchytraeus bigeminus* (Enchytraeidae). Oikos, 24: 287–294.

Christensen, B., Jelne, F. and Berg, I. 1978. Long term isozyme variation in parthenogenetic polyploidy form of *Lumbricillus lineatus* (Enchytraeidae: Oligochaeta) in recently established environments. Hereditas, 88: 65–73.

Christensen, B. 1980. Constant differential forms of *Lumbricillus lineatus* (Enchytraeidae) (Oligochaeta). Heriditas, 92: 193–198.

Christensen, B. 1984. Asexual propogation and reproductive strategies in aquatic oligochaeta. Hydrobiologia, 115: 91–95.

Christensen, B., Hvilsom, M.M. and Pedersen, B.V. 1989. On the origin of clonal diversity in parthenogentic *Fridericia striata* (Enchytraeidae, Oligochaeta). Hereditas, 110: 89–91.

Christensen, B., Hvilson, M. and Pedersen, B.V. 1992. Genetic variation in co-existing sexual diploid and parthenogenetic triploid forms *Fridericia galba* (Enchytraeidae, Oligochaeta). Heriditas, 110: 89–91.

Chu, J.W. and Levin, L.A. 1989. Photoperiod and temperature regulation of growth and reproduction in *Streblospio benedicti* (Polychaeta, Spionidae). Invert Reprod Dev, 15: 131–142.

Chu, J. and Pai, S. 1944. The relations between natural fission and regeneration in *Stylaria fossularis* (Annelida). Physiol Zool, 17: 159–166.

Cikutova, M.A., Fitzpartrick, L.C., Venables, B.J. and Goven, A.J. 1993. Sperm count in earthworm (*Lumbricus terrestris*) as a biomarker for environmental toxicology: effects of cadmium and chlordane. Env Pollution, 81: 123–125.

Clark, R.B. and Bonney, D.G. 1960. Influence of supra-oesophageal ganglion on posterior regeneration in *Nereis diversicolor*. J Embryol Exp Morph, 8: 112–118.

Clark, R.B., Clark, M.E. and Ruston, R.J.G. 1962. The endocrinology of regeneration in some errant polychaetes. In: *Neurosecretion*. (eds) Heller, H. and Clarke, R.B. Academic Press, New York, 12: 275–286.

Clavier, J. 1984. Production due to regeneration by *Euclymene oerstedi* (Claparède) (Polychaeta: Maldanidae) in the maritime basin of the Rance (Northern Britanny). J Exp Mar Biol Ecol, 75: 97–106.

Coates, K.A. 1995. Widespread polyploidy forms of *Lumbricillus lineatus* (Muller) (Enchytraeidae: Oligochaeta): comments on polyploidism in the enchytraeids. Can J Zool, 73: 1727–1734.

Collado, R., Hass-Cordes, E. and Schmelz, R. 2011. Microtaxonomy of fragmenting *Enchytraeus* species using molecular markers, with a comment on species complexes in enchytraeids. Turk J Zool, 36: 85–94.

Cornec, J.P., Cresp, J., Delye, P. et al. 1987. Tissue responses and organogenesis during regeneration in the oliogochaete *Limnodrilus hoffmeisteri* (Clap.). Can J Zool, 65: 403–414.

Corpas, I., Gaspar, I., Martinez, S. et al. 1995. Testicular alterations in rats due to gestational and early lactational administration of lead. Reprod Toxicol. 14: 57–62.

Costopulos, J.J., Stephens, G.C. and Wright, S.H. 1979. Uptake of amino acids by marine polychaetes under the anoxic conditions. Biol Bull, 157: 434–444.

Cowart, J.D., Fielman, K.T., Woodin, S.A. and Lincoln, D.E. 2000. Halogenated metabolites in two marine polychaetes and their planktotrophic and lecithotrophic larvae. Mar Biol, 136: 993–1002.

Creaser, E.P., Jr. 1973. Reproduction of the bloodworm (*Glycera dibranchiate*) in the Sheepscot estuary. Mar J Fish Res Bd Can, 30: 161–166.

Creaser, P. and Clifford, D.A. 1982. Life history studies of the sandworm, *Nereis virens* Sars, in the Sheepscot estuary, Maine. Fish Bull, 80: 735–743.

Cuomo, M.C. 1985. Sulphide as a larval settlement cue for *Capitella* sp I. Biogeochemistry, 1: 169–181.

Currie, D.R., McArthur, M.A. and Cohen, B.F. 2000. Reproduction and distribution of the invasive European fanworm *Sabella spallanzanii* (Polychaeta: Sabellidae) in Port Phillip Bay, Victoria, Australia. Mar Biol, 136: 645–656.

Curry, J.P. and Bolger, T. 1984. Growth, reproduction and litter and soil consumption by *Lumbricus terrestris* L. in reclaimed peat. Soil Biol Biochem, 16: 253–257.

Daisy, N.P., Subramanian, E.R., Christyraj, J.D.S. et al. 2016. Studies on regeneration of central nervous system and social ability of the earthworm *Eudrilus eugeniae*. Invert Neurosci, 16: 1–13.

Daly, J.M. 1972. The maturation and breeding biology of *Harmothoe imbricata*. Mar Biol, 12: 53–66.

Daly, J.M. 1973. Segmentation, autotomy and regeneration of lost posterior segments in *Harmothoe imbricata* (L.) (Polychaeta: Polynoidae). Mauri Ora, 1: 17–28.

Daly, J.M. 1975. Reversible epitoky in the life history of the polychaete *Odontosyllis polycera* (Schmarda, 1861). J Mar Biol Ass UK, 55: 327–344.

Daly, J.M. 1978a. The annual cycle and the short term periodicity of breeding in a Northumberland population of *Spirorbis spirorbis* (Polychaeta: Serpulidae). J Mar Biol Ass UK, 58: 161–176.

Daly, J.M. 1978b. Growth and fecundity in a Northumberland population of *Spirorbis spirorbis* (Polychaeta: Serpulidae). J Mar Biol Ass UK, 58: 177–190.

Dash, M.C. and Patra, U.C. 1977. Density, biomass and energy budget of a tropical earthworm population in Orissa, India. Rev Ecol Biol Sol, 14: 461–471.

Dash, M.C. and Patra, U.C. 1979. Worm cost production and nitrogen concentration to soil by a tropical earthworm population in Orissa, India. Rev Ecol Biol Sol, 16: 79–83.

Dash, M.C., Satpathy, B., Behera, N. and Dei, C. 1984. Gut load and turnover of soil, plant and fungal material by *Drawida calebi*, a tropical earthworm. Rev Ecol Biol Sol, 21: 387–393.

Dash, M.C. 1987. The other annelids. In: *Animal Energetics*. (eds) Pandian, T.J. and Vernberg, F.J. Academic Press, San Diego, 1: 261–299.

Dash, H.K., Beura, B.N. and Dash, M.C. 1986. Gut load and transit time in some tropical earthworm. Pedobiologia, 29: 13–20.

David, A.A. and Williams, J.D. 2011. Asexual reproduction and anterior regeneration under high and low temperatures in the sponge associate *Polydora colonia* (Polychaeta: Spionidae). Invert Reprod Dev: 1–10.

David, A.A. and Williams, J.D. 2016. The influence of hypoosmatic stress on the regenerative capacity of the invasive polychaete *Marenzelleria viridis* (Annelids: Spionidae) from its native range. Mar Ecol, 37: 821–830.

Davies, R.W. and Singhal, R.N. 1987. The chromosome numbers of five North American and European leech species. Can J Zool, 68: 681–684.

Davies, R.W. and McLoughin, N.J. 1996. The effects of feeding regime on the growth and reproduction of the medicinal leech *Hirudo medicinalis*. Fresh Biol, 36: 563–568.

Davila-Jimenez, Y., Tovar-Hernandez, M.A. and Simoes, N. 2017. The social feather duster worm *Bispira brunnea* (Polychaeta: Sabellidae): aggregations, morphology and reproduction. Mar Biol Res, 13: 782–796.

Davison, A. 2006. The ovotestis: an underdeveloped organ of evolution. Bioessays, 28: 642–650.

De Vlas, J. 1979. Secondary production by tail regeneration in a tidal flat population of lugworms (*Arenicola marina*) cropped by flatfish. Neth J Sea Res, 13: 362–393.

Dean, H.K. 1995. A new species of *Raricirrus* (Polychaeta: Ctenodrilidae) from wood collected in the Tongue of the Ocean, Virgin Islands. Proc Biol Soc Washington, 108: 169–179.

Degraer, S., Wittoeck, J., Appeltans, W. et al. 2006. The macrobenthos atlas of the Belgian part of the North Sea. Belgian Science Policy. D/2005/1191/3. ISBN 90-810081-6-1, 164 pp.

Dhainaut, A. 1970. Etude cytochimique et ultrastructurale de l'evolution ovocytaire de *Nereis pelagica* L. (Annelide Polychete). I. Ovogenese naturelle. Z Zellforsch Mikrosk Anat, 104: 375–389.

Diaz Cosin, D.J., Novo, M. and Fernandez, R. 2011. Reproduction in earthworms: Sexual selection and parthenogenesis. In: *Biology of Earthworms*. (ed) Karaca, A., Springer Verlag, Berlin, pp 69–86.

Dixon, D.R. 1976. The energetic of growth and reproduction in the brackish water serpulid polychaete, *Mercierella enigmatica* Fauvel. In: *Population Dynamics*. Proc 10th Euro Symp Mar Biol. (eds) Persoone, G. and Jastpers, E., Universa Press, Wetteren, Belgium, pp 197–209.

Dixon, D.R. 1980. The energetic of tube production by *Mercierella enigmatica* (Polychaeta: Serpulidae). J Mar Biol Ass UK, 60: 655–659.

Domínguez, J., Edwards, C.A. and Ashby, J. 2001. The biology and ecology of *Eudrilus eugeniae* (Kinberg) (Oligochaeta) bred in cattle wastes. Pedobiologia, 45: 341–353.

Dominguez, J., Velando, A., Aira, M. and Monroy, F. 2003. Uniparental reproduction of *Eisenia foetida* and *E.andrei* (Oligochaeta: Lumbricidae): evidence of self-insemination. Pedobologia, 47: 530–534.

Drewes, C.D. and Fourtner, C.R. 1990. Morphallaxis in an aquatic oligochaete, *Lumbriculus variegatus*: Reorganization of escape reflexes in regenerating body fragments. Dev Biol, 138: 94–103.

Drewes, C.D. and Fourtner, C.R. 1991. Reorganization of escape reflexes during asexual fission in an aquatic oligochaete, *Dero digitata*. J Exp Zool, 260: 170–180.

Dualan, I.V. and Williams, J.D. 2011. Palp growth, regeneration, and longevity of the obligate hermit crab symbiont *Dipolydora commensalis* (Annelida: Spionidae). Invert Biol, 130: 264–276.

Dubilier, N. 1988. H2S—A settlement cue or a toxic substance for *Capitella* sp I larvae. Biol Bull, 174: 30–38.

Dubilier, N., Giere, O., Distel, D.L. and Cavanaugh, C.M. 1995. Characterization of chemoautotrophic bacterial symbionts in a gutless marine worm (Oligochaeta, Annelida) by phylogenetic 16S rRNA sequence analysis and *in situ* hybridization. Appl Env Microbiol, 61: 2346–2350.

Durchon, M. 1952. Recherches expérimentales sur deux aspects de la reproduction chez les Annélides Polychètes: l'épitoquie et la stolonisation. Ann Sci Nat Zool Biol Anim, 14: 119–206.

Durchon, M. 1975. Sex reversal in the Syllinae (Polychaeta: Annelida). In: *Intersexuality in the Animal Kingdom.* (ed) Reinboth, R. Springer-Verlag, Heidelbeg, pp 41–47.

E Costa, P.F., Oliveira, R.F. and da Fonseca, L.C. 2006. Feeding ecology of *Nereis diversicolor* (O.F. Muller) (Annelida: Polychaeta) on estuarine and lagoon environments in the southwest Coast of Portugal. PANAMJAS, 1: 114–126.

Eckelbarger, K.J. 1973. The reproductive biology and development of the terebellid polychaete, *Nicolea zostericola* (Oersted: Grube, 1860) and the origin and role of coelomic cells in oogenesis, Ph.D. thesis, Northeastern University, Boston.

Eckelbarger, K.J. 1975. A light and electron microscope investigation of gametogenesis in *Nicolea zostericola* (Polychaeta: Terebellidae). Mar Biol, 30: 353–370.

Eckelbarger, K.J. 1983. Evolutionary radiation in polychaete ovaries and vitellogenic mechanisms and their role in life history patterns. Can J Zool, 61: 487–504.

Eckelbarger, K.J. 1984. Comparative aspects of oogenesis in polychaetes. In: *Polychaete Reproduction. Progress in Comparative Reproductive Biology.* (eds) Fischer, A. and Pfannenstiel, H.D., Fortschr Zool, 29: 123–148.

Eckelbarger, K.J. 1986. Vitellogenic mechanisms and the allocation of energy to offspring in polychaetes. Bull Mar Sci, 39: 426–443.

Eckelbarger, K.J. 2005. Oogenesis and oocytes. Hydrobiologia, 535/536: 179–198.

Eckelbarger, K.J. 2006. Oogenesis. In: *Reproductive Biology and Phylogeny of Annelida.* (eds) Rouse, G.W. and Pleijel, F., Science Publishers, Enfield, New Hampshire, pp 23–44.

Engelstadter, J. 2008. Constraints on the evolution of asexual reproduction. Bio Essays, 30: 1138–1150.

Ernst, W. and Goerke, H. 1969. Aufnahme und Umwandlung geloster Glucose-^{14}C durch *Lanice conchilega* (Polychaeta, Terebellidae). Veroff Inst Meeresforsch Bremerh, 11: 313.

Erséus, C. 1994. The Oligochaeta. In: *Taxonomic Atlas of the Santa Maria Basin and Western Santa Barbara Channel.* (eds) Blake, J.A. and Hilbig, B. Mus Nat Hist, Santa Barbara, California, 4: 5–38.

Erseus, C. 1999. Sperm types and their use for a phylogenetic analysis of aquatic clitellates. Hydrobiologia, 402: 225–237.

Erseus, C. 2003. The gutless Tubificidae (Annelida: Oligochaeta) of the Bahamas. Meiofauna Marina, 12: 59–84.

Evjemo, J.O. and Olsen, Y. 1997. Lipid and fatty acid content in cultivated organisms compared to marine copepods. Hydrobiologia, 358: 159–162.

Fage, L. and Legendre, R. 1927. Peches planctoniques a la lumiere, effectuees it Banyuls-sur-Mer et 11 Concarneau. I. Annelides Polychetes. Arch Zool, 67: 23–222.

Fairchild, E.A., Bergman, A.M. and Trushenski, J.T. 2017. Production and nutritional composition of white worms *Enchytraeus albidus* fed different low-cost feeds. Aquaculture, 481: 16–24.

Falconi, R., Renzulli, T. and Zaccanti, F. 2006. Survival and reproduction in *Aeolosoma viride* (Annelida, Aphanoneura). Hydrobiologia, 564: 95–99.

Falconi, R., Gugnali, A. and Zaccanti, F. 2015. Quantitative observations on asexual reproduction of *Aeolosoma viride* (Annelida, Aphanoneura). Invert Biol, 134: 151–161.

Faulkner, G.H. 1930. The anatomy and the histology of bud-formation in the serpulid *Filograna implexa*, together with some cytological observations on the nuclei of the neoblasts. J Linn Soc Zool, 37: 109–190.

Faulwetter, S., Markantonatou, V., Pavloudi, C. et al. 2014. Polytraits: A database on biological traits of marine polychaetes. Biodiversity Data J, 2: e1024, doi: 10.3897/BDJ.2.e1024.

Fauvel, P. 2010. Annelida polychaeta of the Indian Museum, Calcutta. Mem Ind Museum, 12: 1–262.

Felbeck, H. 1981. Chemoautotrophic potential of the hydrothermal vent tube worm, *Riftia pachyptila* Jones (Vestimentifera). Science, 213: 336–338.

Ferguson, J.C. 1971. Uptake and release of free amino acids by starfishes. Biol Bull, 141: 22–129.

Ferguson, J.C. 1982. A comparative study of the net metabolic benefits derived from the uptake and release of free amino acids by marine invertebrates. Biol Bull, 162: 1–7.

Fernández, R., Novo, M., Gutiérrez, M. et al. 2010. Life cycle and reproductive traits of the earthworm *Aporrectodea trapezoides* (Dugès, 1828) in laboratory cultures. Pedobiologia, 53: 295–299.

Ferraguti, M. 1984. Cytological aspects of oligochaete spermiogenesis. Hydrobiologia, 115: 59–64.

Ferraguti, M. and Erseus, E. 1999. Sperm types and their use for a phylogenetic analysis of aquatic clitellates. Hydrobiologia, 402: 225–237.

Ferraguti, M., Erseus, C., Kaygorodova, I. and Martin, P. 1999. New sperm types in Naididae and Lumbriculidae (Annelida: Oligochaeta) and their possible phylogenetic implications. Hydrobiologia, 406: 213–222.

Fewou, J. and Dhainaut-Courtois, N. 1995 Research on polychaete annelid osmoregulatory peptide(s) by immunocytochemical and physiological approaches. Computer reconstruction of the brain and evidence for a role of angiotension-like molecules in *Nereis* (*Hediste*) *diversicolor* O F Muller. Biol Cell, 85: 21–33.

Fielde, A.M. 1885. Observations on tenacity of life and regeneration of excised parts in *Lumbricus terrestris*. Proc Acad Nat Sci Philadelphia, 37: 20–22.

Finogenova, N.P. and Lobasheva, T.M. 1987. Gorwth of *Tubifex tubifex* Muller (Oligochaeta, Tubificidae) under various trophic conditions. Int Rev Hydrobiol, 72: 709–726.

Fischer, A. 1974. Stages and stage distribution in early oogenesis in the annelid *Platynereis dumerilii*. Cell Tiss Res, 156: 35–45.

Fischer, A. 1985. Reproduction and postembryonic development of the annelid *Platynereis dumerilii*. Film no. 1577 (en). IWF Gottingen.

Fischer, A., Meves, K. and Franke, H-D. 1992. Stolonization and mating behaviour in *Autolytus prolifer* (Annelida, Polychaeta). Film No. C 1799, Institut fur den Wissenschaftlichen Film (IWF), Gottingen, Germany.

Fischer, A. and Fischer, U. 1995. On the life-style and life-cycle of the luminescent polychaete *Odontosyllis enopla* (Annelida: Polychaeta). Invert Biol, 114: 236–247.

Fischer, A. 1999. Reproductive and developmental phenomena in annelids: a source of exemplary research problems. Hydrobiologia, 402: 1–20.

Fischer, E.A. and Petersen, C.W. 1987. The evolution of sexual patterns in the seabasses. Bioscience, 37: 482–489.

Fishelson, L. 1971 Ecology and distribution of the benthic fauna in the shallow waters of the Red Sea. Mar Biol, 10: 113–133.

Fong, P.P. and Pearse, J.S. 1992. Photoperiodic regulation of parturition in the self-fertilizing viviparous polychaete *Neanthes limnicola* from central California. Mar Biol, 112: 81–89.

Franke, H.D. 1976. Experinrentelle Unterschungen uber die Fortplanzung und Entwicklung des Polychaeten *Syllis* (*Typosyllis*) *prolifera* Krobn in Laborzuchten. Diplom-Arbeit, Technische Universitat Braunschweig, pp 1–59.

Franke, H.D. 1977. Die Rolle des Procenvrikels bei der Steuerung von Stolonisariorithmus und sexueller Differenzierurng des Polychaeten *Typosyllis prolifera*. Verh Dtsch Zool Ges, pp 977–324.

Franke, H.D. 1980. Zur Determination der zeitlichen Yerteilung von Fortpflanzungsprozessen in Laborkulturen des Polychaeten *Tvposvllis prolifera*. Helgolander Meeresunters, 34: 61–84.

Franke, H.D. 1985. On a clocklike mechanism timing lunar rhythmic reproduction in *Typosyllis prolifera* (Polychaeta). J Comp Physiol, 156A: 553–561.

Franke, H.D. 1986a. The role of light and endogenous factors in the timing of the reproductive cycle of *Typosyllis prolifera* and some other polychaetes. Am Zool, 26: 433–445.

Franke, H.D. 1986b. Sex ratio and sex change in wild and laboratory populations of *Typosyllis prolifera* (Polychaeta). Mar Biol, 90: 197–208.

Franke, H.D. 1999. Reproduction of the Syllidae (Annelida: Polychaeta). Hydrobiologia, 402: 39–55.

Gambi, M.C., Giangrande, A. and Patti, F.P. 2000. Comparative observations on reproductive biology of four species of *Perkinsiana* (Polychaeta: Sabellidae: Sabellinae). Bull Mar Sci, 67: 299–309.

Gambi, M.C., Patti, F.P., Micaletto, G. and Giangrande, A. 2001. Diversity of reproductive features in some Antarctic polynoid and sabellic polychaetes with a description of *Demonax polarsterni* sp n. (Polychaeta, Sabellidae). Polar Biol, 24: 883–891.

Garcia-Alonso, J., Hoeger, U. and Rebscher, N. 2006. Regulation of vitellogenesis in *Nereis virens* (Annelida: Polychaeta): effect of estradoil-17 beta on eleocytes. Comp Biochem Physiol, 143: 55–61.

Garcia-Alonso, J., Smith, B.D. and Rainbow, P.S. 2013. A compacted culture system for a marine polychaete (*Platynereis dumerilii*). PANAMJAS, 8: 142–146.

Garraffoni, A.R.S., Yokayama, L.Q. and Amaral, A.C.Z. 2014. The gametogenic cycle and life history of *Nicolea uspiana* (Polychaeta: Terebellidae) on the south east coast of Brazil. J Mar Biol Ass UK, 94: 925–933.

Garwood, P.R. and Olive, P.J.W. 1982. The influence of photoperiod on oocyte growth and its role in the control of the reproductive cycle of the polychaete *Harmothoe imbricata*. Int J Invert Reprod, 5: 161–165.

Gaston, G.R. and Hall, J. 2000. Lunar periodicity and bioluminescence of swarming *Odontosyllis luminosa* (Polychaeta: Syllidae) in Belize. Gulf Caribbean Res, 12: 47–51.

Gates, G.E. 1927. Regeneration in a tropical earthworm *Perionyx excavatus* E. Perr. Biol Bull, 53: 351–364

Gates, G.E. 1941. Further notes on regeneration in a tropical earthworm, *Perionyx excavatus* E. Perrier 1872. J Exp Zool, 82: 161–185.

Gates, G.E. 1950. Regeneration in an earthworm *Eisenia foetida*. II posterior regeneration. Biol Bull, 98: 36–45.

Gates, G.E. 1971. On reversions to former ancestral conditions in megadrile oligochaetes. Ecolution, 25: 245–248.

Gazave, E., Behague, J., Laplane, L. et al. 2013. Posterior elongation in the annelid *Platynereis dumerilii* involves stem cells molecularly related to primordial germ cells. Dev Biol, 382: 246–267.

Ghiselin, M.T. 1969. The evolution of hermaphroditism among animals. Q Rev Biol, 44: 189–208.

Giangrande, A. 1997. Polychaete reproductive patterns, life cycles and life histories: an overview. Oceanograph Mar Biol Ann Rev, 35: 323–86.

Giangrande, A., Licciano, M., Schirosi, R. et al. 2014a. Chemical and structural defensive external strategies in six sabellid worms (Annelida). Mar Ecol, 35: 36–45.

Giangrande, A., Pierri, C., Fanelli, G. et al. 2014b. Rearing experiences of the polychaete *Sabella spallanzanii* in the Gulf of Taranto (Mediterranean Sea, Italy). Aquacult Int, 22: 1677–1688.

Gibbs, P.E. 1978. Macrofauna of the intertidal sand flats on low wooded islands, Northern Great Barrier Reef. Phil Trans R Soc London, 284B: 81–97.

Gibson, G.D. 1997. Variable development in the spionid *Boccardia proboscidea* (Polychaeta) is linked to nurse egg production and larval trophic mode. Invert Biol, 116: 213–226.

Gibson, G.D., Paterson, I., Taylor, H. and Woolridge, B. 1999. Molecular and morphological evidence of a single species, *Boccardia proboscidea* (Polychaeta) with multiple development modes. Mar Biol, 134: 743–751.

Gibson, G.D. and Harvey, J.M.L. 2000. Morphogenesis during asexual reproduction in *Pygospio elegans* Claparede (Annelida, Polychaeta). Biol Bull, 199: 41–49.

Gibson, G.D. and Paterson, I.G. 2003. Morphogenesis during sexual and asexual reproduction in *Amphipolydora vestalis* (Polychaeta: Spionidae). NZJ Mar Freshwat Res, 37: 741–752.

Gibson, P.H. 1977. Reproduction in the cirratulid polychaete, *Dodecaceria concharum* and *D. pulchra*. J Zool, 182: 89–102.

Gibson, P.H. 1978. Systematics of *Dodecaceria* (Annelida: Polychaeta) and its relation to the reproduction of its species. J Lin Soc Zool, 63: 275–287.

Gibson, P.H. 1979. The specific status of two cirratulid polychaetes, *Dodecaceria fimbriata* and *D. caulleryi*, compared by their morphology and methods of reproduction. Can J Zool, 57: 1443–1451.

Gibson, P.H. 1981. Gametogenesis in the cirratulid polychaetes *Dodecaceria concharum* and *D. caulleryi*. Proc Zool Soc Lond, 193: 355–370.

Giere, O. 1979. Studies on marine Oligochaeta from Bermuda, with emphasis on new *Phallodrilus*-species (Tubificidae). Cah Biol Mar, 20: 301–314.

Giere, O. 1981. The gutless marine oligochaete *Phallodrilus leukodermatus*. Structural studies on an aberrant tubificid associated with bacteria. Mar Ecol Prog Ser, 5: 353–357.

Giere, O., Felbeck, H., Dawson, R. and Liebezeit, G. 1984. The gutless oligochaete *Phallodrilus leukodermatus* Giere, a tubificid of structural, ecological and physiological relevance. Hydrobiologia, 115: 83–89.

Giere, O., Conway, N.M., Gastrock, G. and Schmidt, C. 1991. 'Regulation' of gutless annelid ecology by endosymbiotic bacteria. Mar Ecol Prog Ser, 68: 287–299.

Giere, O. and Erseus, C. 2002. Taxonomy and new bacterial symbioses of gutless marine Tubificidae (Annelida, Oligochaeta) from the Island of Elba (Italy). Org Div Evol, 2: 289–297.

Giere, O. 2006. Ecology and biology of marine oligochaeta—an inventory rather than another review. Hydrobiologia, 564: 103–116.

Gillis, P.L., Diener, L.C., Reynoldson, T.B. and Dixon, D.G. 2002. Cadmium-induced production of a metallothionein like protein in *Tubifex tubifex* (Oligochaeta) and *Chironomus riparius* (Diptera): Correlation with reproduction and growth. Env Taxicol Chem, 21: 1836–1844.

Gilpin-Brown, J.B. 1959. The reproduction and larval development of *Nereis fucata* (Savigny). J Mar Biol Ass UK, 38: 65–80.

Goerke, H. 1971. Nahrungsaufnahme, Nahrungsausnutzung und Wachstum von *Nereis viridens* (Polychaeta, Nereidae). Veroeff Inst Meersforsch Bremerh, 13: 51–78.

Golding, D.W. 1967. Endocrinology, regeneration and maturation in *Nereis*. Biol Bull, 133: 567–577.

Golding, D.W. 1983. Endocrine programmed development and reproduction in *Nereis*. Gen Comp Endocrinol, 52: 456–466.

Golding, D.W. and Yuwono, E. 1994. Latent capacities for gametogenic cycling in the semelparous invertebrate *Nereis*. Proc Natl Acad Sci USA, 91: 11777–11781.

Goncalves, M.F.M., Gomes, S.I.L. and Soares, A.M.V.M. 2017. *Enchytraeus crypticus* fitness: effect of density on a two-generation study. Ecotoxicology, doi 10.1007/s10646-017-1785-4.

Gordon, D.C. 1966. The effects of the deposit feeding polychaete *Pectinaria gouldii* on the intertidal sediments of Barnstable harbor1. Limnol Oceanogr, 11: 327–332.

Goss, R.J. 1974. *Regeneration*. Georg Thieme, Stuttgart.

Govedich, F.R and Bain, B.A. 2005. All about Leeches. www.leechtherapy/health.

Gremare, A. 1986. A comparative study of reproductive energetic in two populations of the terebellid polychaete *Eupolymnia nebulosa* Montagu with different reproductive modes. J Exp Mar Biol Ecol, 96: 287–302.

Gremare, A. and Olive, P.J.W. 1986. A preliminary study of fecundity and reproductive effort in two polychaetous annelids with contrasting reproductive strategies. Int J Invert Reprod Dev, 9: 1–16.

Gremare, A., Amouroux, J.M. and Vetion, G. 1998. Long-term comparison of macrobenthos within the soft bottoms of the Bay of Banylus-sur-mer (Northwestern Mediterranean Sea). J Sea Res, 40: 281–302.

Greve, W. 1974. Planktonic spermatophores found in a culture device with spionid polychaetes. Helgolander wiss Meeresunters, 26: 370–374.

Grimm, R. 1987. Contributions towards the taxonomy of the African naididae (Oligochaeta). IV. Zoogeographical and taxonomical considerations on African Naididae. Dev Hydrobiol Aquat Oligochaet, 40: 27–37.

Guerin, J.P. 1991. Elevage de Spionides (Annelides, Polychetes) en cycle complet. 3. Description du developpement larvaire de *Boccardia semibranchiata*. Ann Inst Oceanogr Monneo, 67: 145–154.

Guerrero III, R.D. 2009. Earthworm culture for vermicompost and vermimeal production and for vermiceutical application in the Phliippines (1978–2008)—A review. Dyn Soil Dyna Plant, 3: 28–31.

Gusso, C.C., Gravina, M.F. and Maggiore, F.R. 2001. Temporal variations in soft bottom benthic communities in central Tyrrhenian Sea (Italy). Archo Oceanogr Limnol, 22: 175–182.

Hadfield, M.G., Nedved, B.T., Wilbur, S. and Koehl, M.A.R. 2014. Biofilm cue for larval settlement in *Hydroides elegans* (Polychaeta): is contact necessary. Mar Biol, 161: 2577–2587.

Hafer, J., Fischer, A. and Frenz, H.J. 1992. Identification of the yolk receptor protein in oocytes of *Nereis virens* (Annelids, Polychaeta) and comparison with the locust vitellogenin receptor. J Comp Physiol, 162B: 148–152.

Halt, M.N., Rouse, G.W., Petersen, M.E. and Pleijel, F. 2006. Cirratuliformia. In: *Reproductive Biology and Phylogeny of Annelida*. (eds) Rouse, G. and Pleijel, F., Science Publishers, Enfield, New Hampshire, pp 497–520.

Hamond, R. 1969. On the preferred foods of some autolytoids (Polychaeta, Syllidae). Cah Biol Mar, 10: 439–445.

Hanafiah, Z., Sato, M., Nakashima, H. and Tosuji, H. 2006. Reproductive swarming of sympatric nereidid polychaetes in an Estuary of the Omuta-gawa river in Kyushu, Japan, with special reference to simultaneous swarming of two *Hediste* species. Zool Sci, 23: 205–217.

Hardege, J.D., Bentley, M.G., Beckmann, M. and Muller, C. 1996. Sex pheromones in marine polychaetes: volatile organic substances (VOS) isolated from *Arenicola marina*. Mar Ecol Prog Ser, 139: 157–166.

Hardege, J.D., Muller, C., Beckmann, M. and Hardege, H.D.B. 1998. Timing of reproduction in marine polychaetes: The role of sex pheromones. Ecoscience, 5: 395–404.

Hardege, J.D. 1999. Nereidid polychaetes as model organisms for marine chemical ecology. Hydrobiologia, 402: 145–161.

Harder, T. and Qian, P.Y. 1999 Induction of larval attachment and metamorphosis in the serpulid polychaete *Hydroides elegans* by dissolved free amino acids: isolation and identification (Haswell). Mar Ecol Prog Ser, 179: 259–271.

Harder, T., Lam, C. and Qian, P.Y. 2002. Induction of larval settlement in the polychaete *Hydroides elegans* by marine biofilms: an investigation of monospecific diatom films as settlement cues. Mar Ecol Prog Ser, 229: 105–112.

Harley, M.B. 1950. Occurrence of a filter-feeding mechanism in the polychaete *Nereis diversicolor*. Nature, 165: 734–735.

Harms, J. 1993 Check list of species (algae, invertebrates and vertebrates) found in the vicinity of the island of Helgoland (North Sea, German Bight)—a review of recent records. Helgoländer Meeresunters, 47: 1–34.

Harper, E.H. 1904. Notes on regulation in *Stylaria lacustris*. Biol Bull, 6: 173–190.

Hartman, O. 1966. Polychaeta Myzostomidae and Sedentaria of Antarctica. Antarctic Res Ser, Washington, 7: 158.

Hartman, O. 1967. Polychaetous annelids collected by the USNS Eltanin and Staten Island cruises, chiefly from Antarctic seas Allan Hancock Monogr. Mar Biol, 2: 1–387.

Hauenschild, C. 1953. Die phanotypische Geschlechtsbestimmung bei *Grubea clavata* (Clap.) und vergleichende Beobachtungen an anderen Sylliden. Zool Jb Physiol, 64: 14–54.

Hauenschild, C. 1955. Photoperiodizitat als Ursache des von der Mondphase abhangigen Metamorphose. Thythmus bei dem Polychaeten *P. dumerilii*. Z Naturforsch, 10: 658–666.

Hauenschild, C. 1966. Der hormonelle Einfluß des Gehirns auf die sexuelle Entwicklung bei dem Polychaeten *Platynereis dumerilii*. Gen Comp Endocrinol, 6: 26–73.

Hauenschild, D. 1975. Die Beteiligung endokriner mechanischen an der geschlechtlichen Entwicklung einiger Polychaeten. Forsch Zool, 22: 75–92.

Hauenschild, C. and Hauenschild, A. 1951. Untersuchungen uber die stoffiiche Koordination der Paarung des Polychaeten *Grubea clavata*. Zool Jb Physiol, 62: 429–440.

Hauenschild, C. and Fischer, A. 1962. Neurosecretary control of development in *Platynereis dumerilii*. In: *Neurosecretion*. (eds) Heller, H. and Clarke, R.B., Academic Press, New York, 12: 297–312.

Hauenschild, C., Fischer, A. and Hofmann, D.K. 1968. Untersuchungen am pazifischen Palolowurm *Eunice viridis* (Polychaeta) in Samoa. Helgoländer wiss Meeresunters, 18: 254–295.

Heacox, A.E. and Schroeder, P.C. 1982. The effects of prostomium and proventriculus removal on sex determination and gametogenesis in *Typosyllis pulchra* (Polychaeta: Syllidae). Wilhelm Rouxs Arch Dev Biol, 191: 84–90.

Hegde, P.R. and Sreepada, K.S. 2014. Reports on aquatic oligochaetes (Naididae) in paddy fields of Moodabidri Taluk, Dakshina Kannada, South India. J Entonol Zool Stud, 2: 101–107.

Hempelmann, F. 1923. Kausal-analytische Untersuchungen uber das Auftreten vergrosserter Borsten und die Lage der Teilungszone bei *Pristina*. Arch Mikr Anat, 98: 379–445.

Henshaw, J.M., Kokko, H. and Jennions, M.D. 2015. Direct reciprocity stabilizes simultaneous hermaphroditism at high mating rates: A model of sex allocation with egg trading. Evol, 69: 2129–2139.

Herlant-Meewis, H. 1947. Contribution a 1'Ctude de la regeneration chez les Oligochetes. Ann Soc Zool Belg, 77: 5–47.

Herlant-Meewis, H. 1951. Les lois de la scissiparite chez les Aeolosomatidae: *Aeolosoma viride*. Ann Soc Zool Belg, 82: 231–284.

Herpin, R. 1925. Recherches biologiques sur la reproduction et Ie developpement de quelques Annelides Polychaetes. Bull Soc Sci Nat Ouest France, 5: 1–250.

Heuer, C.M. and Loesel, R. 2007. Immunofluorescene analysis of the internal brain anatomy of *Nereis diversicolor* (Polychaeta, Annelida). Cell Tissue Res, 331: 713–724.

Hilario, A., Young, C.M. and Tyler, P.A. 2005. Sperm storage, internal fertilization, and embryonic dispersal in vent and seep tubeworms (Polychaeta: Siboglinidae: Vestimentifera). Biol Bull, 208: 20–28.

Hill, S.D. 1970. Origin of the regeneration blastema in polychaete annelids. Am Zool, 10: 101–112.

Hill, S.D. 1972. Caudal regeneration in the absence of a brain in two species of sedentary polychaetes. J Embryol Exp Morphol, 28: 667–680.

Hirabayashi, K., Oga, K. and Yamamoto, M. 2014. Bathymetric distribution of aquatic oligochaeta in Lake Kizaki, Central Japan, Zoosymposia, 9: 36–43.

Hochachka, P.W., Fields, J. and Mustafa, T. 1973. Animal life without oxygen: Basic biochemical mechanisms. Am Zool, 13: 543–555.

Hodgkin, J. 1990. Sex determinations compared in *Drosophila* and *Caenorhabditis*. Nature, 344: 35–47.

Hofmann, D.K. 1974. Maturation, epitoky and regeneration in the polychaete *Eunice siciliensis* under field and laboratory conditions. Mar Biol, 25: 149–161.

Holbrook, M.J.L. and Grassle, J.P. 1984. The effect of low density on the development of simultaneous hermaphroditism in male *Capitella* species I (Polychaeta). Biol Bull, 166: 103–109.

Holmstrand, L.L. and Collins, H.L. 1985. Reproductive biology of the Erpobdellid leech *Nephelopsis obscura* Verrill, 1872 in culture. Freshwater Invert Biol, 4: 30–34.

Hossain, A., Hasan, M. and Mollah, M.F.A. 2011. Effect of soybean meal and mustard oil cake on the production of fish live food tubificid worms in Bangladesh. World J Fish Mar Sci, 3: 183–189.

Howie, D.I.D. 1963. Experimental evidence for the humoral stimulation of ripening of the gametes and spawning in the polychaete *Arenicola marina* (L.) Gen Comp Endocrinol, 3: 660–668.

Hsieh, H.L. 1997. Self-fertilization: a potential fertilization mode in an estuarine sabellid polychaete. Mar Ecol Prog Ser, 147: 143–48.

Hunt, H.R. 1919. Regenerative phenomeno following removal of a digestive tube and nerve cord of earthworms. Bull Mus Comp Zool Harv, 62: 569–581.

Hurtado, L.A., Lutz, R.A. and Vrijenhoek, R.C. 2004. Distinct patterns of genetic differentiation among annelids of eastern Pacific hydrothermal vents. Mol Ecol, 13: 2603–2615.

Hutchings, P. 1998. Biodiversity and functioning of polychaetes in benthic sediments. Biodiversity Conserv, 7: 1133–1145.

Hyman, L.H. 1916. An analysis of the process of regeneration in certain microdrilous oligochaetes. J Exp Zool, 20: 99–163.

Hyman, L.H. 1938. The fragmentation of *Nais paraguayensis*. Physiol Zool, 11: 126–143.

Hyman, L.H. 1940. Aspects of regeneration in annelids. Am Nat, 74: 513–527.

Inoue, S., Okada, K. and Tanino, H. 1990. 6-Propionyllumazines from the marine polychaete, *Odontosyllis undecimdonta*. Chemistry Letters, 1990: 367–368.

Inoue, S., Okada, K. and Tanino, H. 1993. A new hexagonal cyclic enol phosphate of 6-beta-hydroxy propionyllumazines from the marine swimming polychaete, *Odontosyllis undecimdonta*. Heterocycles, 35: 147–150.

Iori, D., Forti, L., Massamba-N'Siala, G. et al. 2014. Toxicity of the purple mucus of the polychaete *Halla parthenopeia* (Oenonidae) revealed by a battery of ecotoxicological bioassays. Scient Mari, 78: 589–595.

Irvine, S.Q. and Seaver, E.C. 2006. Early annelid development, A molecular perspective. In: *Reproductive Biology and Phylogeny of Annelida*. (eds) Rouse, G. and Pleijel, F., Science Publishers, Enfield, New Hampshire, pp 93–140.

Ivleva, I.V. 1970. The influence of temperature on the transformation of matter in marine invertebrates. In: *Marine Food Chains*. (ed) Steele, J.H. Univ California Press, Berkeley, pp 96–112.

Jablonka, E. and Lamb, M.J. 1990. The evolution of heteromorphic sex chromosomes. Biol Rev, 65: 249–276.

Jaenike, J. and Selander, R.K. 1979. Evolution and ecology of parthenogenesis in earthworms. Am Zool, 19: 729–737.

James, R.J. and Underwood, A.J. 1994. Influence of colour of substratum on recruitment of spirorbid tubeworms to different types of intertidal boulders. J Exp Mar Biol Ecol, 181: 105–115.

Jamieson, B.G.M. 1981. *The Ultrastructure of the Oligochaeta*. Academic Press, London, p 462.

Jamieson, B.G.M. 1986. Some recent studies on the ultrastructure and phylogeny of annelid and uniramian spermatozoa. Dev Growth Diff, 28: 25–26.

Jamieson, B.G.M. and Ferraguti, M. 2006. Non-leech clitellata. In: *Reproductive Biology and Phylogeny of Annelida*. (eds) Rouse, G. and Pleijel, F., Science Publishers, Enfield, New Hampshire, pp 235–392.

Jayachandran, P.R., Prabhakaran, M.P., Asha, C.V. et al. 2015. First report on mass reproductive swarming of a polychaete worm, *Dendronereis aestuarina* (Annelida, Nereididae) Southern 1921, from a freshwater environment in the south west coast of India. Int J Mar Sci, 5: 1–7.

Jenni, K. 2012. Variation in development mode and its effect on divergence and maintenance of population. Jyvaskyla Stud Biol Env Sci, Jyvaskyla Univ, Finland.

Jewel, A.S., Al Masud, A., Amin, R. et al. 2016. Comparative growth of tubificid worms in culture media supplemented with different nutrients. Int J Fish Aquat Stud, 4: 83–87.

Jimenez, J.J. and Decaens, T. 2000. Vertical distribution of earthworms in grassland soils of the Colombian Llanos. Biol Fert Soil, 32: 463–473.

Johnston, A.S.A., Holmstrup, M., Hodson, M.E. et al. 2014. Earthworm distribution and abundance predicted by a process-based model. Appl Soil Ecol, 84: 112–123.

Jollivet, D., Empis, A., Barker, M.C. et al. 2000. Reproductive biology, sexual dimorphism and population structure of the deep sea hydrothermal vent scale-worm *Branchipolynoe seepensis* (Polychaeta: Polynoidae). J Mar Biol Ass UK, 80: 55–68.

Jones, M.L. 1981. *Riftia pachyptila* Jones: observations on the vestimentiferan worm from the Galapagos Rift. Science, 213: 333–336.

Jones, P.C.T. and Mollison, J.E. 1948. A technique for the quantitative estimation for soil microorganisms. J Gen Microbiol, 2: 54–69.

Jorgensen, C.B. 1976. August Putter, August Krogh and modern ideas on the use uf dissolved organic matter in aquatic environments. Biol Rev, 51: 291–328.

Jorgensen, N.O.G. 1979. Uptake of L-valine and other amino acids by the polychaete *Nereis virens*. Mar Biol, 52: 42–52.

Jorgensen, N.O.G. and Kristensen, E. 1980a. Uptake of amino acids by three species of *Nereis* (Annelids: Polychaeta). I. Transport kinetics and net uptake from natural concentrations. Mar Ecol Prog Ser, 3: 329–340.

Jorgensen, N.O.G. and Kristensen, E. 1980b. Uptake of amino acids by three species of *Nereis* (Annelids: Polychaeta). II. Effects of anaerobiosis. Mar Ecol Prog Ser, 3: 341–346.

Joshi, N. and Dabral, M. 2008. Life cycle of earthworms *Drawida nepalensis*, *Metaphire houlleti* and *Perionyx excavatus* under laboratory controlled conditions. Life Sci J, 5: 83–86.

Kahmann, D. and Franke, H-D. 1984. Hormonabhangigkeit der weiblichen Differenzierung bei dem Polychaeten *Brania clavata* (Syllidae). Verh dt zool Ges, 1984: 191.

Kalidas, R.M., Subramanian, E.R., Mydeen, S.A.K.N.M. et al. 2015. Conserved lamin A protein expression in differentiated cells in the earthworm *Eudrilus eugeniae*. Cell Biol Int, 39: 1036–1046.

Karaseva, N.P., Rimskaya-Korsakova, N.N., Galkin, S.V. and Malakhov, V.V. 2016. Taxonomy, geographical and bathymetric distribution of vestimentiferan tubeworms (Annelida, Siboglinidae). Zool Zh, 95: 624–659.

Karmegam, N. and Daniel, T. 2009. Growth, reproductive biology and life cycle of the vermicomposting earthworm, *Perionyx ceylanensis* Mich. (Oligochaeta: Megascolecidae). Bioresour Technol, 100: 4790–4796.

Kaster, J.L., Klump, J.V., Meyer, M. et al. 1984. Comparison of defecation rates of *Limnodrilus hoffmeisteri* Clapare`de (Tubificidae) using two different methods. Hydrobiologia, 111: 181–184.

Kato, K., Orii, H., Wattanabe, K. and Agata, K. 199. The role of dorso-ventral interaction in the onset of planarian regeneration. Development, 126: 1031–1040.

Kaushal, B.R., Bora, S. and Kandpal, B. 1999. Growth and cocoon production by the earthworm *Metaphire houlleti* (Oligochaeta) in different food sources. Biol Ferti Soils, 29: 394–400.

Kavumpuruth, S. 1992. Ploidy induction in ornamental fish. Ph.D. Thesis, Madurai Kamaraj Univerisity, Madurai, India.

Kawamoto, S., Yoshida-Noro, C. and Tochinai, S. 2005. Bipolar head regeneration induced by artificial amputation in *Enchytraeus japonensis* (Annelida, Oligochaeta). J Exp Zool, 303A: 615–627.

Khan, R.A. 1982. Biology of the marine leech *Johanssonia artica* (Johanson) from Newfoundland. Proc Helm Soc Wash, 49: 266–278.

Kharin, A.V., Zagainova, I.V. and Kostyuchenko, R.P. 2006. Formation of the paratomic fission zone in freshwater oligochaetes. Rus J Dev Biol, 37: 354–365.

King, R.S. and Newmark, P.A. 2012. The cell biology of regeneration. J Cell Biol, 196: 553–562.

Kirchman, D., Graham, S., Reish, D. and Mitchell, R. 1982. Bacteria induce settlement and metamorphosis of *JonI/o (Dexiospira) brasiliensis* Grube (Polychaeta: Spirorbidae). J Exp Mar Biol Ecol, 56: 153–163.

Kirk, R.G. 1971. Reproduction of *Lumbricillus rivalis* (Levinsen) in laboratory cultures and in decaying seaweed. An Appl Biol, 67: 255–264.

Kluge, B., Lehmann-Greif, M. and Fischer, A. 1995. Long-lasting exocytosis and massive structural reorganization in the egg periphery during cortical reaction in *Platynereis dumerilii* (Annelida, Polychaeta). Zygote, 3: 141–156.

Knight-Jones, P. and Bowden, N. 1984. Incubation and scissiparty in Sabellidae (Polychaeta). J Mar Biol Ass UK, 64: 809–818.

Knight-Jones, P. and Giangrande, A. 2003. Two new species of an atypical group of *Pseudobranchiomma* Jones (Polychaeta: Sabellidae). Hydrobiologia, 496: 95–103.

Kobayashi, G., Miura, T. and Kojima, S. 2015. *Lamellibrachia sagami* sp nov., a new vestimentiferan tubeworm (Annelida: Siboglinidae) from Sagami Bay and several sites in the northwestern Pacific Ocean. Zootaxa, 4018: 097–108.

Koene, J.M., Pfortner, T. and Michiels, N.K. 2005. Piercing the partner's skin influences sperm uptake in the earthworm *Lumbricus terrestris*. Behav Ecol Sociobiol, 59: 243–249.

Kolbasova, G.D., Tzetlin, A.B. and Kupriyanova, E.K. 2013. Biology of *Pseudopotamilla reniformis* (Müller 1771) in the White Sea, with description of asexual reproduction. Invert Reprod Dev, 57: 264–275.

Korablev, V.P., Radashevsky, V.I. and Manchenko, G.P. 1999. The XX-XY (male heterogametic) sex chromosome system in *Polydora curiosa*. Ophelia, 51: 193–201.

Kosiorek, D. 1974. Development cycle of *Tubifex tubifex* Muller in experimental culture. Pol Arch Hidrobiol, 21: 411–422.

Kostyuchenko, R.P., Kozin, V.V. and Kupriashova, E.E. 2016. Regeneration and asexual reproduction in annelids: Cells, genes, and evolution. Biol Bull, 43: 185–194.

Koya, Y., Onchi, R., Furta, Y. and Yamakuchi, K. 2003. Method for artificial fertilization and observation of the developmental process in Japanese palolo *Typorhynchus heterochaetus* (Annelids: Polychaeta). Sci Rep Fac Edu Gifu Univ (Nat Sci), 27: 85–94.

Kozin, V.V. and Kostyuchenko, R.P. 2015. *Vasa, PL10* and *Piwi* gene expression during caudal regeneration of the polychaete annelid *Alitta virens*. Dev Genes Evol, 225: 129–138.

Kudenov, J.D. 1974. The reproductive biology of *Eurythoe complanata* (Pallas, 1766) (Polychaeta: Amphinomidae). Ph.D. Thesis, University of Arizona, p 128.

Kupriyanova, E.K., Nishi, E., ten Hove, H.A. and Rzhavsky, A.V. 2001. Life history patterns in serpulimorph polychaetes: ecological and evolutionary perspectives. Oceanograph Mar Biol Ann Rev, 39: 1–101.

Kupriyanova, E.K. 2006. Fertilization success in *Galeolaria caespitosa* (Polychaeta, Serpulidae): gamete characteristics, role of sperm dilution, gamete age and contact time. Scient Mari, 70: 309–317.

Kutschera, U. and Wirtz, P. 1986. Reproductive behaviour and parental care of *Helobdella striata* (Hirudinea, Glossiphoniidae): a leech that feeds its young. Ethology, 72: 132–142.

Kutschera, U. 1989. Reproductive behaviour and parental care of *Helobdella striata* (Hirudinea, Glossiphoniidae). Zool Anz, 222: 122–128.

Kutschera, U. and Wirtz, P. 2001. The evolution of parental care in fresh water leeches. Theory Biosci, 120: 115–137.

Lamb, A., Gibbs, D. and Gibbs, C. 2011. Strait of Georgia biodiversity in relation to bull kelp abundance. Pacific Fisheries Resource Conservation Council, 111 p.

Lanfranco, M. and Rolando, A. 1981. Sexual races and reproductive isolation in *Ophryotrocha labronica* la Greca and Bacci (Annelida Polychaeta). Boll Zool, 48: 291–294.

Langhammer, H. 1928. Teilungs- und Regenerations-Vorginge bei *Procerasteu halleziana* undihre Beziehung zu der Stolonisation von *Autolytus prolifer*. Thesis, University of Marburg.

Lardicci, C., Ceccherelli, G. and Rossi, F. 1997. *Streblospio shrubsolii* (Polychaeta: Spionidae): temporal fluctuations in size and reproductive activity. Cah Biol Mar, 38: 207–214.

Lassarre, P. 1971. Oligochaete from the marine neiobenthos: Taxonomy and ecology. In: Proc I Int Conf Meiofauna. (ed) Hulings, N.C. Smithsonian Contribution to Zoology, 76: 71–86.

Lau, K.K. and Qian, P.Y. 1997. Phlorotannis and their analogs as larval settlement inhibitors of a tube-building polychaete *Hydroides elegans* (Hawell). Mar Ecol Prog Ser, 159: 219–227.

Lawrence, A.J. and Olive, P.J.W. 1995. Gonadotrophic hormone in *Eulalia viridis* (Polychaea, Annelida). Stimulation of vitellogenesis. Invert Reprod Dev, 28: 43–52.

Lawrence, A.J. and Soame, J.M. 2009. The endocrine control of reproduction in Nereidae: a new multi-hormonal model with implications for their functional role in a changing environment. Phil Trans R Soc, 364B: 3363–3376.

Learner, M.A. 1972. Laboratory studies on the life-histories of four enchytraeid worms (Oligochaeta) which inhabit sewage percolating filters. Ann Appl Biol, 70: 251–266.

Leelatanawit, R., Uawisetwathana, U., Khudet, J. and Karoonuthaisiri, N. 2014. Effects of polychaetes (*Perinereis nuntia*) on sperm performance of the domesticated black tiger shrimp (*Penaeus monodon*). Aquaculture, 433: 266–275.

Lietz, D.M. 1987. Potential for aquatic oligochaetes as live food in commercial aquaculture. Hydrobiologia, 155: 309–310.

Lesiuk, N.M. and Drewes, C.D. 1999. Autotomy reflex in a freshwater oligochaete, *Lumbriculus variegatus* (Clitellata: Lumbriculidae). Hydrobiologia, 406: 253–261.

Levin, L.A. 1984. Life history and dispersal patterns in a dense infaunal polychaete assemblage: community structure and response to disturbance. Ecology, 65: 1185–1200.

Levin, L.A. and Huggett, D.V. 1990. Implications of alternate reproductive modes for seasonality and demography in an estuarine polychaete. Ecology, 71: 2191–2208.

Levin, L.A., Zhu, J. and Creed, E. 1991. The genetic basis of life history characters in a polychaete exhibiting planktotrophy and lecithotrophy. Evolution, 45: 380–397.

Levin, L.A. and Bridges, T.S. 1994. Control and consequences of alternative developmental modes in a poecilogonous polychaete. Am Zool, 34: 323–332.

Licciano, M., Murray, J.M., Watson, G.J. and Giangrande, A. 2012. Morphological comparison of the regeneration process in *Sabella spallanzanii* and *Branchiomma luctuosum* (Annelida, Sabellida). Invert Biol, 131: 40–51.

Liebmann, E. 1942. The coelomocytes of Lumbricidae. J Morph, 71: 221–245.

Liebmann, E. 1946. The correlation between sexual reproduction and regeneration in a series of Oligochaeta. J Exp Zool, 91: 373–389.

Lindegaard, C., Hamburger, K. and Dall, P.C. 1994. Population dynamics and energy budget of *Marionina southerni* Cernosvitov (Enchytraeidae, Oligochaeta) in the shallow littoral of Lake Ersom, Denmark. Hydrobiologia, 278: 291–301.

Lindsay, S.M. and Woodin, S.A. 1992. The effect of palp loss on feeding behavior of two spionid polychaetes: changes in exposure. Biol Bull, 183: 440–447.

Lindsay, S.M. and Woodin, S.A. 1995. Tissue loss induces switching of feeding mode in spionid polychaetes. Mar Ecol Prog Ser, 125: 159–169.

Lindsay, S.M., Jackson, J.L. and He, S.Q. 2007. Anterior regeneration in the spionid polychaetes *Dipolydora quadrilobata* and *Pygospio elegans*. Mar Biol, 150: 1161–1172.

Lobo, H. and Alves, R.G. 2011. Reproductive cycle of *Branchiura sowerbyi* (Oligochaeta: Naididae: Tubificinae) cultivated under laboratory conditions. Zoologia, 28: 427–431.

Lobo, H. and Espíndola, E.L.G. 2014. *Branchiura sowerbyi* (Oligochaeta: Naididae) as a test species in ecotoxicology bioassays: a review. Zoosymposia, 9: 59–69.

Lohlein, B. 1999. Assessment of secondary production of Naididae (Oligochaeta): an example from a north German lake. Hydrobiologia, 406: 191–198.

Lopez-Jamar, E., Gonzalez, G. and Mejuto, J. 1986. Temporal changes of community structure and biomass in two subtidal macroinfaunal assemblages in La Coruna Bay, NW Spain. Hydrobiologia, 142: 137–150.

Lopez, E., Britayev, T.A., Martin, D. and San Martin, G. 2001. New symbiotic associations involving Syllidae (Annelida: Polychaeta), with taxonomic and biological remarks on *Pionosyllis magnifica* and *Syllis cf. armillaris*. J Mar Biol Ass UK, 81: 399–409.

Lowe, C.N. and Butt, K.R. 2005. Culture techniques for small soil dwelling earthworms: A review. Pedobiologia, 49: 401–419.

Lowe, C.N. and Butt, K.R. 2008. Life cycle traits of the parthenogenic earthworm *Octolasion cyaneum* (Savigny, 1826). Eur J Soil Biol, 44: 541–544.

Lucey, N.M., Lombardi, C., DeMarchi, L. et al. 2015. To brood or not to brood: Are marine invertebrates that protect their offspring more resilient to ocean acidification. Sci Rep, doi: 10.1038/srep12009.

Lummel, L.A.E. v. 1932. Over lichtende wormpjes in de baai van Bataria. De Tropische Natuur, 21: 85–87.

Lutz, R.A. 1988. Dispersal of organisms at deep-sea hydrothermal vents: A review. Oceanol Acta, 8: 23–30.

Macdonald, T.A., Burd, B.J., Macdonald, V.I. and van Roodselaar, A. 2010. Taxonomic and feeding guild classification for the marine benthic macroinvertebrates of the Strait of Georgia, British Columbia. Can Tech Rep Fish Aquat Sci, 2874: 62 p.

Mackay, J. and Gibson, G.D. 1999. The influence of nurse eggs on variable larval development in *Polydora cornuta* (Polychaeta: Spionidae). Invert Reprod Dev, 35: 167–176.

Mangum, C.P., Lykkeboe, G. and Johansen, K. 1975. Oxgen uptake and the role of hemoglobin in the East African swampworm *Alma emini*. Comp Biochem Physiol, 52A: 477–482.

Mann, K.H. 1957. The breeding, growth and age structure of a population of the leech *Helobdella stagnalis* (L.). J Ani Ecol, 26: 171–177.

Marian, M.P. and Pandian, T.J. 1984. Culturing and harvesting techniques of *Tubifex tubifex*. Aquaculture, 42: 303–315.

Marian, M.P. and Pandian, T.J. 1985. Interference of *Chironomus* in open culture system of *Tubifex tubifex*. Aquaculture, 44: 249–251.

Marian, M.P., Chandran, S. and Pandian, T.J. 1989. A rack culture system for *Tubifex tubifex*. Aquacult Engg, 8: 329–337.

MarineSpecies.org 2050 MarineSpecies.org. http://www.marinespecies.org/index.php.

Markert, R.E., Markert, B.J. and Vertrees, N.J. 1961. Lunar periodicity in spawning and luminescence in *Odontosyllis enopla*. Ecology, 42: 414–415.

Marotta, R., Melone, G., Bright, M. and Ferraguti, M. 2005. Spermatozoa and sperm aggregates in the vestimentiferan *Lamellibrachia luymesi* compared with those of *Riftia pachyptila* (Polychaeta: Siboglinidae: Vestimentifera). Biol Bull, 209: 215–226.

Marotta, R., Crottini, A., Raimondi, E. et al. 2014. Alike but different: the evolution of *Tubifex tubifex* species complex (Annelida, Clitellata) through polyploidization. BMC Evol Biol, 14: 73.

Marsden, J.R., Conlin, B.E. and Hunte, W. 1990. Habitat selection in the tropical polychaete *Spirobranchus giganetus*. 2. Larva. Preferences for corals. Mar Biol, 104: 93–99.

Marsden, J.R. 1992. Reproductive isolation in two forms of the serpulid polychaete *Spirobranchus polycerus* (Schmarda) in Barbados. Bull Mar Sci, 51: 14–18.

Marsh, A.G., Mullineaux, L.S., Young, C.M. and Manahan, D.T. 2001. Larval dispersal potential of the tubeworm *Riftia pachyptila* at deep-sea hydrothermal vents. Nature, 411: 77–80.

Marshall, D.J. and Evans, J.P. 2005. The benefits of polyandry in the free-spawning polychaete *Galeolaria caespitosa*. J Evol Biol, 18: 235–241.

Martin, A.M.S., Gerdes, D. and Arntz, W.E. 2005. Distributional patterns of shallow-water polychaetes in the Magellan region: a zoogeographical and ecological synopsis. Sci Mar, 69: 123–133.

Martin, D. and Gil, J. 2010. Checklist of class Polychaeta (Phylum Annelida). In: *The Biodiversity of the Mediterranean Sea: Estimates, Patterns, and Threats.* (eds) Coll, M. et al., PLoS ONE 5: 36.

Martin, E.A. 1933. Polymorphism and methods of asexual reproduction in the annelid, *Dodecacaria* of Vineyard Sound. Biol Bull, 65: 99–105.

Martin, J.P. and Bastida, R. 2006. Life history and production of *Capitella capitata* (Capitellidae: Polychaeta) in Rio De La Plata Estuary (Argentina). Thalassas, 22: 25–38.

Martin, P., Ferraguti, M. and Kaygorodova, I. 1998. Description of two new *Rhynchelmis* species from Lake Baikal (Russia) using classical morphological and ultrastructural spermatozoa characters. Ann Limnol, 34: 283–293.

Martin, P., Martens, K. and Goddeeris, B. 1999. Oligochaeta from the abyssal zone of Lake Baikal (Siberia, Russia). Hydrobiologia, 406: 165–174.

Martin, P., Martinez-Ansemil, E., Pinder, A. et al. 2008. Global diversity of oligochaetous clitellates ("Oligochaeta": Clitellata) in freshwater. Hydrobiologia, 595: 117–127.

Martinez-Acosta, V.G. and Zoran, M.J. 2015. Evolutionary aspects of annelid regeneration. In: eLs. John Wiley: Chichester. Doi: 10.1002/9780-470015902.a0022103.pub2.

Martinez, D.E. and Levinton, J.S. 1992. Asexual metazoans undergo senescence. Proc Nat Acd Sci USA, 89: 9920–9923.

Martinez, V.G., Menger, G.J. and Zoran, M.J. 2005. Regeneration and asexual reproduction share common molecular changes: upregualtion of a neural glycoepitope during morphallaxis in *Lumbriculus*. Mech Dev, 122: 721–732.

Martinez, V.G., Reddy, P.K. and Zoran, M.J. 2006. Asexual reproduction and segmental regeneration, but not morphallaxis are inhibited by boric acid in *Lumbriculus variegatus* (Annelida: Clitellata: Lumbriculidae). Hydrobiologia, 56: 473–186.

Matsushima, O. et al. 2002 A novel GGNG-related neuropeptide from the polychaete *Perinereis vancaurica*. Peptides, 23: 1379–1390.

Mayer, A.G. 1902. The Atlantic palolo (*Eunice fucata*). Brooklyn Mus Bull, 1: 94–103.

Mba, C.C. 1988. The effects of diet and incubating media on the production and hatchability of the earthworm Eudrilus eugeniae (Kinberg) cocoons. Rev Biol Trop, 36: 89–95.

McEuen, F.S., Wu, B.L. and Chia, F.S. 1983. Reproduction and development of *Sabella* media, a sabellid polychaete with extratubular brooding. Mar Biol, 76: 301–309.

McHugh, D. 1993. A comparative study of reproduction and development in the polychaete family Terebellidae. Biol Bull, 185: 153–167.

McHugh, D. and Tunnicliffe, V. 1994. Ecology and reproductive biology of the hydrothermal vent polychaete *Amphisamytha galapagensis* (Ampharetidae). Mar Ecol Prog Ser, 106: 11–120.

McHugh, D. 1997. Molecular evidence that echiurans and pogonophorans are derived annelids. Proc Natl Acad Sci USA, 94: 8006–8009.

Memis, D., Celikkale, M.S. and Ercan, E. 2004. The effects of different diets on the white worm (*Enchytraeus albidus*, Henle, 1837) reproduction. Turk J Fish Aquat Sci, 4: 5–7.

Mercier, A., Bailon, S. and Hamel, J.F. 2014. Life history and seasonal breeding of the deep sea annelid *Ophryotrocha* sp (Polychaeta: Dorvelleidae). Deep Sea Res, 191: 27–35.

Micaletto, G., Gambi, M.C. and Cantone, G. 2002. A new record of the endosymbiont polychaete *Veneriserva* (Dorvilleidae), with description of a new sub-species, and relationships with its host *Laetmonice producta* (Polychaeta: Aphroditidae) in Southern Ocean waters (Antarctica). Mar Biol, 141: 691–698.

Miller, D.C. and Jumars, P.A. 1986. Pellet accumulation, sediment supply and crowding as determinants of surface deposit feeding rate in *Pseudopolydora kempi japonica* Imajima and Hartman (Polychaeta: Spionidae). J Exp Mar Biol Ecol, 99: 1–17.

Minelli, A. 1993. Biological Systematics: the state of the art. Chapman & Hall, London.

Minetti, C., Sella, G. and Lorenzi, M.C. 2013. Population size, not density, serves as a cue for sex ratio adjustments in polychaete worms. Ital J Zool, 80: 547–551.

Mischke, C.C. and Griffin, M.J. 2011. Laboratory mass culture of the freshwater oligochaetes *Dero digitata*, North Am J Aquacult, 73: 13–16.

Miyamoto, N., Shinozaki, A. and Fujiwara, Y. 2014. Segment regeneration in the vestimentifern tubeworm, *Lamellibrachia satsuma*. Zool Sci, 31: 535–541.

Moment, G.B. 1943. The relation between body level, temperature and nutrients to regeneration growth. J Morph, 73: 108–117.

Moment, G.B. 1946. A study of growth limitation in earthworms. J Exp Zool, 103: 487–506.

Moment, G.B. 1950. A contribution to the anatomy of growth in earthworms. J Morph, 86: 59–72.

Moment, G.B. 1951. Simultaneous anterior and posterior regeneration and other growth phenomena in maldanid polychaetes. J Exp Zool, 117: 1–13.

Monahan, R.K. 1988. Sex ratio and sex change in *Ophryotrocha puerilis puerilis* (Polychaeta, Dorvilleidae). Ph.D. Thesis, State University of New York at Stony Brook.

Monroy, F., Aira, M., Velando, A. and Dominguez, J. 2003. Have spermatophores in *Eisenia fetida* (Oligochaeta, Lumbricidae) any reproductive role? Pedobiologia, 47: 526–529.

Moreno, R.A., Hernandez, C.E., Rivadeneira, M.M. et al. 2006. Patterns of endemism in south-eastern Pacific benthic polychaetes of the Chilean coast. J Biogeogr, 33: 750–759.

Morgan, T.H. 1898. Experimental studies of the regeneration of *Planaria maculata*. Arch Entwm, 7: 364–397.

Morgulis, S. 1907. Observations and experiments on regeneration in *Lumbriculus*. J Exp Zool, 4: 549–574.

Muldal, S. 1952. The chromosomes of the earthworms. I. The evolution of polyploids. Heredity, 6: 55–76.

Muller, C. 1908. Regenerationsversuche an *Lumbriculus variegatus* und *Tubifex rivulorum*. Arch EntwMech Org, 26: 209–77.

Müller, M., Berenzen, A. and Westheide, W. 2003. Experiments on anterior regeneration in *Eurythoe complanata* ("Polychaeta", Amphinomidae): reconfiguration of the nervous system nervous system and its function for regeneration. Zoomorphology, 122: 95–103.

Müller, M. 2004. Nerve development, growth and differentiation during regeneration in *Enchytraeus fragmentosus* and *Stylaria lacustris* (Oligochaeta). Dev Growth Diff, 46: 471–478.

Mullineaux, L.S., Wiebe, P.H. and Baker, E.T. 1995. Larvae of benthic invertebrates in hydrothermal vent plumes over Juan de Fuca Ridge. Mar Biol, 122: 585–596.

Murray, J.M., Watson, G.J., Giangrande, A. et al. 2013. Regeneration as a novel method to culture marine ornamental sabellids. Aquaculture, 410-411: 129–137.

Muruganantham, M., Mohan, P.M., Karunakumari, R. and Ubare, V.V. 2015. First report on *Nereis* (Neanthes) *virens* (Sars) an epitoky polychaete worm from Middle Strait, Baratang, Andaman Island, India. J Res Biol, 5: 1769–1774.

Myohara, M., Yoshida-Noro, C., Kobari, F. and Tochinai, S. 1999. Fragmenting oligochaete *Enchytraeus japonensis*: a new material for regeneration study. Dev Growth Diff, 41: 549–555

Myohara, M. 2004. Differential tissue development during embryogenesis and regeneration in an annelid. Dev Dyn, 231: 349–358.

Myohara, M. 2012. What role do annelid neoblasts play? a comparison of the regeneration patterns in a neoblast-bearing and a neoblast-lacking enchytraeid oligochaete. PLoS ONE, 7: e37319.

Naidu, K.V. 2005. Fauna of India: Aquatic Oligochaeta. Zoological Survey of India, Kolkata, pp 1–294

Narita, T. 2006. Seasonal vertical migration and aestivation of *Rhyacodrilus hiemalis* (Tubificidae, Clitellata) in the sediment of Lake Biwa, Japan. Hydrobiologia, 564: 87–93.

Nascimento, H.L.S. and Alves, R.G. 2009. The effect of temperature on the reproduction of *Limnodrilus hoffmeisteri* (Oligochaeta: Tubificidae). Zoologia, 26: 191–193.

Nayar, K.K. 1966. Neuroendocrinology of annelids. Anim Morph Physiol, 13, 133–143.

Needham, A.E. 1990. Annelida-Clitellata. In: *Reproductive Biology of Invertebrates*. (eds) Adiyodi, K.G. and Adiyodi, R.G., Oxford IBH Publishing, New Delhi, 4B: 1–36.

Negm-Eldin, M.M., Abdraba, M.A. and Benamer, H.E. 2013. First record, population ecology and biology of the leech *Limnatis nilotica* in the Green Mountain, Libya. Trav Inst Scien Rabat Ser Zool, 49: 37–42.

Neuhoff, H.G. 1979. Influence of temperature and salinity on food conversion and growth of different *Nereis* species (Polychaeta: Annelida). Mar Ecol Prog Ser, 1: 255–262.

Newrkla, P. and Mutayoba, S. 1987. Why and where do oligochaetes hide their cocoons? Hydrobiologia, 155: 171–178.

Nielsen, A.M., Eriksen, N.T., Iversen, J.J.L. and Riisgard, H.U. 1995. Feeding, growth and respiration in the polychaetes *Nereis diversicolor* (facultative filter-feeder) and *N. virens* (omnivorous)—a comparative study. Mar Ecol Prog Ser, 125: 149–158.

Nilsson, P., Kurdziel, J.P. and Levinton, J.S. 1997. Heterogeneous population growth, parental effects and genotype–environment interactions of a marine oligochaete. Mar Biol, 130: 181–191.

Nilsson, P., Levinton, J.S. and Kurdziel, J.P. 2000. Migration of a marine oligochaete: induction of dispersal and microhabitat choice. Mar Ecol Prog Ser, 207: 89–96.

Nishi, E. and Nishihira, M. 1994. Colony formation via sexual and asexual reproduction in *Salmacina dysteri* (Huxley) (Polychaeta, Serpulidae). Zool Sci, 11: 589–595.

Nishi, E. 1996. Serpulid polychaetes associated with living and dead coral at Okinawa Island south-west Japan. Publ Seto Mar Biol, 36: 305–318.

Niwa, N., Akimoto-Kato, A., Sakuma, M. et al. 2013. Homeogenetic inductive mechanism of segmentation in polychaete tail regeneration. Dev Biol, 381: 460–470.

Nogueria, J.M.M. and Knight-Jones, P. 2002. A new species of *Pseudobranchiomma* Jones (Polychaeta: Sabellidae) found amongst Brazilian coral, with a redescription of *P. punctata* (Treadwell, 1906) from Hawaii. J Nat Hist, 36: 1661–1670.

Novikova, E.L., Bakalenko, N.L., Nesterenko, A.Y. and Kulakova, M. 2013. Expression of Hox genes during regeneration of nereid polychaete *Alitta* (*Nereis*) *virens* (Annelida, Lophotrochozoa). Evol Dev, 4: 14.

Novo, M., Almodóvar, A., Fernández, R. et al. 2010. Mate choice of an endogeic earthworm revealed by microsatellite markers. Pedobiologia, 53: 375–379.

Nygren, A. 1999. Phylogeny and reproduction in Syllidae (Polychaeta). Zool J Linn Soc, 126: 365–386.

Nyholm, S.V., Robidart, J. and Girguis, P.R. 2008. Coupling metabolite flux to transcriptomics: Insights into the molecular mechanisms underlying primary productivity by the hydrothermal vent tubeworm *Ridgeia piscesae*. Biol Bull, 214: 255–265.

O'Brien, J.P. 1946. Studies on the cellular basis of regeneration in *Nais paraguayensis*, and the effects of x-rays thereon. Growth, 10: 25–44.

Obenat, S.M. and Pezzani, S.E. 1994. Life cycle and population structure of the polychaete Ficopomatus enigmaticus (Serpulidae) in Mar Chiquiba coastal lagoon, Argentina. Estuaries, 17: 263–270.

Ockelmann, K.W. and Akesson, B. 1990. *Ophryotrocha socialis* n. sp, a link between two groups of simultaneous hermaphrodites within the genus (Polychaeta, Dorvilleidae). Ophelia, 31: 145–162.

O'Connor, F.B. 1957. An ecological study of the enchytraeid worm population of a coniferous forest soil. Oikos, 8: 161–169.

O'Connor, R.J. and Lamont, P. 1978. The spatial organization of intertidal *Spirorhis* community. J Exp Mar Biol Ecol, 32: 143–169.

Okada, Y.K. 1929. Regeneration and fragmentation in the syllidian polychaetes. Arch Entw Mech Org, 115: 542–600.

Okada, Y.K. 1933. Two interesting syllids, with remarks on their asexual reproduction. Mem Coil Sci, Kyoto Imp Univ, 7B: 325–338.

Okrzesik, J., Kachamakova-Trojanowska, N., Jozkkowicz, A. et al. 2013. Reversible inhibition of reproduction during regeneration of cereberal ganglia and coelomocytes in the earthworm *Dendrobaena veneta*. Invert Sur J, 10: 151–161.

Olive, P.J.W. 1970. Reproduction of a Northumberland and population of the polychaete *Cirratulus cirratus*. Mar Biol, 5: 259–273.

Olive, P.J.W. 1975. A vitellogenesis promoting influence of the prostomium in the polychaete *Eulalia viridis* (Muller) (Polychaete: Phyllodocidae). Gen Comp Endocrinol, 26: 266–273.

Olive, P.J.W. 1976. Further evidence of a vitellogenesis-promoting hormone and its activity in *Eulalia viridis* (L.) (Polychaeta: Phyllodocidae). Gen Comp Endocrinol, 30: 397–403.

Olive, P.J.W. 1978. Reproduction and annual gametogenic cycle in *Nephyts hombergii* and *N. caeca* (Polychaeta: Nephytidae). Mar Biol, 46: 89–90.

Olive, P.J.W. and Bentley, M.G. 1980. Hormonal control of oogenesis, ovulation and spawning in the annual reproductive cycle of the polychaete *Nephtys hombergi* Sav. (Nephtyidae). Int J Invert Reprod, 2: 205–221.

Olive, P.J.W. and Pillai, G. 1983. Reproductive biology of the polychaete *Kefersteinia cirrata* Keferstein (Hesionidae). II. The gametogenic cycle and evidence for photoperiodc control of oogenesis. Int J Invert Reprod, 6: 307–315.

Olive, P.J.W., Morgan, P.J., Wright, N.H. and Zhang, S.L. 1984. Variable reproductive output in Polychaeta: Options and design constraints. Adv Invert Reprod, 3: 399–408.

Olive, P.J.W., Bentley, M.G., Wright, N.H. and Morgan, P.J. 1985. Reproductive energetic, endocrinology and population dynamics of *Nephtys caeaca* and *N. hombergi*. Mar Biol, 88: 235–246.

Olive, P.J.W. and Lawrence, A.J. 1990. Gonadotrophic hormone in Nephtyidae (Polychaeta, Annelida): stimulation of ovarian protein synthesis. Int J Inv Reprod Dev, 18: 189–195.

Olive, P.J.W. 1999. Polychaete aquaculture and polychaete science: a mutual synergism. Hydrobiologia, 402: 175–183.

Oliver, J.S. 1984. Selection for asexual reproduction in an Antarctic polychaete worm. Mar Ecol Prog Ser Oldendorf, 19: 33–38.

Oplinger, R.W., Bartley, M. and Wagner, E.J. 2011. Culture of *Tubifex tubifex*: Effect of feed type, ration, temperature and density on juvenile recruitment, production and adult survival. North Am J Aquacult, 73: 68–75.

Oshida, P.S., Word, L.S. and Mearns, A.J. 1981. Effects of hexavalent and trivalent chromium on the reproduction of *Neanthes arenaceodentata* (polychaeta). Mar Env Res, 5: 41–49.

Osman, I.H., Gabr, H.R., Saito, H. and El-Etryby, S.G.H. 2010. Reproductive biology of the highly commercial polychaetes in the Suez Canal. J Mar Biol Ass UK, 90: 281–290.

Ozpolat, D. and Bely, A.E. 2015. Gonad establishment during asexual reproduction in the annelid *Pristina leidyi*. Dev Biol, 405: 123–136.

Paavo, B., Bailey-Brock, J.H. and Akesson, B. 2000. Morphology and life history of *Ophryotrocha adherens* sp nov. (Polychaeta, Dorvilleidae). Sarsia, 85: 251–264.

Palmer, P.J. 2010. Polychaete-assisted sand filters. Aquaculture, 306: 369–377.

Pandian, T.J. 1975. Mechanism of heterotrophy. In: *Marine Ecology*. (ed) Kinne, O., John Wiley, London, 3 Part 1: 61–249.

Pandian, T.J. and Marian, M.P. 1985a. Techniques for culture and harvest of *Tubifex*. Privately circulated pamphlet.

Pandian, T.J. and Marian, M.P. 1985b. Estimation of absorption efficiency in polychaetes using nitrogen content of food. J Exp Mar Biol Ecol, 90: 289–295.

Pandian, T.J. and Marian, M.P. 1985c. Nitrogen content of food as an index of absorption efficiency in fishes. Mar Biol, 85: 301–311.

Pandian, T.J. and Marian, M.P. 1985d. An indirect procedure for the estimation of assimilation efliciency of aquatic insects. Freshwat Biol, 16: 93–98.

Pandian, T.J. and Marian, M.P. 1985e. Prediction of assimilation efficiency in reptiles. Nat Acad Sci Let, 7: 351–354.

Pandian, T.J. 1987. Fish. In: *Animal Energetics*. (eds) Pandian, T.J. and Vernberg, F.J., Academic Press, San Diego, 2: 357–465.

Pandian, T.J. 2010. *Sexuality in Fishes*. Science Publishers/CRC Press, USA, p 208.

Pandian, T.J. 2011. *Sex Determination in Fish*. Science Publishers/CRC Press, USA, p 270.

Pandian, T.J. 2012. *Genetic Sex Differentiation in Fish*. CRC Press, USA, p 214.

Pandian, T.J. 2013. *Endocrine Sex Differentiation in Fish*. CRC Press, USA, p 303.

Pandian, T.J. 2015. *Environmental Sex Determination in Fish*. CRC Press, USA, p 299.

Pandian, T.J. 2016. *Reproduction and Development in Crustacea*. CRC Press, USA, p 301.

Pandian, T.J. 2017. *Reproduction and Development in Mollusca*. CRC Press, USA, p 299.

Pandian, T.J. 2018. *Reproduction and Development in Echinodermata and Prochordata*. CRC Press, USA, p.

Parandavar, H., Kim, K.H. and Kim, C.H. 2015. Effects of rearing density on growth of the polychaete rockworm *Marphysa sanguinea*. Fish Aquat Sci, 18: 57–63.

Pasteris, A., Bonomi, G. and Bonacina, C. 1996. Age, stage, and size structure as population state variables for *Tubifex tubifex* (Oligochaeta, Tuibificidae). Hydrobiologia 334: 125–132.

Paulus, T. and Muller, M.C.M. 2006. Cell proliferation dynamics and morphological differentiation during regeneration in *Dorvillea bermudensis* (Polychaeta, Dorvilleidae), J Morphol, 267: 393–403.

Pawlik, J.R. 1986. Chemical induction of larval settlement and metamorphosis in the reef-building tube worm *Phragmatopoma cali/ornica* (Sabellariidae: Polychaeta). Mar Biol, 91: 59–68.

Pelegri, S.P. and Blackburn, T.H. 1995. Effects of *Tubifex tubifex* (Oligochaeta: Tubificidae) on N-mineralization in freshwater sediments measured with ^{15}N isotopes. Aquat Microb Ecol, 9: 289–294.

Penners, A. 1922. Die Furchung von *Tubifex rivulorum*. Zool Jb Anat, 43: 323–367.

Pesch, G.G. and Pesch, C.E. 1980. Chromosome complement of the marine worm *Neanthes arenaceodendata* (Polychaeta: Annelida). Can Fish Aquat Sci, 37: 286–288.

Pesch, C.E., Zajac, R.N., Whitlatch, R.B. and Balboni, M.A. 1987. Effect of intraspecific density on life history traits and poplation growth rate of *Neanthes arenaceodentata* (Polychaeta: Nereidae) in the laboratory. Mar Biol, 96: 545–554.

Petersen, M.E. 1999. Reproduction and development in Cirratulidae (Annelida: Polychaeta). Hydrobiologia, 402: 107–128.

Petersen, S., Arlt, G., Faubel, A. and Carman, K.R. 1998. On the nutritive significance of dissolved free amino acids uptake for the cosmopolitan oligochaete *Nais elinguis* Muller (Naididae). Estu Coas Shelf Sci, 46: 85–91.

Peterson, D.L. 1983. Life cycle and reproduction of *Nephelopsis obscura* Verrill (Hirudinea: Erpobdellidae) in permanent ponds of Northwestern Minnesota. Freshwat Invert Biol, 2: 165–172.

Petraitis, P.S. 1985a. Females inhibit males propensity to develop into simultaneous hermaphrodites in *Capitella* species I (Capitellidae). Biol Bull, 168: 395–402.

Petraitis, P.S. 1985b. Digametic sex determination in the marine polychaete, *Capitella capitata* (species type I). Heredity, 55: 151–156.

Petraitis, P.S. 1988. Occurrence and reproductive success of feminized males in the polychaete *Capitella capitata* (species type I). Mar Biol, 97: 403–412.

Petraitis, P.S. 1991. The effects of sex ratio and density on the expression of gender in the polychaete *Capitella capitata*. Evol Ecol, 5: 393–404.

Pfannenstiel, L.D. 1971. Zur sexuellen Differerizierung des Borstentuurtns *Ophryotrocha puerilis*. Naturwiss, 58: 367.

Pfannenstiel, L.D. 1973. Zur sexuellen Differerizierung von *Ophryotrocha puerilis*. (Polychaetan: Eunicidae). Hlar Biol, 20: 245–258.

Pfannenstiel, L.D. 1974. The effect of starvation and decapitation on the sexual differentiation in the gonochoristic polychaete *Ophryotrocha notoglandulata*. Wilhelm Roux Arch, 175: 52–53.

Pfannenstiel, L.D. 1975. Mutual influence on the sexual differentiation in the protandric polychaete *Ophryotrocha puerilis*. In: *Intersexuality in the Animal Kingdom*. (ed) Reintboth, R., Springer-Verlag, Heidelberg, pp 48–56.

Pfannenstiel, L.D. 1976. Ist der Polychaet *Ophryotrocha labronica* ein proterandrischer Hermaphrodit? Mar Biol, 38: 169–178.

Pfannenstiel, L.D. 1977a. Endokrinologische und genetische Untersuchungen einer proterandrischer Population des Polychaeten *Ophryotrocha labronica*. Hlar Biol, 39: 319–329.

Pfannenstiel, L.D. 1977b. Experimental analysis of the 'Paarkultzrreffekf' in the protandric polychaete, *Ophryotrocha puerilis*. Clap Mecz J Exp Mar Biol Ecol, 28: 31–40.

Pfannenstiel, L.D. 1977c. Unterschiedlicbe Hormoniabhangigkeit der Oogenese bei den Weibchen-Typen des proterandrischen Polychaeten *Ophryotrocha labronica*. Verh Dtsch Zool Ges, 1977: 325.

Pfannenstiel, L.D. 1978a. Die Entwicklung der Kontaktstruktur von Ei und Nahrzelle im Zuge der Oogenese von *Ophryotrocha puerilis* Claparede and Mecznikow (Polychaeta, Dorvilleidae) Zomorphologie, 90: 181–196.

Pfannenstiel, L.D. 1978b. Endocrinology of polychaete reproduction and sexual development. Boll Zool, 45: 171–188.

Pfannenstiel, H.D. and Grueing, C. 1987. Gametogenesis and reproduction in nereidid sibling species (*Platynereis dumerilii* and *P. massiliensis*). Bull Biol Soc Wash, 7: 272–279.

Pfannenstiel, H.-D. and Grunig, C. 1990. Spermatogenesis and sperm ultrastructure in the polychaete genus *Ophryotrocha* (Dorvilleidae). Helgol Meeresunt, 44: 159–171.

Pires, A., Freitas, R., Quintino, V. and Rodrigues, A.M. 2012. Can *Diopatra neapolitana* (Annelida: Onuphidae) regenerate body damage caused by bait digging or predation? Estuar Coast Shelf Sci, 110: 36–42.

Pires, A., Figueira, E., Moreira, A. et al. 2015. The effects of water acidification, temperature and salinity on the regenerative capacity of the polychaete *Diopatra neapolitana*. Mar Environ Res, 106: 30–41.

Plyuscheva, M., Martin, D. and Britayev, T. 2004. Population ecology of two sympatric polychaetes *Lepidonotus squamatus* and *Harmothoe imbricata* (Polychaeta, Polynoidea) in the White Sea. Invert Zool, 1: 65–73.

Pocklington, P. and Hutcheson, M.S. 1983. New record of viviparity for the dominant benthic invertebrate *Exogone hebes* (Polychaeta: Syllidae) from the Grand Banks of New Foundland. Mar Ecol Prog Ser, 11: 239–244.

Poddubnaya, T.L. 1984. Parthenogenesis in Tubificidae. Hydrobiologia, 115: 97–99.

Policansky, D. 1982. Sex change in plants and animals. Annu Rev Ecol Syst, 13: 471–495.

Pollack, H. 1979. Populationsdynamik, Produktivitlt und Energiehaushalt des Wattwurms *Arenicola marina* (Annelida, Polychaeta). Helgolander Wiss Meeresunters, 32: 313–358.

Porchet, M. 1967. Role des ovocytes submatures dans l'arret de l'inhibition cerebrale chez *P. cultrifera*. Acad Sci, 265: 1394–1396.

Porchet, M. and Cardon, C. 1976. The inhibitory feed-back mechanism coming from oocytes and acting on brain endocrine activity in *Nereis* (Polychaetes, Annelids). Gen Comp Endocrinol, 30: 378–390.

Porchet, M., Baert, J-L. and Dhainaut, A. 1989. Evolution of the concepts of vitellogenesis in polychaete annelids. J Invert Reprod Dev, 16: 53–61.

Porto, P.G., Velando, A. and Dominguez, J. 2012. Multiple mating increases cocoon hatching success in the earthworm *Eisenia andrei* (Oligochaeta: Lumbricidae). Biol J Linn Sic, 107: 175–181.

Pradillon, F., Shillito, B., Young, C.M. and Gaill, F. 2001. Developmental arrest in vent worm embryos. Nature, 413: 698–699.

Premoli, M.C. and Sella, G. 1995. Sex economy in benthic polychaetes. Ethol Ecol Evol, 7: 27–48.

Premoli, M.C., Sella, G. and Berra, G.P. 1996. Heritable variation of sex ratio in a polychaete worm. J Evol Biol, 9: 845–854.

Preston, R.L. and Stevens, V.R. 1982. Kinetic and thermodynamic aspects of sodium-coupled amino acid transport by marine invertebrates. Am Zool, 22: 709–721.

Prevedelli, D. 1992. Growth rates of *Perinereis rullieri* (Polychaeta, Nereididae) under different conditions of temperature and diet. Ital J Zool, 59: 261–265.

Prevedelli, D. and Vandini, R.Z. 1999. Survival, fecundity and sex ratio of *Dinophilus gyrociliatus* (Polychaeta: Dinophilidae) under different dietary conditions. Mar Biol, 132: 163–170.

Prevedelli, D. and Simonini, R. 2000. Effects of salinity and two food regimes on survival, fecundity and sex ratio in two groups of *Dinophilus gyrociliatus* (Polychaeta: Dinophilidae). Mar Biol, 137: 23–29.

Prevedelli, D. and Simonini, R. 2003. Life cycles in brackish habitats: Adaptive strategies of some polychaete from the Venice lagoon. Oceanol Acta, 26: 77–84.

Probst, G. 1931. Beiträge zur Regeneration der anneliden. Wilhelm Roux Arch Entwicklungs Org, 124: 369–403.

Purschke, G. 2006. Problematic annelid groups. In: *Reproductive Biology and Phylogeny of Annelida*. (eds) Rouse, G. and Pleijel, F., Science Publishers, Enfield, New Hampshire, pp 639–668.

Putter, A. 1909. Die Ernhrung der Wassertiere und der Stoffhaushalt der Gewasser. G. Fischer, Jena.

Qian, P.Y. and Chia, F.S. 1993. Larval growth and development as influenced by food limitation in two polychaetes: *Capitella* sp and *Polydora ligni*. J Exp Mar Biol Ecol, 166: 93–105.

Qian, P.Y. and Pechenik, J.A. 1998. Effects of larval starvation and delayed metamorphosis on juvenile survival and growth of the tube-dwelling polychaete *Hydroides elegans* (Haswell). J Exp Mar Biol Ecol, 227: 169–185.

Qian, P.Y. 1999. Larval settlement of polychaetes. Hydrobiologia, 402: 239–253.

Qian, P.Y. and Uwe-Dahms, H. 2006. Larval ecology of the annelids. In: *Reproductive Biology and Phylogeny of Annelida*. (eds) Rouse, G. and Pleijel, F., Science Publishers, Enfield, New Hampshire, pp 179–234.

Qiu, J.W. and Qian, P.Y. 1997. Combined effects of salinity, ternperature and food concentration on the early development of the polychaete *Hydroides elegans* (Haswell, 1883). Mar Ecol Prog Ser, 152: 79–88.

Quiroz-Martinez, B., Schmitt, F.G. and Dauvin, J.C. 2012. Statistical analysis of polychaete population density: dynamics of dominant species and scaling properties in relative abundance fluctuations. Nonlin Processes Geophys, 19: 45–52.

Radashevsky, V.I. 1996. Morphology, ecology and asexual reproduction of a new *Polydorella* species (Polychaeta: Spionidae) from the South China Sea. Bull Mar Sci, 58: 684–693.

Radashevsky, V.I. and Nogueira, J.M.de M. 2003. Life history, morphology and distribution of *Dipolydora armata* (Polychaeta: Spionidae). J Mar Biol Ass UK, 83: 375–384.

Ram, J.L., Müller, C.T., Beckmann, M. and Hardege, J.D. 1999. The spawning pheromone cysteine-glutathione disulfide (nereithionine) arouses a multicomponent nuptial behavior and electrophysiological activity in *Nereis succinea* males. FASEB J, 13: 945–952.

Ramskov, T. and Forbes, V.E. 2008. Life history and population dynamics of the opportunistic polychaete *Capitella* sp I in relation to sediment organic matter. Mar Ecol Prog Ser, 369: 181–192

Randolph, H. 1891. The regeneration of the tail in *Lumbriculus*. Zool Anz, 14: 154–156.

Randolph, H. 1892. The regeneration of the tail in *Lumbriculus*. J Morph, 7: 317–44.

Rasmussen, E. 1953. Asexual reproduction in *Pygospio elegans* Claparede (Polychaeta Sedentaria). Nature, 171: 1161–1162.

Ratsak, C.H. and Verkuijlen, J. 2006. Sludge reduction by predatory activity of aquatic oligochaetes in wastewater treatment plants: science or fiction? A review. Hydrobiologia, 564: 197–211.

Read, G.B. 1974. Egg masses and larvae of the polychaete *Nereis falcaria* (Note). New Zealand J Mar Freshwat Res, 8: 557–561.

Read, G.B. 1975. Systematics and biology of polydorid species (Polychaeta: Spionidae) from Wellington Harbour. J R Soc New Zealand, 5: 395–419.

Read, G.B. 2004. Guide to New Zealand Shore Polychaetes. Web publication. Http://www.annelida.net/nz/Polychaeta/References/NZPolySpeciesListV2.htm.

Rebscher, N., Lidke, A.K. and Ackermann, C.F. 2012. Hidden in the crowd: primordial germ cells and somatic stem cells in the mesodermal posterior growth zone of the polychaete *Platynereis dumerillii* are two distict cell populations. Ecol Dev, 3: 9 http://www.evodevjournal.com/content/3/1/9.

Reinecke, S.A. and Reinecke, A.J. 1997. The influence of lead and manganese on spermatozoa of *Eisenia fetida* (Oligochaeta). Soil Biol Biochem, 29: 737–742.

Reinecke, A.J. and Venter, J.M. 1987. Moisture preferences, growth and reproduction of the compost worm *Eisenia foetida* (Oligochaeta). Biol Fert Soil, 3: 135–141.

Reish, D.J. and Stephens, G.C. 1969. Uptake of organic material by aquatic invertebrates. V. The influence of age on the uptake of glycine-^{14}C by the polychaete *Neanthes arenaceodentata*. Mar Biol, 3: 352–355.

Rettob, M. 2012. Fecundity and body length of rag worm *Pernereis cultifera* (Grube, 1840) from wearlilir beach waters, small Kei Island, Southeast Maluku District. J Coast Dev, 16: 84–88.

Reynolds, J.W. 1974. Are oligochaetes really hermaphroditic amphimictic organisms? The Biologists, 56: 98–99.

Reynoldson, J.B. 1943. A comparative account ofthe life cycles of *Lumbricillus lineatus* Mull. and *Enchytraeus albidus* Henle in relation to temperature. Ann Appl Biol, 30: 60–66.

Rhoden, C. 2015. *Aporrectodea longa* (Clitellata, Lumbricidae)—one or two species in Scandinavia. B.Sc Disseration, University of Gothenburg.

Rice, S.A. 1980. Ultrastructure of the male nephridium and its role in spermatophore formation in spionid polychaetes (Annelids). Zoomorphologia, 95: 181–194.

Rioja, E. 1929. Un caso de reproducción asexual en un sabélido (*Branchiomma linaresi* Rioja). Bol Real Soc Esp Hist Nat Biológica, 29: 33–36.

Rodriguez-Romero, A., Jarrold, M.D., Masamba-N'Siala et al. 2015. Multi-generational responses of a marine polychaete to rapid change in seawater pCO_2. Evol Appli, doi: 10.1111/eva.12344.

Rossi, A.M., Saidel, W.M., Marotta, R. et al. 2013. Operculum ultrastructure in leech cocoons. J Morphol, 274: 940–946.

Rossi, A.M., Saidel, W.M., Gravante, C. et al. 2016. Mechanics of cocoon secretion in a segmented worm (Annelida: Hirudinidae). Micron, 86: 30–35.

Rota, E., Martin, P. and Erseus, C. 2001. Soil-dwelling polychaetes: enigmatic as ever? Some hints on their phylogenetic relationships as suggested by a maximum parsimony analysis of 18S rRNA gene sequences. Cont Zool, 70: 127–138.

Rouse, G. and Fitzhugh, K. 1994. Broadcasting fables: Is external fertilization really primitive? Sex, size, and larvae in sabellid polychaetes. Zool Script, 23: 271–312.

Rouse, G.W. 1999. Polychaete sperm: phylogenetic and functional considerations. Hydrobiologia, 402: 215–224.

Rouse, G.W. 2006. Annelid larval morphology. In: *Reproductive Biology and Phylogeny of Annelida*. (eds) Rouse, G. and Pleijel, F., Science Publishers, Enfield, New Hampshire, pp 141–178.

Rouse, G.W. and Pleijel, F. 2006. Annelid phyology and systematics. In: *Reproductive Biology and Phylogeny of Annelida*. (eds) Rouse, G.W. and Pleijel, F., Science Publishers Enfield, NH, pp 3–22.

Rouse, G.W., Wilson, N.G., Goffredi, S.K. and Vrijenhoek, R.C. 2008. Spawning and development in *Osedax* boneworms (Siboglinidae, Annelida). Mar Biol, 156: 395–405.

Rower, G. 2010 A provisional guide to the family Opheliidae (Polychaeta) from the shallow waters of the British Isles. Report to the NMBAQC 2008 taxonomic workshop participants—Dove Marine Laboratory. EMU Report, 12 pp, June 2010.

Ruby, E.G. and Fox, D.L. 1976. Anaerobic respiration in the polychaete *Euzonus* (*Thoracophelia*) *mucronata*. Mar Biol, 35: 149–153.

Ruppert, E.E., Fox, R.S. and Barnes, R.D. 2004. Invertebrate Zoology. *A Functional Evolutionary Approach*. 7th Ed. Brooks/Cole, Thomson Learning, Inc. 990 p.

Rychel, A.L. and Swalla, B. 2009. Regeneration in hemichordates and echinoderms. In: *Stem Cells in Marine Organisms*. (eds) Rinkevich, B. and Matranga, V., Springer Verlag, Dordrecht, pp 245–266.

Safarik, M., Redden, A.M. and Schneider, M.J. 2006. Density-dependent growth of the polychaete *Diopatra aciculata*. Scient Mari, 70: 337–341.

Sahin, G.K. and Cinar, M.E. 2012. A check-list of polychaete species (Annelida-Polychaeta) from the Black Sea. J Black Sea/Medit Environ, 18: 10–48.

Salazar-Vallejo, S.I. 1996. Lista de species y bibliografía de Poliquetos (Polychaeta) del Gran Caribe. Anales Inst Biol Univ nac Autón, México, Ser Zool, 67: 11–50.

Salazar-Vallejo, S.I. and Londaño-Mesa, M.H. 2004. Lista de especies y bibliografía de poliquetos (Polychaeta) del Pacífico Oriental Tropical. Anales del Instituto de Biología, Universidad Nacional Autónoma de México, Serie Zoología, 75: 9–97.

Samuel, S.C.J.R., Subramanian, E.R., Vedha, Y.B. et al. 2012. Autofluorescence in BrdU-positive cells and augmentation of regeneration kinetics by riboflavin. Stem Cells Dev, 21: 2071–2083.

Sato, M., Tsuchiya, M. and Nishihira, M. 1982. Ecological aspect of the development of the polychaete, *Lumbrinereis latreilli* (Audouin et Milne-Edwards): signification of direct development and non simultaneous emergence of the young from the jelly mass. Bull Mar Biol Stn, 17: 71–85.

Sato, M. and Ikeda, M. 1992. Chromosomal complements of two forms of *Neanthes japonica* (Polychaeta, Nereididae) with evidence of male-heterogametic sex chromosomes. Mar Biol, 112: 299–307.

Sato, M. and Osanai, K. 1996. Role of jelly matrix of egg masses in fertilization of the polychaete *Lumbrinereis latreilli*. Inv Reprod Develop, 29: 185–191.

Sato, M. 1999. Divergence of reproductive and developmental characteristics in *Hediste* (Polychaeta: Nereididae). Hydrobiologia, 402: 129–143.

Sawyer, R.T., Lepont, F., Stuart, D.K. and Kramer, A.P. 1981. Growth and reproduction of the giant glossiphoniid leech *Haementeria ghilianii*. Biol Bull, 160: 322–331.

Sayers, C.W., Coleman, J. and Shain, D.H. 2009. Cell dynamics during cocoon secretion in the aquatic leech *Theromyzon tessulatum* (Annelids: Clitellata: Glossiphoniidae). Tissue Cell, 41: 35–42.

Sayles, L.P. 1932. External features of regeneration in *Clymenella torquata*. J Exp Zool, 62: 237–258.

Sayles, L.P. 1936. Regeneration in the polychæte *Clymenella torquata*. III Effect of level of cut on type of new structures in anterior regeneration. Biol Bull, 70: 441–459.

Schiedges, K.L. 1979. Reproductive biology and ontogenesis in the polychete genus *Autolytus* (Annelida: Syllidae): observation on laboratory-cultured individuals. Mar Biol, 54: 239–250.

Schiedges, K.L. 1980. Morphological and systematic studies of an *Autolytus* population (Polychaeta, Syllidae, Autolytinae) from the Oosterscheilde Estuary. Netherlands J Sea Res, 14: 208–219.

Schierwater, B. and Hauenschild, C. 1990. A photoperiod determined life-cycle in an oligochaete worm. Biol Bull, 178: 111–117.

Schleicherova, D., Lorenzi, M.C. and Sella, G. 2006. How outcrossing hermaphrodites sense the presence of conspecifics and suppress female allocation. Behav Ecol, 17: 1–5.

Schleicherová, D., Lorenzi, M.C., Sella, G. and Michiels, N.K. 2010. Gender expression and group size: a test in a hermaphroditic and a gonochoric congeneric species of *Ophryotrocha* (Polychaeta). J Exp Biol, 213: 1586–1590.

Schmelz, R.M. and Collado, R. 2012. An updated checklist of currently accepted species of Enchytraeidae (Oligochaeta). Agri Forest Res, 357: 67–87.

Schmelz, R.M., Niva, C.C., Römbke, J. and Collado, R. 2013. Diversity of terrestrial Enchytraeidae (Oligochaeta) in Latin America: Current knowledge and future research potential. Appl Soil Ecol, 69: 13–20.

Schmelz, R.M. and Collado, R. 2015. Checklist of taxa of Enchytraeidae (Oligochaeta): an update. Soil Organisms, 87: 149–152.

Schneider, S., Fischer, A. and Dorresteijn, A.W.C. 1992. A morphometric comparison of dissimilar early development in sibling species of *Platynereis* (Annelida, Polychaeta). Roux's Arch Dev Biol, 201: 243–256.

Schottler, U. 1979. On the anaerobic metabolism of three species of *Nereis* (Annelida). Mar Ecol Prog Ser, 1: 249–254.

Schotter, U., Wienhausen, G. and Zebe, E. 1983. The mode of energy production in the lugworm *Arenicola marina* at different oxygen concentrations. J Comp Physiol, 149: 547–555.

Schroeder, P.C. 1967. Morphogenesis of epitokous setae during normal and induced metamorphosis in the polychaete annelid *Nereis grubei*. Biol Bull, 133: 426–37.

Schroeder, P.C. and Hermans, C.O. 1975. Annelids-Polychaeta. In: *Reproduction of Marine Invertebrates*. (eds) Giese, A.C. and Pearse, J.S., Academic Press, New York, 3: 1–213.

Schroeder, P.C. 1989. Annelids-Polychaeta. In: *Reproductive Biology of Invertebrates*. (eds) Adiyodi, K.G. and Adiyodi, R.G., Oxford and IBH Publishing, New Delhi, 4a: 383–442.

Sella, G. 1988. Reciprocation, reproductive success and safeguards against cheating in a hermaphroditic polychaete worm *Ophryotrocha diadema*, Åkesson 1976. Biol Bull, 175: 212–217.

Sella, G., Premoli, M.C. and Turri, F. 1997. Egg trading in simultaneously hermaphroditic polychaete worm *Ophryotrocha gracilis* (Huth). Behav Ecol, 8: 83–86.

Sella, G. and Ramella, L. 1999. Sexual conflict and mating systems in the dorvilleid genus *Ophryotrocha* and the dinophilid genus *Dinophilus*. Hydrobiologia, 402: 203–213.

Senapati, B.K. and Dash, M.C. 1983. Energetics of earthworm populations in tropical pastures from India. Proc Ind Acad Sci, 92: 315–321.

Serebiah, J.S. 2015. Culture of marine polychaetes. In: *Advances in Marine Brackishwater Aquaculture*. (eds) Santhanam, P., Thirunavukarasu, A.R. and Perumal, P., Springer, India, New Delhi, pp 43–49.

Shanks, A.L. 2001. An identification guide to the larval marine invertebrates of the Pacific Northwest. Oregon State University Press, Corvallis, Oregon.

Siddique, J., Amir Khan, A., Hussain, I. and Akhther, S. 2005. Growth and reproduction of earthworm (*Eisenia fetida*) in different organic media. Pak J Zool, 37: 211–214.

Siebers, D. 1976. Absorption of neutral and basic amino acids across the body surface of two annelid species. Helgolander wiss. Meeresunters, 28: 456–466.

Siebers, D. 1982. Bacterial invertebrate interactions in uptake of dissolved organic matter. Am Zool, 22: 723–733.

Simonini, R. and Prevedelli, D. 2003. Effects of temperature on two Mediterranean populations of *Dinophilus gyrociliatus* (Polychaeta: Dinophilidae): I. Effects on life history and sex ratio. J Exp Mar Biol Ecol, 291: 79–93.

Simpson, I.C., Roger, P.A., Oficial, R. and Grant, I.F. 1993. Density and composition of aquatic oligochaete populations in different farmer's ricefields. Biol Fertil Soils, 16: 34–40.

Singhal, R.N. and Davies, R.W. 1996. Effects of an organophosphorus insecticide (Temephos) on gametogenesis in the leech *Hirudinaria manillensis* (Hirudinidae). J Inv Pathol 67: 100–101.

Sket, B. and Trontelj, P. 2008. Global diversity of leeches (Hirudinea) in freshwater. Hydrobiologia, 595: 129–137.

Smith, H.L. and Gibson, G.D. 1999. Nurse egg origin in the polychaete *Boccardia proboscidea* (Spionidae). Invert Reprod Dev, 35: 177–185.

Smith, C.R., Drazen, J. and Mincks, S.L. 2006. Deep sea biodiversity and biogeography: Perspectives from abyss. Int Seabed Seamount Biodiversity Symp, Hawaii.

Smith, D.S., del Castillo, J. and Anderson, M. 1973. Fine structure and innervation of an annelid muscle with the longest recorded sarcomere. Tiss Cell, 5: 281–302.

Southward, A.J. and Southward, E.C. 1972. Observations on the role of dissolved organic compounds in the nutrition of benthic invertebrates. II. Uptake by other animals living in the same habitat as pogonophores and by some littoral polychaetes. Sarsia, 48: 61–70.

Southward, A.J. and Southward, E.C. 1980. The significance of dissolved organic compounds in the nutrition of *Siboglinum ekmani* and other small species of Pogonophora. J Mar Biol Ass UK, 60: 1005–134.

Springett, J.A. 1970. The distribution and life histories of some moorland Enchytraeidae (Ohgochaeta). J Anim Ecol, 39: 725–737.

Sruthy, P.B., Anjana, J.C., Rathinamala, J. and Jayashree, S. 2013. Screening of earthworm (*Eudrilus eugeniae*) gut as a transient microbial habitat. Adv Zool Bot, 1: 53–56.

Stephens, G.C. 1963. Uptake of organic material by aquatic invertebrates. II. Accumulation of amino acids by the bamboo worm *Clymenella torquata*. Comp Biochem Physiol, 10: 191–202.

Stephens, G.C. 1975. Uptake of naturally occurring primary amines by marine annelids. Biol Bull, 149: 397–407.

Stevens, B.R. and Preston, R.L. 1980. A transport of L-alanine by the integument of the marine polychaete *Glycera dibranchiata*. J Exp Zool, 2012: 119–127.

Stevens, R.B., Kerans, B.L., Lemmon, J.C. and Rasmussen, C. 2001. The effects of *Myxobolus cerebralis* myxospore dose on triactinomyxon production and biology of *Tubifex tubifex* from two geographic regions. J Parasitol, 87: 315–321.

Stolc, A. 1903. Uber den Lebenszyclus der nidrigen Stisswasserannulaten und tiber einige anschliessende biologische Fragen (Aeolosoma-Arten). Bull Int Acad Tcheque Sci, 7: 74–130.

Stone, R.G. 1932. The effects of x-rays on regeneration in *Tubifex tubifex*. J Morphol, 53: 389–431.

Stone, R.G. 1933. The effects of X-rays on anterior regeneration in *Tubifex tubifex*. J Morph, 54: 303–320.

Strathmann, M.F. 1987. *Reproduction and Development of Marine Invertebrates of the Northern Pacific Coast*. University of Washington Press, Seattle, WA.

Strecher, H.J. 1968. Zur Organization und Fortpflanzung von *Pisione remota* (Southern) (Polychaeta, Pisionidae). Z Morphol Okol Tiere, 61: 347–410.

Sturzenbaum, S.R., Andre, J., Kille, P. and Morgan, A.J. 2009. Earthworm genomes, genes and proteins: the (re)discovery of Darwin's worms. Proc R Soc, 276B: 789–797.

Styan, C.A., Kupriyanova, E. and Havenhand, J.N. 2008. Barriers to cross fertilization between populations of widely dispersed polychaete species are unlikely to have arisen through gametic compatibility arms-races. Evolution, 62-12: 3041–3051.

Subramanian, E.R., Sudalaimani, D.K., Samuel, S.C.J.R. et al. 2017. Studies on oragnogenesis during regeneration in the earthworm *Eudrilus eugeniae* in support of symbiotic association with *Bacillus endophyticus*. Turk J Biol, 41: 113–126.

Sugio, M., Yoshida-Noro, C., Ozawa, K. and Tochinai, S. 2012. Stem cells in asexual reproduction of *Enchytraeus japonensis* (Oligochaeta, Annelida): Proliferation and migration of neoblasts. Dev Growth Diff, 54: 439–450.

Suomalainen, E. 1950. Parthenogenesis in animals. Adv Genet, 3: 193–253.

Suomalainen, E., Saura, A. and Lokki, J. 1987. *Cytology and Evolution in Parthenogenesis*. CRC Press, Boca Raton, p 216.

Surholt, B. 1977. The influence of oxygen deficiency and electrical stimulation on the concentrations of ATP, ADP, AMP and phosphotaurocyamine in the body-wall musculature of *Arenicola marina*. Hoppe Seyler's Z. Physiol Chem, 358: 1455–1461.

Tadokoro, R., Sugio, M., Kutsuna, J. et al. 2006. Early segregation of germ and somatic lineages during gonadal regeneration in the annelid *Enchytraeus japonensis*. Curr Biol, 16: 1012–1017.

Takeo, M., Yoshida-Noro, C. and Tochinai, S. 2008. Morphallactic regeneration as revealed by region specific gene expression in the digestive tract *Enchytraeus japonensis* (Oligochaeta, Annelida). Dev Dyn, 237: 1284–1294.

Takeo, M., Yoshida-Noro, C. and Tochinai, S. 2010. Functional analysis of grimp: a novel gene required for mesodermal cell proliferation at an initial stage of regeneration in *Enchytraeus japonensis* (Enchytraeidae, Oligochaete). Int J Dev Biol, 54: 151–160.

Tenore, K.R. 1983. What controls the availability to animals of detritus derived from vascular plants: Organic nitrogen enrichment or caloric availability? Mar Ecol Prog Ser, 10: 307–309.

Tessmar-Raible, K., Raible, F., Christodoulou, F. et al. 2007. Conserved sensory-neurosecretory cell types in annelid and fish forebrain: insights into hypothalamus evolution. Cell, 129: 1389–1400.

Testerman, J.K. 1972. Accumulation of free fatty acids from sea water by marine invertebrates. Biol Bull, 142: 160–177.

Tettamanti, G., Grimaldi, A., Congiu, T. et al. 2005. Collagen reorganization in leech wound healing. Biol Cell, 97: 557–568.

Theede, H. 1973. Comparative studies on the influence of oxygen deficiency and hydrogen sulfide on marine bottom invertebrates. Neth J Sea Res, 7: 244–252.

Theede, H., Schaudinn, J. and Saffe, F. 1973. Ecophysiological studies on four *Nereis* species of the Kiel Bay. Oikos, 15: 246–252.

Timm, T. 1984. Potential age of aquatic Oligochaeta. Hydrobiologia, 115: 101–104.

Tandoh, E.J. and Lavelle, P. 1997. Effect of two organic residues on growth and cocoon production in the earthworms *Hyperiodrilus africanus* (Eudrillidae). Eur J Soil Biol, 33: 13–18.

Tondoh, E.J. 1998. Effect of coffee residues on growth and reproduction of *Hyperiodrilus africanus* (Oligochaeta, Eudrilidae) in Ivory Coast. Biol Fert Soil, 26: 336–340.

Tosuji, H., Miyamoto, J., Hayata, Y. and Sato, M. 2004. Karyotyping of female and male *Hediste japonica* (Polychaeta: Annelida) in comparison with those of two closely related species *H. diadroma* and *H. atoka*. Zool Sci, 21: 147–152.

Tosuji, H. and Sato, M. 2006. Salinity favorable for early development and gamete compatibility in two sympatric estuarine species of the genus *Hediste* (Polychaeta: Nereididae) in the Ariake Sea, Japan. Mar Biol, 148: 529–539.

Tovar-Hernández, M.A. and Knight-Jones, P. 2006. Species of *Branchiomma* (Polychaeta: Sabellidae) from the Caribbean Sea and Pacific coast of Panama. Zootaxa, 1189: 1–37.

Tovar-Hernández, M.A., Méndez, N. and Salgado-Barragán, J. 2009. *Branchiomma bairdi*: a Caribbean hermaphrodite fan worm in the south-eastern Gulf of California (Polychaeta: Sabellidae). Mar Biodiver Rec, 2: e43.8 pages.

Tovar-Hernández, M.A. and Dean, H. 2014. A new gregarious sabellid worm from the Gulf of California reproducing by spontaneous fission (Polychaeta, Sabellidae). J Mar Biol Ass UK, 94: 935–946.

Traut, W. 1969a. Zur Sexualität von *Dinophilus gyrociliatus* (Archiannelidae) I. Der Einfluss von Aussenbedingungen und genetischen Factoren auf das Geschlechtsverhaetnis. Biol Zentralbl, 88: 467–695.

Traut, W. 1969b. Zur Sexualität von *Dinophilus gyrociliatus* (Archiannelidae) II. Der Aufbau des Ovars und die Oogenese. Biol Zentralbl, 88: 695–714.

Traut, W. 1970. Zur Sexualitat von *Dinophilus gyrociliatus* (Archiannelida). III. Die Geschlechtsbestimmung. Biol Zentralbl, 89: 137–161.

Tsai, C.F., Shen, H.P., Tsai, S.C. and Lee, H.H. 2007. Four new species of terrestrial earthworms belonging to the genus *Amynthas* (Megascolecidae: Oligochaeta) from Taiwan with discussion on speculative synonyms and species delimitation in oligochaete taxonomy. J Natl Hist, 41: 357–379.

Tsuji, F.I. and Hill, E. 1983. Repetitive cycles of bioluminescence and spawning in the polychaete *Odontosyllis phosphorea*. Biol Bull, 165: 444–449.

Tzetlin, A.B. and Britayev, T.A. 1985. A new species of the Spionidae (Polychaeta) with asexual reproduction associated with sponges. Zool Scr, 14: 177–181.

Urcuyo, I.A., Bergquist, D.C., MacDonald, I.R. et al. 2007. Growth and longevity of the tubeworm *Ridgeia piscesae* in the variable diffuse flow habitats of the Juan de Fuca Ridge. Mar Ecol Prog Ser, 344: 143–157.

Utvesky, S., Kovalenko, N., Doroshenko, K. et al. 2009. Chromosome numbers of three species of medicinal leeches (*Hirudo* spp). Syst Parasitol, 74: 95–102.

Van Cleave, C.D. 1937. A study of the process of fission in the naid *Pristina longiseta*. Physiol Zool, 10: 299–314.

van der Have, T.M., Broeekx, P.B. and Kersbergen, A. 2015. Risk assessment of live bait. Bureau of Waardenburg, Gulemborg, The Netherlands, p 62.

Van Dover, C.L. 1994. *In situ* spawning of hydrothermal vent tubeworms (*Riftia pachyptila*). Biol Bull, 186: 134–135.

Velando, A., Dominguez, J. and Ferreiro, A. 2006. Inbreeding and outbreeding reduces cocoon production in the earthworm *Eisenia andrei*. Eur J Soil Biol, 42: 354–357.

Velando, A., Eiroa, J. and Dominguez, J. 2008. Brainless but not clueless:earthworms boost their ejaculates when they detect fecund non-virgin partners. Proc R Soc, 275: 1267–1072.

Viljoen, S.A. and Reinecke, A.J. 1989. The life-cycle of the African Nightcrawler, *Eudrilus eugeniae* (Oligochaeta). South Afri J Zool, 24: 27–32.

Vittor, B.A. 2002. Gray's Reef benthic macroinvertebrate community assessment, April 2001. Final Report to NOAA under contract No. 50-DGNC-0-90024, NOAA NOS, Charleston, SC.

Von Haffner, K. 1928. Uber die Regeneration der vordersten Segmente von *Lumbriculus* und ihre Fahigkeit, ein Hinterende zu regenerieren. Z Wk Zool, 132: 37–72.

Watson, A.T. 1906. A case of regeneration in polychaete worms. Proc R Soc, 77B: 332–336.

Watson, G.J. and Bentley, M.G. 1997. Evidence for a coelomic maturation factor controlling oocyte maturation in the polychaete *Arenicola marina* (L.). Invert Reprod Dev, 31: 297–306.

Watson, G.J., Cadman, P.S., Paterson, L.A. et al. 1998. Control of oocyte maturation, sperm activation and spawning in two lugworm species: *Arenicola marina* and *A. defodiens*. Mar Ecol Prog Ser, 175: 167–176.

Wehe, T. and Fiege, D. 2002. Annotated checklist of the polychaete species of the seas surrounding the Arabian Peninsula: Red Sea, Gulf of Aden, Arabian Sea, Gulf of Oman, Arabian Gulf. Fauna of Arabia, 19: 7–238.

Weidhase, M., Bleidorn, C. and Helm, C. 2014. Structure and anterior regeneration of musculature and nervous system in *Cirratulus cf cirratus* (Cirratulidae, Annelida). J Morphol, 275: 1418–1430.

Weidhase, M., Beckers, P., Bleidorn, C. and Aguado, M.T. 2016. On the role of the proventricle region in reproduction and regeneration in *Typosyllis antoni* (Annelida: Syllidae). BMC Evol Biol, 16: 196.

Weinberg, J.R., Starczak, V.R., Mueller, C. et al. 1990. Divergence between populations of a monogamous polychaete with male parental care: premating isolation and chromosome variation. Mar Biol, 107: 205–213.

Westheide, W. 1967. Monographie der Gattung Hesionides Frederich und *Microphthalmus* Polychaeten. Z Morph Tiere, 61: 1–159.

Westheide, W. 1971. Interstitial Polychaeta (excluding Archiannelida). In: *Proceedings of the First International Conference on Meiofauna.* (ed) Hulings, N.C., Smithsonian. Contributions to Zoolgy, 76: 57–70.

Westheide, W. 1979. Zur Ultrastruktur der Genitalorgane in interstitiellen Polychaeten. II. Mannliche Kopulationsorgane mit intracellulären Skelettstaben in einer *Microphthalmus*-Art. Zool Scr, 8: 111–118.

Westheide, W. 1984. The concept of reproduction in polychaetes with small body size: adaptations in interstitial species. Fortschr Zool, 29: 265–287.

Westheide, W. 1988. Genital organs. In: *The Ultrastructure of Polychaeta.* (eds) Westheide, W. and Hermans, C.O., Microfauna Mar, 4: 263–279.

Westheide, W. and Muller, M.C.M. 1996. Cinematographic documentation of enchytraeid morphology and reproductive biology. Hydrobiologia, 334: 263–267.

Westheide, W. and Purschke, G. 2013. Annelida. In: *Lehrbuch der Speziellen Zoologie.* (eds) Westheide, W. and Rieger, G.E., Springer, Heidelberg, 1: 357–415.

Whitford, T.A. and Williams, J.D. 2016. Anterior regeneration in the polychaete *Marenzelleria viridis* (Annelids: Spionidae). Invert Biol, 135: 357–369.

Williams, G.B. 1964. The effect of extracts of *Fucus serratus* in promoting the settlement of larvae of *Spirobis borealis* (Polychaeta). J Mar Biol Ass UK, 45: 397–414.

Williams, J.D. 2004. Reproduction and morphology of *Polydorella* (Polychaeta: Spionidae), including the description of a new species from the Philippines. J Natl Hist, 38: 1339–1358.

Williams, M.E. and Bentley, M.G. 2002. Fertilization success in marine invertebrates: The influence of gamete age. Biol Bull, 202: 34–42.

Wilson, W.H. 1985. Food limitation of asexual reproduction in a spionid polychaete. Int J Invert Reprod Dev, 8: 61–65.

Wilson, W.H. 1991. Sexual reproductive modes in polychates: Classification and diversity. Bull Mar Sci, 48: 500–516.

Won, S., Park, B.K., Kim, B.J. et al. 2014. Molecular identification of *Haemadispa rjukuana* (Hirudiniformes: Haemadipsidae) in Gageo Island, Korea. Korean J Parasitol, 52: 169–175.

Woodin, S.A. 1982. Browsing important in marine sedimentary environments? Spionid polychaete example. J Exp Mar Biol Ecol, 60: 35–45.

Xiao, N., Ge, F. and Edwars, C.A. 2011. The regeneration capacity of an earthworm *Eisenia fetida* in relation to the site of amputation along the body. Acta Ecol Sinica, 31: 197–204.

Yang, T., Wang, T-h. and Guo, Z-k. 1997. Chromosome number and characteristics of a hirudinidae leech *Hirudo nipponia.* Zool Res, 18: 26, 32, 50.

Yasmin, S. and D'Souza, D. 2010. Effects of pesticides on the growth and reproduction of earthworm: A review. Appl Env Soil Sci, 2010: 1–9.

Yesudhason, B.V., Jegathambigai, J., Thangaswamy, P.A. et al. 2012. Unique phenotypes in the sperm of the earthworm *Eudrilus eugeniae* for assessing radiation hazards. Emb Monit Assess, 185: 4745–4752.

Yokota, H. and Kaneko, N. 2002. Naidid worms (Oligochaeta, Naididae) in paddy soils as affected by the application of legume mulch and/or tillage practice. Biol Fertil Soils, 35: 122–127.

Yoshida-Noro, C. and Tochinai, S. 2010. Stem cell system in asexual and sexual reproduction of *Enchytraeus japonensis* (Oligochaeta: Annelida). Dev Growth Diff, 52: 43–65.

Young, J.O. 1988. Intra- and interspecific predation on the cocoons of *Erpobdella octoculata* (L.) (Annelids: Hirudinea). Hydrobiologia, 169: 85–89.

Yuan, S.L. 1992. Reproductive Biology of *Megalomma* sp (Polychaeta: Sabellidae) in Hsiao-Liu-Chiu. M.S Thesis, National Sun Yat-sen University, Taiwan, 41 p.

Zajac, R.N. 1985. The effects of sublethal predation on reproduction in the spionid polychaete *Polydora ligni* Webster. J Exp Mar Biol Ecol, 88: 1–19.

Zajac, R.N. 1986. The effects of intra specific density and food supply on growth and reproduction in an infaunal polychaete *Polydora ligni* Webster. J Mar Res, 44: 339–359.

Zajac, R.N. 1991. Population ecology of *Polydora ligni* (Polychaeta: Spionidae). I. Seasonal variation in population charactersistics and reproductive activity. Mar Ecol Prog Ser, 77: 197–206.

Zajac, R.N. 1995. Sublethal predation on *Polydora cornuta* (Polychaeta: Spionidae): patterns of tissue loss in a field population, predator functional response and potential demographic impacts. Mar Biol, 123: 531–541.

Zattara, E.E. and Bely, A.E. 2011. Ecolution of a novel developmental trajectory: Fission is distinct from regeneration in the annelid *Pristina leidyi*. Evol Dev, 13: 80–95.

Zattara, E.E. 2012. Regeneration, fission and the evolution of developmental novelty in naid annelids. Ph.D. Thesis, University of Maryland, College Park, p 209.

Zattara, E.E. and Bely, A.E. 2013. Investment choices in post-embryonic development: quantifying interactions among growth, regeneration, and asexual reproduction in the annelid *Pristina leidyi*. J Exp Zool, 320B: 471–488.

Zattara, E.E. and Bely, A.E. 2016. Phylogenetic distribution of regeneration and asexual reproduction in annelids: regeneration is ancestral and fission evolves in regenerative clades. Invert Biol, 135: 400–414.

Zeeck, E., Hardege, J., Bartels-Hardege, H. and Wesselmann, G. 1988. Sex pheromone in a marine polychaete: Determination of the chemical structure. J Exp Zool, 246: 285–292.

Zeeck, E., Harder, T., Beckmann, M. and Muller, C.T. 1996. Marine gamete release pheromones. Nature, 382: 214.

Zeeck, E., Harder, T. and Beckmann, M. 1998. Uric acid: The sperm release pheromone of the marine polychaete *Platynereis dumerilii*. J Chem Ecol, 24: 13–22.

Author Index

Species Index

Subject Index

Author's Biography

Recipient of the S.S. Bhatnagar Prize, the highest Indian award for scientists, one of the ten National Professorships, T.J. Pandian has served as editor/member of editorial boards of many international journals. His books on Animal Energetics (Academic Press) identify him as a prolific but precise writer. His five volumes on Sexuality, Sex Determination and Differentiation in Fishes, published by CRC Press, are ranked with five stars. He is presently authoring a multi-volume series on Reproduction and Development of Aquatic Invertebrates, of which the volumes on Crustacea, Mollusca, and Echinodermata and Prochordata, have already been published. The present book is on Annelida. The next one, Platyhelminthes is being prepared.